FLETCHER CHRISTIAN
BOUNTY MUTINEER

His Life. His Fate. The Repercussions

GLYNN CHRISTIAN

PRAISE FOR *FRAGILE PARADISE*

The Discovery of Fletcher Christian

`Tells a fascinating tale.' *New York Times Book Review*

'. . . abounds in the action and atmosphere the film versions sorely lacked ... enriched by a thorough accounting of Tahitian mores and the British seafaring life-style.' *The Smithsonian*

`Glynn Christian's book has all the ingredients of success. It is an epic story written with the first-hand passion of the hero's descendant, who followed in his ancestor's footsteps and so came closer to the original. This deeply researched and highly readable book must become the classic account [of the story of Fletcher Christian]'

Robin Hanbury-Tenison OBE: Explorer,
Gold Medallist, Royal Geographical Society

`It is exclusively due to Glynn Christian's investigative efforts that we know Charles Christian (his brother) participated in a mutiny and that Fletcher must have known about it ; a major contribution to *Bounty* research.'

Dr Sven Wahlroos, Bounty historian,
author of Mutiny and Romance in the South Seas:
A Companion to the Bounty Adventure

`. . . undoubtedly the best written account of *Bounty* for many years . . . And for once this is a book which has something new to offer readers. The second part of the book . . . the author's journey to Pitcairn . . . is a masterly story, honest and frank and beautifully told.'

Gavin Kennedy: author of Bligh and William Bligh: The Man and his Mutinies

No one has ever delved so deeply into Christian's background . . . sources are skillfully woven into the text without diverting the reader's attention from the facts . . . one shares the author's thrill of discovery.'

Oceans Magazine

`there are still more books to be written about *Bounty;* but none can ever be so entertaining, nor so informative as this.' *Carlisle Evening News and Star*

`sweeps away . . . inaccurate and fanciful notions of previous writers . . . packed with intrigue and conflict, it combines history with biography, travel with detection.'

Keswick Reminder

Fletcher Christian *Bounty* Mutineer

ISBN 978-1-9162984-4-6

First published as FRAGILE PARADISE – The Discovery of Fletcher Christian, BOUNTY Mutineer

Hamish Hamilton, London: 1982
Atlantic, Little Brown, Boston: 1982
Book Club Associates, London: 1983
Doubleday, Sydney (revised): 1999
Long Riders Guild Press: 2005

Cover design by Peter Groves

Revised edition FLETCHER CHRISTIAN *BOUNTY* MUTINEER
Hendon Books London: 2019
www.glynnchristian.com

FLETCHER CHRISTIAN
BOUNTY MUTINEER

Glynn Christian

CHRISTIAN FAMILY OF MILNTOWN, ISLE OF MAN & UNERIGG/EWANRIGG, CUMBERLAND

John MacCrysten I, Deemster 1408, born ca 1368
William MacCrysten II, Deemster 1417, Member House of Keys,
John McCrysten III, Deemster 1448-98
John McChristen IV, Deemster, died c.1511
John McChristen V Deemster, died c 1533: bought Milntown 1511
William McChristen VI Deemster, died 1535
Ewan McChristen VII MHK died 1539
William McChristen VIII Deemster died 1568
William McChristen IX Deemster died c. 1593
Ewan Christian X Deemster 1605-53 Deputy Governor 1634-7
John Christian XI Deemster 1602 -73
Edward Christian XII Deemster 1628-93
Ewan XIII 1651-1719 1st to live permanently at Unerigg/Ewanrigg
John XIV 1688-1745 = Bridget Senhouse of Netherhall
Charles Christian, brother of Ewan XV and John XVI = Ann Dixon
Fletcher Christian = Mauatua
Charles 'Hoppa' = Sully
Isaac = Miriam Young
Godfrey = Frances Edwards
William = Evelyn Smith
Royce = Enid Pitman (Colin=Nola Rowlands: Keith=Patricia Barber)
Glynn Bruce Faye Ross

See also: MILNTOWN by Derek Winterbottom, Alondra Books, Isle of Man

THE BEGINNINGS

April 28, 1789

Aboard *Bounty* off Tofua, Friendly Islands

September 1764 – December 1787

Cumberland, Isle of Man, India and the Caribbean

Is this what Fletcher Christian looked like?
Based on the many known portraits of Fletcher's ancestors and 18th century uncles and cousins, it was painted by the late John Lockett, an expert in facial form and structure, as a commission by Adrian Teal to illustrate his thesis on the many ways Fletcher Christian is portrayed in literature

CHAPTER 1

IT'S WHO THEY WERE

The atmosphere aboard *Bounty* on April 27 1789 was never worse and none was more oppressed or affected than 24-year old acting-lieutenant Fletcher Christian. Darkly tanned and tattooed after months on Tahiti, he ran forward on deck with great tears welling from his eyes. The carpenter William Purcell asked what had happened. *'Can you ask me and hear the treatment I receive?'*, Fletcher answered. Purcell suggested he received the same treatment but Christian pointed out a difference. As a warrant officer Purcell could not be flogged *'... but if I should speak to him as you do he would probably break me, turn me before the mast and perhaps flog me, and if he did it would be the death of us both, for I am sure I should take him in my arms and jump overboard with him.'*

As young Fletcher wept, he added: *'I would rather die ten thousand deaths than bear this treatment. I always do my duty as an officer and a man ought to do, yet I receive this scandalous usage. Flesh and blood cannot bear this treatment.'*

On December 22 1787, the night before *Bounty* sailed from Portsmouth's Spithead to Tahiti, Fletcher's mutineering older brother Charles had assured him that each man has a breaking point. Fletcher Christian was daily being pushed by Captain Bligh to reach his.

When Fletcher was woken for his 4am watch, he had gone to bed less than an hour before, having chosen not abandon the ship on a raft. Confused by lack of sleep, exhausted by the emotions of the last twenty-four hours, he left his hammock to command his watch. On *Bounty's* almost empty, early-morning deck a resolution to his conflicts suddenly became simple and practical. *'Why should I go? Why shouldn't Bligh go?'*

That way it didn't even sound like mutiny.

Gathered around the arms chest, Christian and his supporters agreed the cutlasses, pistols, muskets and bayonets were only to be deterrents. There would be bloody threats but there was to be no bloodshed.

At about 4.30am on April 28th, 1789, Fletcher Christian descended *Bounty's* aft ladderway. Followed by Burkitt, Mills and Churchill, he burst through the habitually open door of Bligh's cabin, waking his victim with a flourish of naked steel and shouting, *'Bligh, you are my prisoner!'*

• • •

Thousands of 18th-century men and women in Great Britain and the South Seas were directly affected by the repercussions of Fletcher's mutiny. Even now there are lives pained daily by defence or loathing of him or of William Bligh and others who were aboard *Bounty.*

Yet for over 200 years proper discussion was not possible, because no-one knew who Fletcher Christian was. Books about *Bounty* and William Bligh were based on supposition, prejudice or unfounded ideals, as many still are today. 1982 changed this, when the first edition of *Fragile Paradise - The Discovery of Fletcher Christian* was published. The last words I wrote in that book were: *'I look forward to the day when there is no longer the urge to cast Bligh or Christian as black or white. They are men who are remembered. Few men who are remembered can have been wholly one or the other'.* I believe this just as passionately today. The 1789 mutiny on *Bounty* proves fact can be stranger and more thrilling than fiction. What drove Fletcher Christian and William Bligh from passionate friendship to personal betrayal in the world's most famous mutiny has generated five feature films, uncountable tv documentaries and over 2500 books and major articles, with more published every year.

This renamed and thoroughly revised 21st-century edition, still the only biography of Fletcher Christian, adds to the facts I earlier discovered and proudly rights the age-old truism that most history is written by men about men. The story of Fletcher and his mutiny is indeed about men and their challenges. The story of Pitcairn Island is about women and their challenges to protect their children by mutineering against the drunken brutality of its men.

We would know little about Christian, Bligh or *Bounty* if it were not for these women of Tahiti, Huahine and Raiatea, *te tupuna vahine*, the Ma'ohi foremothers of Pitcairn Island. Their bravery in defence of their children led to Pitcairn being the first community in the world where women had the permanent right to vote at 18 and girls' education was mandatory. Both were direct inheritances of Fletcher Christian's views on social reform, something his relatives led. It was 80 years before a few UK women over 30 were franchised.

By introducing more about the life led by women in Tahiti, victims of daily bullying, exclusion and abuse, this book also makes it quite clear that charges are utterly unfounded that Fletcher Christian had to kidnap women to sail with him on *Bounty*. Pitcairn's 12 *tupuna vahine* were given two opportunities to leave the ship but didn't when others did.

Those 12 women who landed on Pitcairn did so willingly and proved this by abandoning almost everything to do with their native islands including tattooing, face, hand and skull reshaping, lewd dancing, food restrictions and infanticide. Of the many things of which Fletcher Christian has been falsely accused, kidnap is the most undeserved.

Fletcher Christian is best known for one terrible, criminal act against William Bligh aboard *Bounty.* I hope you discover in this book much more about the black and the white of both Fletcher Christian and William Bligh and will then judge neither with rancour.

It's who they were and 1789 is a long time ago.

William Bligh painted after his return from the South Pacific and his courageous open-boat journey; Bounty's launch and erupting Tofua are in the background.

CHAPTER 2

BRAVERY IN RETROSPECT

I was nine when I learned I was descended from Fletcher Christian. My teacher that year at Auckland's Owairaka Primary School was Mr Jackson, a rare man, able to conjure up in the minds of children the reality of times past. He revealed that history is simply the lives of other people. By exciting those untrammelled senses of romance and adventure that the young enjoy before the afflictions of puberty, Mr Jackson helped me realise that I, marooned in the suburbs of Auckland, was linked to men and women who had been alive at the time Drake was first to sail around the globe, when William the Conqueror won the Battle of Hastings, during the rise and fall of Julius Caesar or the Inca or every Chinese Emperor.

I became so stimulated that I desperately wanted to know just where the men and women who made me had been when, say, Queen Victoria was crowned or when Columbus reported back to Ferdinand and Isabella. I ran home after school, so I could ask: 'Are we . . . am I descended from anyone famous?' By remarkable coincidence, MGM's 1930s *Mutiny on the Bounty* with Clark Gable as Fletcher Christian had been re-released that week in local cinemas. Grandma Christian took me to see it at Auckland's Embassy Theatre in Wellesley Street and then Mr Jackson let me tell the class about it two mornings in a row.

There was a dramatic improvement in my playground status, which I did everything to encourage. Being the smallest and most easily bullied kid in school became marginally more bearable, except when I repeated what the papers said about Fletcher Christian, that Pitcairn Island was a major concentration of the blood of Plantagenet kings because of his direct descent through his grandmother from King Edward I and his wife Eleanor of Castile. It's true but I learned you do not claim to have royal blood and live in Owairaka and say so.

The story of the man who stole one of the king's ships from William Bligh and then went to live with semi-naked women on a remote South Pacific island was too much properly to absorb, too dramatic for me really to feel part of it. Pieces did help explain some of the confusion that discoloured my childhood, as until then I had been troubled about why my grandfather looked so different and why he was so tall. When he talked with his sisters it was a language that even my father, his eldest son, could little understand. At first, I attributed his extraordinarily idiosyncratic English to ignorance. He made singular words plural and vice versa, saying he had lost his 'glass' when I knew he meant his spectacles. He was decidedly different and children hate being associated with difference.

So, for all the status it offered, talking about Fletcher Christian had to be tempered until I was able to understand that difference and appreciate the specialness of my grandfather and his equally over-powering sisters, my great-aunts. It took me another five years to learn that almost universally other people found them fascinating.

Like most men whose grandparents are dead I deeply regret I did not know my grandfather better. I should have learned how to speak his special language from Norfolk Island, a mixture of 18th-century English and 18th-century Tahitian, the language with which Fletcher Christian and his Tahitian consort Mauatua and their Pitcairn colleagues created. I should now love to hear more of those lilting peculiarities, peppered with wonderful words and accented softly, like men of England's West Country, but I was discomforted by his size and learned what I could from Aunt Renee, the least tall of his sisters. She told me what she would, not what she could, and explained without knowing it some of the extraordinary aspects of myself that had made me a shy and frightened child.

From my earliest days I was aware of the ease with which ecstasy could overtake me as soon as I was alone under the sun or in the thick, evergreen New Zealand bush. I could lie for hours on hot grass or under trees or beside flowers, entertained to the point of bewildered fulfilment just by being there, by the heat and the smells. Sometimes I took off my clothes to walk freely and secretly in the bush or along a beach, and had done so since I was seven. It was physical and sexual and I knew it but it was imperative.

The eventual absorption of the fact that Fletcher Christian's wife Mauatua, my great-great-great-great-grandmother, was Tahitian, and that my family had lived on Pacific islands for generations, helped to explain that my instinctive fascination with the sun, the tropics and the fruits and flowers made by them was perfectly natural. It was my heritage and from that time I developed an enormous but very private sense of pride in what made me different. It reinforced what Mr Jackson had taught, that links with the past are more than names on paper and faded photographs.

I learned that Fletcher Christian had hidden from justice on Pitcairn Island. He and Mauatua had two sons and a daughter, Mary, who was supposedly the inspiration for Mary Russell Mitford's poem *Christina: The Maid of the South Seas* but who died a crabby old spinster. His sons were Thursday October and Charles and the only sons of Pitcairn's forefathers who married full-blooded Tahitian women, both older foremothers who had arrived with the ship many years before. Thursday married Teraura the older ex-consort of Edward Young and widow of Matthew Quintal. Charles married Sully, who had arrived on Pitcairn Island as an infant with *Bounty* in 1790. Thus, Fletcher's grandchildren were more Tahitian than Caucasian, explaining why, of all *Bounty* families, the Christians are most likely to show or feel their Tahitian ancestry.

Fletcher's second son Charles was my great-great-great grandfather and his son Isaac was born on Pitcairn in 1825 but by 1856 the island was considered unsuitable to support its population of 187. Queen Victoria, who had always taken an interest in the Pitcairners and who had sent them gifts including an organ, agreed that the *Bounty* descendants, forgiven the sins of their forefathers, should be given Norfolk Island, which sits temperately between Australia and New Zealand and hence Isaac went, too.

Norfolk Island had been the British Devil's Island, the ultimate punishment for men transported from Great Britain for stealing bread or killing others. For decades it was a cesspool of sodomy, massacre and exploitation but now, deserted apart from a few caretaker ex-convicts, it was to be settled by the world's most fascinating and God-fearing community. Most Pitcairners hated it on sight. The sparse, green, rolling hills, spiked by lofty Norfolk pines, peppered with the sedate Georgian masonry of the prison and surrounding village, were no match for the lush ruggedness of Pitcairn. There were no coconuts for milk and the stuff that came out of cows was considered unhealthy as well as being difficult to obtain. Several families returned to the tropic remoteness of Pitcairn as soon as they could, a retreat based neatly upon blood lines. Broadly, Christians descended from Thursday October returned, the family of Charles stayed on Norfolk Island.

Great-great grandfather Isaac had migrated to Norfolk Island when he was 31 and there married Miriam Young. Their son was also named Isaac but to avoid confusion was known by his second name, Godfrey, a tradition that has been continued in the family up to this day; I am William Glynn. Isaac-called-Godfrey grew to manhood at a time when Norfolk Island was an important and busy whaling centre. Eventually he owned his own ship and married an outsider, Frances Edwards, the daughter of an American whaler. Her mother was a Flowerday and so she was a granddaughter of the Hansens, the first white children born and brought up in New Zealand. Their marriage linked the blood of two of the South Pacific's earliest pioneering European families.

Great-grandfather Godfrey and Frances had the typical large family of the time, one of whom was my grandfather William, a favourite son and brother known through his youth as Sunny for his good nature and smile. His sister Aunt Renee told me stories that I first disbelieved but I heard them from other sources, too.

Grandfather William first tried to stowaway on his father's whaler when he was eight. He was tall and muscular enough to pass for 12 or 13, by which time you were certainly expected to earn your living in those tough times. He managed to get aboard undetected but made the great mistake of trying too hard and lit up a pipe of tobacco, his first. Nausea exposed what his light boy's voice had not and he was returned.

The stories of his bravery and individuality included William being a whaler's hand-harpoonist by the time he really was 12. Later he sailed with the Bishop of Melanesia to South America and was aboard the cable ship *Iris* during the First World War when Count von Lucknow was arrested. When he went to New Zealand in 1916, he married Evelyn Smith, descended from William Davern, an Irishman who came to NZ as a soldier of the 58th Foot in 1852 to fight in what are known today as the Land Wars. William and Evelyn's eldest son Royce was born in 1918 and became my father in 1942.

My mother Enid's grandfather Alexander Bow was another adventurer, who came to NZ in 1852 with land grants but chose instead to teach Maori children in remote settlements. In retrospect, I can see that with adventuring pioneers in so many bloodlines I had no choice but to be filled with curiosity about other people, different places, new ways to live.

Once I had these descents clear, I was able to work out that Fletcher Christian was my great-great-great-great-grandfather but that was all I knew. After I left school, I thought of Fletcher only in terms of Clark Gable. When Marlon Brando announced he was making a new version of the film in the early '60s, I was reassured to hear how much research he said he had done on Fletcher Christian. It was all a terrible disappointment. The dandified gentleman portrayed by Brando was possibly more accurate than Gable's character (he had first refused to play the part, feeling that pigtails and knee breeches would damage his masculine screen image) but the fiery immolation immediately after arriving at Pitcairn Island with which Brando ended the character's life caused me infinite pain and trouble. Too many people believed it to be the way Fletcher Christian had really died and thus that my claim to descent was bogus.

There seemed to be no book that would tell me the truth, or a consistent version of any part of the story. Was Bligh the tyrant that my great-aunts said? Who would tell me if Fletcher Christian was from the Isle of Man or Cumberland? Who knew what had become of him? It became increasingly important that I should know all these things.

By September 1965, already successful as one of New Zealand's first full-time radio and TV scriptwriters in an advertising agency, I was frustrated, feeling I could go no further in that country. I decided to leave and six weeks later I was aboard RHMS *Ellinis*, bound for Southampton. Apart from £10 my parents had given me, the money I carried was borrowed. I knew only one person in England and hardly knew why I was going. I had made one of those unpremeditated decisions that irrevocably changes a life, a bit like mutiny I supposed.

It was brave and ambitious move for a 23-year old man who still blushed easily and who could not go to a pub by himself until over 30-years old. At the time, I was aware only of being fired by absolute fear of failure or ridicule. With each battle I fought to survive in England I was even more determined not to return to New Zealand until I had achieved something that made my leaving worthwhile. My thoughts turned to Fletcher Christian often as I realised his decision to take *Bounty* had put him into the same situation. More and more I wondered if it might be possible to discover why he had made that one decision that so irretrievably changed his life and then had repercussions for so many others for so very long.

On the surface I could not compare myself with Fletcher Christian. He apparently came from a privileged background and was physically strong and athletic. Yet when I visited Tahiti on my way to Europe, I knew at once that my intoxication with nature came from my Tahitian ancestry and that first day on Tahiti remains one of the most precious experiences of my life. Having established one connection with the past, I hoped that knowing more about Fletcher Christian might explain some of my other, wilder ways and the paradox of self-doubt and ambition. If this were possible it had to follow that there were other men and women in my background whose traits and personalities would allow me to see myself simply as another link in a continuing chain of adventurers. That is precisely what has happened.

My discovery of Fletcher Christian has proved far more rewarding than I dared imagine. In unearthing his background and motivations I discovered many of my own.

MY PITCAIRN DESCENT

Fletcher Christian = Mauatua

Charles = Sully, daughter of Teio and unknown Tahitian male

Isaac = Miriam Young: granddaughter of Edward Young and Toofaiti and daughter of Elizabeth Mills, thus granddaughter of John Mills and Vahineatua

Godfrey = Frances Edwards

William = Evelyn Smith

Royce = Enid Pitman (Colin = Nola Rowlands: Keith = Patricia Barber)

Glynn: Bruce: Faye: Ross

CHAPTER 3

THE NEW GENERATION

There never was a mutiny of *Bounty*. Rather there was a revolt of one man against another, Fletcher Christian against William Bligh, with some of *Bounty*'s men joining him. Logically that clash can only be understood if the passions and perversities of both men are chronicled.

Fletcher Christian is recognised as a very fine navigator, a shining pupil of William Bligh, who in turn reflected the brilliance of Captain James Cook. None of these men's success was a matter of native skill. When *Bounty* sailed off to the South Seas in 1787, she carried Larcum Kendall's K2, one of the earliest chronometers and that had already been to the South Pacific with Captain James Cook and Bligh, who was his sailing master. This invention made it possible to keep accurate time at sea, which made it possible to discern longitude accurately for the first time, that is knowing how far west or east you were from the Greenwich meridian. Not knowing this had meant the loss of hundreds of ships and tens of thousands of sailors, sometimes in sight of land. K2 can be seen working today in Greenwich's Royal Observatory.

After his mutiny, Fletcher Christian safely sailed *Bounty* on an epic 8,000 mile/12,875 km mid-Pacific voyage in search of a hideaway. He journeyed as far west as Fiji and *Bounty* was the first European ship to discover Rarotonga. Early in 1790, the K2 chronometer helped him find invisibility on Pitcairn Island, which had been incorrectly charted.

Pitcairn Island was possibly the first closed community in which white men could choose women only of another colour; even missionaries in Africa or South America could take a wife of their own heritage or send for one. Pitcairn was a volatile mixture of men and women who knew little of each other's culture or language.

The descendants of these ill-matched pioneers became the most God-fearing and admired community on earth, an inspiration both to Queen Victoria and, fatally for Pitcairn, to the Anglican Church's interfering missionaries.

Fletcher Christian has been called many things, a mutineer, a pirate, a murderer, a gentleman, a blackguard, a kidnapper and more. Factually, he was an adventuring explorer, the founding father of a unique people and, like so many of his powerful family and others in the 18th century, a courageous social pioneer. What Rousseau and the Age of Enlightenment proposed, Christian tried to do.

The books and pamphlets of the proselytising Victorian Church and Hollywood made Fletcher Christian a hero of protest and escapism, whom I have discussed on the white sands of the Gulf of Siam and in the Palace of Westminster, in the U.S. Capitol, at National Geographic and on the stage of the Royal Geographical Society, at Teotihuacan, in the Australian Outback, on QE2 and countless other ships and more recently at the Nantucket Whaling Museum. But who was he, what was the truth about him? Could he be discovered after so long, when books couldn't even agree if he was from Cumberland or the Isle of Man?

My quest to discover Fletcher Christian began in 1978 and took me three years, without the aid of the internet, before the days even of faxes. Once I knew who he was, a sailing expedition to Pitcairn Island helped me answer the mystery that had occupied minds for almost two centuries. What happened to Fletcher Christian?

• • •

Fletcher Christian was born to Charles and Ann Christian in the stone-built farmstead of the substantial landholding of Moorland Close on September 25 1764 and was carried over the hill to be baptised in St Mungo's, Brigham, the same day, something that happened only when a baby was not expected to survive. Either Fletcher was born sickly or there was an epidemic of some kind in the purlieus of close-by Cockermouth, all of which used river water to power leather-tanning, hat making and wood or metal workshops.

Not far from Cumberland's Lake District, Moorland Close belongs to the small community of Eaglesfield. The farm's ancient walls and buildings sit snug on the long gentle brow of a great green hill that slopes for 2 ½ miles / 4kms to the market town of Cockermouth, built where the Cocker joins the River Derwent. It surrounds the remains of a romantic Norman castle that is still occupied, except for the areas wasted by Robert the Bruce and the Civil War.

When dialect was common the property was called Mairlandclere and was *'a quadrangular pile of buildings in the style of a medieval manor house, half castle and half farmstead'*, originally fortified against Lowland reivers who came to plunder, wreck and, time permitting, rape. The peace brought to the Borders by the accession of the Scot King James VI to the throne of England as James I had existed for well over a century when Charles Christian and Ann Dixon began their married life at the Close on May 2 1751 and they had every expectation of comfort and peaceful prosperity. The marriage was quite in keeping with their station in Georgian society and both were in possession of their inheritances. Ann's was Moorland Close. Her father, who was a dyer but described as a gentleman, was her co-heir and lived with them.

Moorland Close homestead was built in 1709 outside a defensive enclosure of ancient brick. In Fletcher Christian's time the walls and towers enclosed the remains of a medieval 'part-castle' and other buildings, an evocative adventure playground for a boy.

Fletcher's father Charles was the fifth son of John Christian, the 14th documented head of the Christians of Milntown on the Isle of Man and of Ewanrigg Hall in Cumberland. They were an extraordinarily prosperous and powerful clan, with fingers firmly in every temporal, pastoral or spiritual pie at home and abroad. The strictures of primogeniture meant that younger sons, even of the richest families, had to marry an heiress or to embrace the equally rewarding bosom of the church, the comradeship of the armed forces or, *in extremis*, such professions as the law. Christians, who were landed gentry rather than aristocrats, had done this for centuries.

Charles had two sources of income, one from inherited shares in his family's mining and shipping interests, the other from his practice as an attorney in Cockermouth. In the Christian family, the pursuit of a legal career was the proud following of a tradition that had its roots deep in centuries as First Deemsters on the Isle of Man. With financial security and scion status, Charles was seriously eligible. One generally married for the comfort of a wife's fortune and looked elsewhere for the exercise of the libido. In Ann Dixon, Charles found a woman who offered both.

Ann's mother Mary was a Fletcher and, though not as rich as the Christians, they had been in Cumberland far longer. Her grandfather was descended from William Fletcher, the builder of Cockermouth Hall (now a parking lot), who had succoured Queen Mary of Scotland when she fled her mutinous realm. He gave her clothes and cloth of red velvet out of pity for her abandoned state.

Ann seems not to have had a fortune in cash but was co-heiress of Moorland Close, that her late mother owned. It offered a roof and an income to a potential husband and gave her distinct market value. Jane Austen's later female characters are not untypical for being introduced by their fortunes rather than their faces; in such circles the latter was valueless without the former.

Once, the Close's house and outbuildings had been clustered inside a tall and ancient red-brick wall iced with crenellations. A thin, rutted road ran under the long east side. In this wall was the main entrance, guarded by a high, square watch-tower with a pointed roof. To enter you had to dismount and walk through a succession of low narrow doors under the tower, a lesser version of the defences of a medieval castle.

During Queen Anne's safer times in 1709, the Fletchers built a new house outside the battlements, a simple solid building of elegant proportions in honey-coloured, dressed stone, now a dull grey stucco. The partly-eviscerated quadrangle was planted as an orchard and garden, emulating the walled wonders of espalier and asparagus seen on the estates of the noble and rich, a startling conceit for a modest moorland farm. Still, the battlements did provide protection from biting northerly winds that damaged young plants quite as badly as invaders' boots had once done. They also made the most seductive setting imaginable for the adventurous fantasies of children and for the illusions of their parents because from certain angles the property looked positively stately. In the 1860s there were still those who remembered seeing some of the medieval 'castle' and other buildings within the wall, but they are all gone now.

As a son of John XIV and Bridget Senhouse, Charles had been brought up in Ewanrigg Hall, the Christian's mainland mansion above Maryport that had been re-built in 1640. Called Unerigg for most of its history and constantly improved, it was refaced in red brick in sober Georgian style in 1777-8, something Fletcher would have seen in his early teens. In 1782-3 a wing was added on either side; one was a new library and dancing room, the other a kitchen. Ewanrigg was now 13 bays wide, three and four storeys high and had 42 rooms. For Charles, any disadvantage of size or facility at Moorland Close was outweighed by its being an easy ride from his family's seat of Ewanrigg.

Fletcher Christian's paternal grandmother Bridget Senhouse was directly descended from Edward I and Queen Eleanor of Castile

Ewanrigg Hall was the Christian family seat in Cumberland since the mid-1600s. First little more than a fortified Pele tower, it was extended in the 1680s. In 1777-8, new rooms and a new facade doubled its size; the two low wings were added 1782-3. The rear views show what a mixture of styles it had become. This is where Fletcher Christian's father Charles was born and brought up and it became the first seat of Fletcher's 1st cousin John Christian XVII, head of the family. The last vestiges have gone and the grounds are a housing estate.

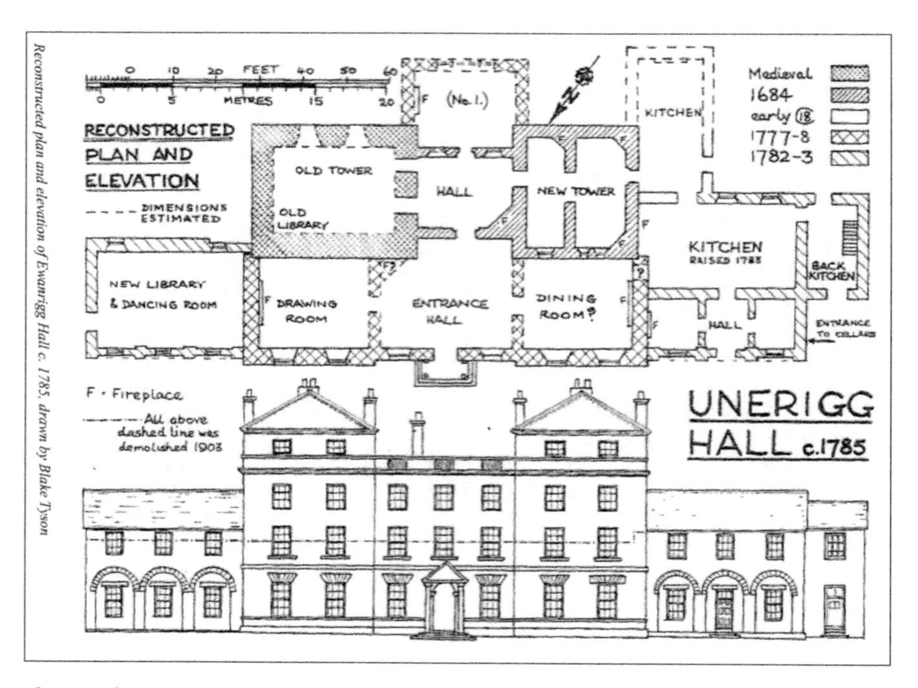

Reconstructed plan and elevation of Ewanrigg Hall c. 1785, drawn by Blake Tyson

through Joan of Acre, born when the couple was on Crusade. Her brother Edward II married Joan's daughter Margaret de Clare to his lover Sir Piers Gaveston, said to have worn more jewels than any woman at his boyfriend's coronation but she later become Duchess of Gloucester. The Senhouse family seat of Netherhall in what became Maryport was even bigger than Ewanrigg.

Charles' youngest brother Humphrey had to move to Docking, Norfolk, to marry Elizabeth Brett. Through complicated genealogical arrangements their eldest son, just eight years older than Fletcher, inherited Docking Hall provided he changed his name to Hare. When I visited in the early 1980s, Hares still lived in the fine old Hall, with portraits of Christians as their ancestors and were considering changing their family name back to Christian.

Charles and Ann were 22 and 20 respectively when they married and had ten children, six of whom survived to adulthood, a good record for the time. John, Ewan, Jacob, Edward, Mary, Charles, Fletcher, Frances, Ann and Humphrey were all named for brothers, uncles, aunts, sisters, and grandparents, mainly names that the Christian family had used since the 14th century. Ewan and Jacob died within weeks of each other early in 1757. Frances and Ann were twins, who survived but a short time. There might have been even more children, but three months after Humphrey was born, Charles senior died and was buried in Brigham churchyard on March 13 1768, beside his wife's grandfather. Fletcher was three and a half.

Charles's tomb shows the importance attached to being a Christian, alive or dead. The inscription says he was not only of Moorland

Close but was also the son of John Christian of Ewanrigg. It is an elegant table grave with turned legs and, at its head, a massive, moulded rendering of the Christian's coat of arms with its unicorn-head crest.

When the Gentleman's Magazine of March 1768 reported Charles Christian's death, they described him as Coroner for Cumberland. This can't be confirmed. Cumberland's Coroners were in the direct appointment of Lord Egremont, incumbent of Cockermouth Castle and no Christian appears in the list of his appointees.

Charles had made his Will on Valentine's Day 1768 declaring himself weak of body and leaving everything to Ann. With nice feeling for that saint's day he pays Ann pretty compliments. To her he committed *'the custody and tuition of the children with whom it hath pleased God to bless us during their respective minorities, not doubting but as she has approved herself to me the best of wives she will also prove the best of mothers'.*

Ann was a widow at 36. On May 3 1768 she and her father Jacob Dixon, both giving Moorland Close as their address, swore that *'the Paper Writing to you now shown'* was the last Will and Testament of Charles and paid £800 to the Keeper General of the Exchequer and Prorogative Court of Robert, Lord Archbishop of York. Throughout I have used the guidance of the National Archives website to translate money to its approximate value today; £800 then is about

Netherhall was the family seat of the Senhouses. Fletcher's paternal grandmother Bridget Senhouse was directly descended from King Edward I and Queen Eleanor of Castile, via their daughter Princess Joan of Acre; this is why it is said Pitcairn is one of the greatest concentrations of Plantagenet blood.

When Fletcher was 3 ½ years old, his father Charles was buried in St Mungo's churchyard Brigham, half way between Moorland Close and Ewanrigg Hall at Dearham. His unusual table-grave is topped with a bold Christian coat of arms and the inscription points out his connection with Unerigg/Ewanrigg.

£70,000 today. The money was a bond to ensure they carried out the provisions of the Will, paying creditors and funeral expenses and gives an indication of the prosperity of Moorland Close. Ann was also expected to show *'a true and perfect inventory of Goods, Rights, Credits, Cattles and Chattles'*; this so that there could be no pleading of false poverty to avoid payment of just debts.

The fatherless children, aged from 16 to three months, were reasonably provided for and the Close was a going concern. They had a resident maternal grandfather and there seem to have been plenty of Dixons ready to turn a hand on the farm. Ann took the children's tuition seriously, for her children had been left more than possessions. They were the new faces of an extraordinary heritage, part of the newest generation of the Christian family, which had prospered in Cumberland and on the Isle of Man for centuries.

Most families point proudly to men of power or wealth or influence in one generation or another. Some might even trace one or all of these attributes for several successive lives or in related family branches. There are few, royalty and nobility not excepted, that can claim the combination of all three in ever-growing strength and in an unbroken line of direct male-only descent, including brother to brother, from the mid-14th century, before which most records are suspect anyway. The Christians could. It is so unusual for a progression to have so much strength for so long a time that it had marked relevance to anyone born into that family.

It would have been impossible for Fletcher, Charles's and Ann Christian's seventh child and sixth son, not to know about or be influenced by this heritage.

Milntown, in the parish of Lezayre, Ramsey, in the north of the Isle of Man became the Christian family seat in 1511. The original buildings were replaced by a severe five-bay Georgian house that Fletcher would have known. It was enlarged in the 1830s, when fashionable Gothic battlements were added. The library is said to be panelled in bog oak from the original house.

CHAPTER 4

'HE IS TO BE WONDER'D AT WHO WONDERS AT IT ...'

Although born in Cumberland, Fletcher Christian would early have learned that the Christians prospered on the Isle of Man for 500 years or more. The name has variously been prefixed with Mc or Mac, a Celtic device, but the family is unquestionably of Scandinavian origin. There are conflicting theories, but the most likely is descent from Gillocrist Mackerthac, a Norwegian noble, sent to Man in 1238 as a trouble-shooter. His descendants would have inherited the power and the land with which he was invested, explaining the Christians early emergence as major landowners and politicians in the Manx north, the only pocket of noticeably Scandinavian clanship, names and customs.

Fletcher's ancestry has also been linked to Gillocrist, foster brother of King Godred II of Man. In 1154 land was exchanged between that king and the Priory of St Bees on the Cumberland coast, which might explain how Fletcher Christian's forefathers owned Staffland and the Barony of Maughold, names identifying them as having once been ecclesiastical holdings. When Icelandic fugitive Godred Crovan conquered Man in 1077, he landed in the north of the island and fought a famous battle at Skyehill, now Scafell. The island already had its Viking-style open-air Tynwald parliament, which celebrated its millennium in 1979, making it the most ancient and continuous in the world.

Without the dedicated struggle of the Christian family over the centuries, the Isle of Man and Tynwald would not be the unique things they are. Today, all laws passed by the legislators, a freely elected House of Keys, must still be proclaimed from the Hill of Tynwald at St Johns on July 5th, the old Midsummers Day.

The round, stepped Tynwald Hill, 80 feet/24.5 meters in diameter and twelve feet/3.65 meters high, is reputed to be constructed of earth from all parts of the Scandinavian kingdom to which Man once belonged. It could equally be of Druidic origin, for their legal and civil offices and officers are as clearly reflected on the island as those of Vikings. From the steps of this hill, a stirring thing whatever its origins, the First Deemster as the island's chief judge reads the new laws in Manx and English, that no man can plead ignorance of them and no official can secretly impose his own rule. The earliest Christian I could positively trace as an ancestor of Fletcher Christian was such a First Deemster in 1408.

As First Deemsters, generations of Fletcher Christian's grandfathers mounted ancient Tynwald Hill and followed the centuries-old tradition of reading out Manx laws on Mid-Summer Day, so no citizen could plead ignorance of them.

There is such a plethora of Williams, Johns, and Ewans among the successive Deemsters of the Christian pedigree that they are numbered according to succession rather than by how many had preceded them with the same forename. Deemster John Christian I, who with his son Deemster William II, signed a declaration against the Scrope family claim to the island in 1408, held the post if not by accepted hereditary right, then by tradition too strong to break. Christians whose forenames are lost doubtless performed the Tynwald duties even earlier.

The Isle of Man requires two Deemsters and the First, from the north, was always the more powerful. As Chief Judges Deemsters are responsible for the judicial side of life. The office has always been of the greatest dignity and importance and its inherent Manxness is nowhere better illustrated than in the last part of the Deemster's Oath, '. . . *I . . . do swear that I will, without respect of favour or friendship, love or gain, consanguinity or affinity, envy or malice, execute the laws . . . betwixt party and party, as indifferently as the herring backbone doth lie in the midst of the fish.'*

As well as their own Deemster Courts, which were truly the backbone of daily life and illustrative of its strife, the Deemsters sat on every other court. Perhaps their most difficult role was that of being the main buffer between the people and the Lord or King of Man, usually via his representative the Governor.

Neither Manx laws nor their interpretation began to be written until 1594. The Deemsters practised Breast Law, often from the steps of their house, their announcement, *'We give for the Law'* being based solely on unwritten precedent to which they alone had full access. It would have been a dangerous thing to give such power to an untrained man, perhaps one who did not clearly understand that even juries (Enquests) went to great lengths to avoid imposing the death penalty, though ears were lopped off with abandon. The Manx allowed the post to become hereditary, at least in the case of the Christians as First Deemsters.

There used to be doubt about when the Christians first owned their ancestral seat of Milntown at Lezayre in the north of the island, which they retained until 1886, although family members rented it into the 20th century. New research by Derek Winterbottom for his book *MILNTOWN – The story of an historic Manx house and the Christian and Edwards families* Alondra Books, IOM) finally establishes the facts. At the time known as John McChristen, John V bought Milntown in 15ll. It was a rambling, fortified property with the distinct advantage of including one of the island's few licensed mills. The main domestic building included a hall 30ft by 18 ft – about 9.2m x 5.5m – and a series of parlours, a large kitchen and separate bakery, both of which on the mainland would have been detached, for fire safety. There was every type of farm building for cattle and poultry and grounds that included a large kitchen garden called the Scotch Garden. Bed chambers must have been on an upper floor, although the parlours might have been used. Milntown remained in this state as the seat of the First Deemsters until 1750.

When James, seventh Earl of Derby and uncrowned king of Man went to the island in 1627, he found the Christians firmly holding their inherited lands and influence, impervious to the politics of the rest of the world. Later, concerned for the future of his family, who had lost their English estates when Civil War broke out in 1642, the earl wrote at length to his son Charles: *There be many Christians in this country—that is Christins, the true name; But they have made themselves chief here, wherefore if a better Name could be found they would likely pretend unto it . . . But it is not so much that so many be called Christians, as that by Policie they are crept into the principall Places of Power; and they be seated round about the Countrey and in the Heart of it; they are matched with the best Families; have the best Livings [that is farms] and must not be neglected.*

There were three 'pretenders' that the earl and his son could not neglect. At that time the First Deemster was Ewan Christian X. He was 26 when he was appointed in 1612 and remained in office for fifty years.

Ewan held power during the troubled time of the Commonwealth. Although cousins, nephews, and sons were punished, imprisoned and put to death, he escaped all such retribution. Yet he was by no means cool of temper. The records are full of tales of litigations, charges and counter-charges, with the Deemster often admitting guilt and paying a bloodwite for having struck someone with whom he disagreed.

Ewan Christian X's private life seems to have been fully as passionate as his public one. Lord Derby also wrote: *Someone in a pleasant humour says he thought the Deemster did not get so many Bastards for Lust's Sake, as in Policie, to make the Name of the Christians flourish . . . It is very true that there be many Bastards here in this Isle: and he is to be wonder'd at who wonders at it. Be sure it would be very well if that Law were here as in other Places that all knowne Bastards were called after their Mother's names . . . they are subject to make Factions . . . Men of the same Name will side with one another against any Body.*

If the Christians didn't own your land, or marry into your family, then they had children by your daughters, bastard children who took the surname Christian and who then stuck by the family in times of strife. When Fletcher Christian visited Man, he was sure at some stage to be asked if he were a real Christian, that is one who had his surname through the marriage of his parents. The same question can still be asked on Man and on Pitcairn and Norfolk Islands, for the same reason.

Lord Derby quickly got the Christians on his side by appointing another 'pretender' governor. This was Edward Christian, first cousin of Ewan X. At 22 he had sailed in 1612 with a flotilla of the East India Company, charged with capturing the trade of Western India from the Portuguese. After remarkable adventures during which he had been kidnapped and threatened with death, he returned three years later as commander of one of the vessels and an extremely successful man. Back in London he was arraigned for having made a private fortune and for having *'carried himself too proudlie'*.

Yet, it suited both Derbys and Christians to have Edward as Governor. They could run the island virtually as they wished but Edward's love of profit and status became troublesome and the earl removed him from the governorship. Then, hearing rumours that Parliament supporters might attack Man, he gave Edward charge of the militia. Edward took over Peel Castle and set up a training camp. At the same time there were riots throughout the island about the tithing system. The new Governor, John Greenhalgh, wanted to impose martial law, but Edward refused his troops. Judging his own situation more dangerous than that of the Crown, Lord Derby sailed back with English troops to quell the island's 12,000 people. Edward Christian was charged with sedition and brought to trial. With the help of his cousin, Deemster Ewan, he got off reasonably lightly. Instead of being executed, he was fined 1,000 marks, getting on for £70,000 today, *'and his bodie to perpetual imprisonment or untill hee shall be relapsed by the said Lord'.*

Ewan X, aware of the delicacy of his position, now made over Milntown to his son John. The small estate of Ronaldsway (the present Manx airport) went to William, his youngest son, commonly known as Illiam Dhone (Brown-haired William). William was the third 'pretender' and had been a member of the House of Keys and steward of the Abbey. By 1648 he was Receiver-General, responsible for the Derbys' finances. In 1649, after the execution of King Charles, Parliament asked Derby to surrender the island and, when he refused, gave it to Lord Fairfax, who did nothing to claim it.

In August 1651, Derby moved to the mainland to continue fighting, leaving the government in the hands of his wife, and the remaining militia in the command of William Christian. Then, in September, Derby was captured in Chester. Knowing the island could not hold out against Cromwell, Lady Derby wrote offering to surrender it in return for the release of her husband. The Manxmen were not pleased. They had never been part of the British Isles and never wanted to be; they would consider surrendering, but on their own terms. On October 19 1651, 1,800 Manxmen met at Ronaldsway and agreed to oppose Lady Derby. Within ten days they had taken over all the forts on the island except her Castle Rushen. Meanwhile, Lord Derby had been executed in Bolton, on the 15th.

The Commonwealth fleet, under the command of Colonel Duckenfield, shortly arrived at Ramsey, close to Milntown. A deputation of four men, led by John Christian acting as Deemster on behalf of his father Ewan X offered surrender without bloodshed in return for 'laws and liberties as formerly they had'. There was no immediate answer. Instead William Christian of Ronaldsway was enlisted by Duckenfield to present a request for surrender to Lady Derby. Was he a traitor to his liege, or was he a patriot, behaving in the manner best suited to secure what he could on behalf of the people of Man? Did Duckenfield enlist him because Illiam Dhone controlled the forts and the people and had made some kind of deal at the expense of the Derbys? This is but part of the controversy which surrounds Illiam Dhone. Lady Derby was arrested and the Isle of Man became last to join the Commonwealth.

England wanted to dispense with the office of Governor but found that several courts would be unconstitutional, so William was appointed to the position, retaining that of Receiver-General.

Rose Castle was the official residence of the Bishop of Carlisle, Edmund Law and his wife, Fletcher's aunt Mary Christian. Fletcher's Law 1st cousins were great successes in many fields. John and George became bishops, Edward rose to be Lord Chief Justice and 1st Baron Ellenborough; Joan became Lady Rumbold, wife of a notorious Governor of Madras. (Rose Castle Foundation)

Ewan X had recently died but the two new Deemsters were William's brother John XI and his nephew Edward XII, John's eldest son. Once more the Christians of Milntown had a firm grip on the island.

When Charles II was restored to the throne, Charles, eighth Earl of Derby, tried to tempt him to return confiscated Derby properties. The king was persuaded that to do this would be a breach of the Act of Indemnity. In a revenge by proxy, Derby accused William Christian of plotting against his father and signed a mandate `*to all his officers, both civil and military, to bring William Christian to account for all his illegal actions and rebellions on or before the year of our Lord 1651'*. That is, Illiam Dhone was to be tried for his actions during the time of the Commonwealth, something King Charles ruled only the regicides should suffer.

William denied the right of the court to try him but Derby argued that the Act of Indemnity pardoned only crimes against the English Crown and that it was possible separately to be guilty of treason against the King and Lord of Man. Derby had said he was happier to be known as a major English lord than as a minor king but was prepared to don this latter rank if it gave him revenge.

Neither William's Deemster brother nor his Deemster son could save him and Illiam Dhone was shot on January 2 1663 at Hango Hall within sight of his home. The last speech he made began with haunting words I have never forgotten: *There is but a thin veil between life and death . . .*

King Charles was incensed at this formalised murder and Derby was ordered to pay damages and to restore Ronaldsway to William's sons. Everyone else involved was punitively fined and Deemsters John and Edward were reinstated. But the Derbys' malevolence continued and in 1770 Illiam Dhone's descendants left Man to flourish in Ireland and Virginia. One is on a stamp with Daniel Boone and there is a Christiansburg in West Virginia.

Political expediency in the difficult aftermath of the Commonwealth and Restoration saw Ewan XIII (1651-1719) permanently settle at Unerigg Hall, close to the coast of Cumberland and the fishing village Elnefoot, later to become Maryport. While keeping all their land and most of their influence on Man, they were safe from the bloody thrusts of the island's politics and could concentrate on filling coffers rather than coffins.

When Ewan's son John XIV was 29, he married Bridget Senhouse of Netherhall House, and there was great prosperity at Unerigg during the life of John and Bridget. They built the kennels and huntsmen's quarters of beautifully dressed and pointed limestone when hunting became formalised and *de rigueur* for gentlemen. Men commented that the Christians' kennels were better than most people's houses and when I saw the dressed stone ruins of the building I understood why.

John XIV and Bridget's children and grandchildren, who included son Charles and grandson Fletcher were remarkable for their talents and attainments. In 1740 Charles' sister Mary married the Reverend Edmund Law DD taking a fortune of £3,000, today close to £350,000, which made her more than an ordinary bride and is a clear indication of the Christians' financial comfort. Law became Master of Peterhouse in November 1754, a position he seemed to keep even when elevated to the Bishopric of Carlisle and tenancy of magnificent Rose Castle from 1768-87. Within the family, Mary is remembered for always referring to her 13 children as her baker's dozen and also for the resolutions she made on the April evening before she was wed, aged 18.

RESOLUTIONS OF MARY CHRISTIAN UPON THE DAY OF HER MARRIAGE

I resolve:

Never to contradict my dear Husband without it be quite necessary and then with the greatest good nature I am mistress of.

To serve God more sincerely than I have done in the state I am now about to leave and to lead a life suitable to the Blessed calling of my Husband.

Never to fret or fall into a Passion about small matters but to have always a cheerful heart, knowing my blessings much exceed any troubles that can possibly befall me, and in all dangers to commit myself and family to an All-Wise Providence, and then to be easy about the event.

I likewise resolve to lay aside all fondness for dress but to be always neat and clean.

I resolve to be very active and never for the sake of saving myself a walk to neglect anything, though it be never so great a trifle.

I resolve to be very frugal and never to put my Husband to any needless expense.

I resolve to be very kind to my servants, as well to their souls as to their bodies, and always to give exact orders and never to be in a passion if they not be executed.

I resolve to treat my friends kindly, but never extravagantly, and to be full as glad to see his relations as my own.

Thus would I live
Thus would I die
And when this world I leave
To Heaven I'd fly.

Mary Christian 28 April 1740

On the same date almost half a century later her nephew Fletcher would enact something totally inimical to her piety and morality, deep in the South Pacific.

Mary's children, Fletcher's first cousins, were illustrious. The eldest surviving son, John, became Chaplain to the Duke of Portland when he was Lord Lieutenant of Ireland. When 37 he was appointed Bishop of Clonfert and was subsequently Bishop of Killala and then of Elphin. Edward became Chief Justice of England, the last to have automatic cabinet rank and elevation to the peerage, as first Baron Ellenborough. All his life he made fun of his younger cousin Edward Christian of Moorland Close, an older brother of Fletcher who was also a lawyer and ultimately his greatest defender.

Mary's sons Ewan, Joseph and Thomas Law all started their careers in India. Ewan made a fortune, Joseph died there, and Thomas left the East India Company for America where he was befriended by George Washington and married his granddaughter. George Law became yet another bishop, eventually occupying the see of Bath and Wells; you can see his portrait in the Pump Room of Bath's Roman baths. The youngest of the Law's daughters was Joanna, who kept up the Indian connection begun in the 16th century. In 1772 she became the second wife of Thomas Rumbold, who had been President of the British East India Company and Governor of Madras in that town's Fort St George for three years from 1777, during which he was made a baronet.

Sir Thomas was famed as one of the most rapacious of the 'nabobs' and when previously serving the East India Company in Madras is said regularly to have sent home amounts three times that of his salary. Overall it was believed he had taken £600,000 illicitly, over £52 million in 2019. His parliamentary connections and bribes stymied any prosecution and he and Lady Rumbold lived luxuriously at their new Palladian mansion, Woodhall Park in Hertfordshire, where she gave birth to three sons and three daughters. After his death in November 1791 the house was sold to benefit the children of this second marriage. Bligh was back telling of his travails in early 1790 and so Lady Rumbold would have known about the *Bounty* mutiny by her first cousin Fletcher. There's bound to be astonished correspondence between her and her barrister, bishop and merchant brothers. A task for another time.

John XIV and Bridget Christian's eldest son was Ewan XV. He worked with his mother's family the Senhouses of Netherhall to extend their coal mining interests and continued the expansion of the tiny seaside hamlet of Elnefoot into Maryport, named for his Senhouse grandmother, soon to become an important port for trade with the New World, especially in sugar and rum.

Winterbottom's new research shows that around 1750 Ewan XV demolished old Milntown and built a fashionable five-bay Geor-

John Christian Curwen was Fletcher's 1st cousin and head of the Christian family. Isabella Curwen was the second wife. A great agriculturalist and social reformer, John is credited as first to introduce the system where employee, employer and state each contributes to a Social Security system and twice refused a peerage. Isabella Curwen, the very rich cousin after whom Fletcher Christian named his Tahitian wife Mauatua. The last member of one of England's 10th oldest families, Isabella eloped with her cousin John Christian XVII in 1782 when she was only 17. He changed his name to John Christian Curwen weeks before news of Fletcher's mutiny reached England in 1790.

gian brick mansion, reputably incorporating materials from the old house including bog oak and timbers from wrecked ships of the Spanish Armada. This is the Manx family seat Fletcher, his mother, sister and brothers would have known. It was not until 1829 that John XVIII added the showy Gothic-revival towers and crenellations seen today. Milntown is open to the public and welcomes Christian descendants.

Ewan died unmarried and was succeeded by his brother, John XVI, who had married into the rich Curwen family, one of the ten oldest names in the country. They had two sons and six daughters, the eldest of whom was only 16 when Mrs Christian died in 1762. Four years later John himself died in Petty France in Gloucestershire and 11-year-old John XVII inherited. He was a ward of his uncle Henry Curwen and would one day opt to become John Christian Curwen, a name change erroneously associated with news of Fletcher's mutiny on *Bounty.*

John Christian XVII was well cast as head of this generation al-

Isabella Curwen's Workington Hall, to which she and John Christian XVII moved. Even though he had recently spent a great deal to improve Ewanrigg, their joint mining fortunes meant they significantly extended and rebuilt Workington into something so grand and imposing it was also known as Curwen Castle. Is this what Fletcher imagined his life would be like?

though only eight years older than his first cousin, Fletcher. He was perhaps the most influential of all Christians and certainly made the greatest contribution to national life. At 19 he married Margaret Taubman in Kirk Malew on the Isle of Man. Margaret was seven years older and was also his sister-in-law, for her brother had married John's eldest sister, Dorothy. They lived on Man for a while at the Bowling Green, Margaret's parents' house, where their only child, John XVIII, was born in 1776. She died after two and a half years of marriage in 1778. John remained on Man a while as he was elected to the Manx House of Keys in 1777.

John left the island in 1779, saying it had been created solely to plague anyone who had dealings with it but his son was mainly brought up there by his Taubman grandparents. During the next two years he spent wildly to rebuild the main façade of Unerigg and change its name to Ewanrigg. Soon after his wife had died, his old guardian Henry Curwen also died, leaving his only daughter Isabella, then aged 13, to the care of John's spinster sisters, Bridget and Jane.

Isabella was splendidly rich, the last of the Curwens, one of England's 10 oldest names and sole heiress to Workington Hall and huge mining interests. She attracted titled and rich suitors as she grew older and undoubtedly was noticed by her cousin Fletcher Christian on family occasions at Ewanrigg and elsewhere. He was just a year older and could have seen her as a way of reversing his fortune because, after his mother was bankrupted and had to sell the family house in 1780, he was essentially a man with no fortune at 16. He had little chance against his rich, glamorous and older first cousin John Christian XVII.

Isabella's trustees humoured her with such gifts as the island in Lake Windermere, eventually named Belle Isle after her. A rich merchant had shown much nerve by commissioning a round house to be built on it but he was so mocked for his 'pepper-pot' folly that it was advertised for auction and bought in the spring of 1781 for Isabella at 1,640 guineas, around £150,000 today and less than half the amount expended upon it.

Isabella always had an affection for her *'prodigious'* cousin John and in spite of every distraction, including smart schools in London, her wards could not cope with her. John and she fell in love, he returning at intervals from his Grand Tour of Europe to see her. She was a Ward of Chancery and thus legally could not marry until 18 but in practice as long as both parties were over 16 a marriage was not necessarily invalid. That was enough for such a spirited and financially independent couple. They sped secretly to Edinburgh and were married at the Gaelic Chapel on October 9 1782, and, whatever the law thought, there was a high old time at Ewanrigg on the 10th, when John's sisters danced with the servants. The Lord

Chancellor seeming to have no objections, they married again on November 4, at the late 17th-century Parish Church of St Mary Magdalene, Bermondsey. Isabella was 17 and John was 26. He gave her a diamond costing £1,000, about £85,000 today, and they went to Paris and the South of France with her spinster carers and now sisters-in-law Bridget and Jane still in attendance. On September 25th 1782 Fletcher turned 18 and six months later he joined the Royal Navy and sailed to India. With no fortune and no prospect of one, it was the best, perhaps the only thing he could do to gain dignity and to regain social position in 18th-century England.

Belle Isle was named for Isabella Curwen and her unique round Georgian House on an island in Lake Windemere. It became a great social centre after she married John Christian XVII. Not everyone admired the house; Dorothy Wordsworth was scathing.

John XVII now abandoned Ewanrigg, which he had extended and given an entrance hallway of marble into which it was said he could, and did, drive a coach and four although looking at surviving pictures and the floor plan this seems unlikely. He went to live at Isabella's much bigger and older Workington Hall and they turned into a striking castle, as well as making Belle Isle into a blissful summer retreat; descendants lived here until 1993. Apart from leading a brilliant social life in London and the provinces with his wife, John energetically developed the mining interests of the Christians and Curwens and was one of the first to put steam pumps into mines. He pioneered the distribution of free milk to the poor and providing enhanced winter feed to cattle, began the first Agricultural Society and was MP for Carlisle for many years. When the British Parliament first legislated to introduce Social Security in 1912, John was given fulsome credit for having been first to devise and to operate the equivalent of National Insurance, having said.

'The first person to subscribe, must be the worker who was going to benefit, the second must be the employer and the third, the state'. This was exactly the scheme that was adopted and that John had created as Friendly Societies on his estates. John Christian XVII twice refused a peerage and was the ideal end product of the enlightenment of the last two centuries, when equality was an important objective. It is little wonder that his cousin Fletcher grew up with revolutionary ideas; it was the Christian way.

Like so many of his predecessors, John fought for those less fortunate than himself and took his pleasures among them, too. At one political meeting in Workington he reminded his audience that he was like a father to them all. There was much laughter when a woman called out that he was indeed father to at least half the town. John and his sisters knew and cared for the family at Moorland Close and Fletcher saw his Manx heritage in the large number of family portraits that hung in Ewanrigg Hall. He could not help seeing his cousin's benevolence, industry and charity and of course he envied the wilful and very rich Isabella, the obvious object of a young man's affections, especially those of someone with ambition.

So, for uncountable generations, the Viking-rooted Christians had been First Deemsters of the Isle of Man. They owned the best land there and great mining, agricultural and shipping interests in Cumberland. They made the best marriages, wielded judicial as well as financial power, ruled by ancient right, might and an army of bastards. The Christian family was close to their zenith at the end of the 18th century, the last great years of the Age of Enlightenment, when reason, observation and science mixed with ideas of liberty, toleration and the separation of church and state.

Christian family daughters were making the best marriages and sons and were preparing to become memorable men in the corridors of Westminster, the Inns of Court, the Cabinet, the peerage, the palaces of the Church, the governments and commerce of India and the Caribbean, the highest ranks of the Navy and Universities and, most famously, in the little known and thus highly romanticised South Seas.

CHAPTER 5

BROTHERS IN BANKRUPTCY

Being a younger son of a well-to-do family is a pretty good way to grow up. You learn by watching your elders' mistakes and the mistakes made upon them, so you may drift through puberty into adulthood blameless but knowing. As your parents become more financially secure, you enjoy the education and pleasures of their superior indulgence. There is was the pattern Fletcher Christian expected. He had both a mother ambitious for her sons and the example of his relatives to follow, the collected addresses of whom were an almanac of everything rich and grand in Cumberland's landed gentry.

It was suggested in C. S. Wilkinson's *The Wake of the Bounty* that Fletcher's future was always to take over Moorland Close and that this is why he was so named but when his oldest brother John was admitted to an Inn of Court he was described as heir to his father Charles. It's just as likely that any surviving sister was a co-legatee, continuing inheritance through the female line.

Fletcher's childhood world was one of adventure and action. All about were streams to fish, hills to climb, woods, nooks, crannies, glens and valleys to explore on the piebald horse that, until recently, you could see painted on a door of the farmstead. The horizon in every direction is one of mountains, powerful and pastel, as only Cumberland peaks could be and closer there were wonderful battlements and towers of red brick in the backyard.

Fletcher was undoubtedly stimulated by the outdoors and activity, for all who knew him agree he was strongly built and athletic, proud of his muscular and acrobatic abilities. Stories have been published of his playing truant from school to do boyish things but I have never been able to trace these anecdotes to their sources, likely as they seem.

Fletcher Christian's primary school was the small Parish School of Brigham, in a room in Eller Cottage on the High Street. The house stands today in a garden behind a hedge but the name of his Dame is forgotten. Then Fletcher was accepted as a pupil at Cockermouth Free Grammar School, a dour stone building that once stood at the entrance of All Saints Churchyard. It was founded in 1676 and was free to all from Cockermouth, although term fees and other costs were normally paid, which generally meant that the poor did not go after all. Cockermouth was not too badly off, for it was a busy little

Cockermouth Grammar School (now demolished) stood in the grounds of All Saints Church on a knoll at the eastern end of the town centre. Here Fletcher was introduced to the study of navigation.

Eller Cottage was Fletcher's Dame school

place. Main Street was then a cobbled road and the houses mostly low thatched whitewashed cottages. The chief occupations were agriculture, hat making, weaving and noisome tanning. These last three all used plenty of water and the houses of the tradespeople were clustered close to the river banks, off Main Street and Kirkgate. Many of the cottages were primitive enough still to have an outside staircase to the upper floor bedrooms. None of the churches seen now was then built and most of the big buildings were public houses or coaching stations. Except for Cockermouth Castle, the grand blue stone and tile houses of the successful merchants and landowners were spread farther away from the odorous centre.

The route Fletcher rode from Moorland Close to the Grammar School is difficult to follow today. It has been slashed by fences and highways but it meets Cockermouth's Main Street exactly opposite the beautiful house of John Wordsworth, father of the poet William and his limpet-like sister Dorothy. Much has been made of Fletcher Christian and William Wordsworth being contemporaries at the Free Grammar School but Fletcher was six years older than William and schoolboys rarely fraternise over such a wide age gap.

The Grammar School combined instruction in the principles of religion with courses in Greek, Latin, English and a wide range of mathematical subjects, including trigonometry and navigation, giving Fletcher an early start in a subject that would become vital to his later survival. The terms were five shillings/£20 today for entrance and 6 shillings/£25 a quarter. You were expected to pay more for entrance if you were from prosperous stock but children of the poor paid this minimum, known as a cockpenny.

It's long been thought that Fletcher followed his brother Edward to St Bees School, close to Whitehaven. That's what Edward wrote in his Short Reply to Captain Bligh, even inviting doubters to check with other scholars there are the same time. Recent investigations by the St Bees' archivist show this not to be so and it is a new mys-

Fletcher saw the imposing ruins of Cockermouth Castle as he rode to and from school; the old Grammar School abutts the tower of All Saints Church.

tery within the story as to why this mistake was made. Perhaps Edward misremembered his own time there? It always puzzled me why Fletcher was said to have gone to both Cockermouth Grammar and St Bees.

Edward certainly went to St Bees and details about this are interesting as an illustration of the sort of education young men of the Christian family expected, wherever they went. Founded in 1583 by a charter of Queen Elizabeth I, it was a venerable and respected institution by the late 18th century, closely linked to The Queen's College, Oxford, that owned Moorland Close in 1980. In Edward's time in the mid-1770s it was still one building, which now forms the school's dining hall. The ground floor was the schoolroom, the top floor contained the headmaster's quarters and offices for the governors. Boarders were accepted at 13 guineas/now £1250 a year, 15 guineas/£1400 if the long holiday was included. Few boys can actually have boarded at the school itself, there being virtually no extra space, so the majority would have lodged privately at extra expense in the tiny village.

School was from seven in the morning until 11am and from one until five. Unless they had leave to play, the scholars then had to walk home *'two by two together so far as their way lyeth . . .'* On Saturdays they had to spend two hours writing, whilst being examined on the Catechism in *'English, Latin or Greek, according to their capabilities'*. These were young gentlemen from the finest Westmoreland and Cumberland families, so it was not all discipline and examination. On October 13 1778 the Cumberland Pacquet recorded, *'Last week the young gentlemen students at St Bees gave a very elegant ball to the ladies of Egremont and other neighbouring places. Upwards of thirty couples danced country dances, and the whole was conducted with the greatest propriety'*.

When Fletcher Christian's brother Edward was publicly defending him after the mutiny, he pointed out in Fletcher's defence that he had *'stayed at school longer than men generally do who enter the Navy, and being allowed by all to possess extraordinary abilities, is an excellent scholar'*. This is not quite what it seems. Fletcher was certainly as clever as his brother avers but he stayed at school because he had no intention of going into the Navy until such a choice was forced upon him by the mismanagement of the family's financial affairs.

Fletcher's oldest brother was John, born 12 years earlier in 1752. He was admitted to Lincoln's Inn on June 29 1771 as an attorney, the do-it-all lawyer called a solicitor today, when not quite 19. He must have served some sort of apprenticeship to be admitted and in those days he could be so without any view of becoming a barrister. It's rather curious that on February 2nd 1773 he then became a Fellow Commoner at Peterhouse, Cambridge, almost certainly through the influence of his uncle Bishop Law of Carlisle, who was

Master of Peterhouse, the oldest in Cambridge, chartered in 1284 by Edward 1st. I wonder if John knew he was descended from this king via his Senhouse grandmother. A Fellow Commoner was most often a member of or friend of the nobility, admitted with the hope but no real expectation they might take a degree, presumably because the money they paid was more than useful to the College.

Fellow Commoners had superior accommodation and ate at the High Table with the Fellows but wits used to call them 'empty bottles'. It's telling when later bankruptcy is considered that John commissioned William Henshaw, a particularly noted and thus expensive engraver, to produce an intricate rococo-style version of his coat of arms while at Peterhouse.

John was there at least three years but did not get a degree. In the diary of his first cousin Jane Christian she writes on June 19 1776, *'Coz. J. Christian called going to Cambridge next week.'* On January 4th that year there was 'a riot made at the rooms of Mr Christian (a near relative to the Bishop of Carlisle)' that resulted in the rustication of Robert Hopper for 'disorderly behaviour in the Hall and disrespectful behaviour towards the Master and Fellows'.

John Christian practised as an attorney in Cockermouth and in London. In Cockermouth he had very grand premises, a palisaded double-fronted building once known as the Swan Inn on Market Place, complete with seven rooms for lodgers, a variety of parlours with marble chimney pieces, stablings, a dovecote, a dairy, and the right to four seats in a church pew. From surviving letters, it can be inferred that John took some control of the family affairs once he went into business, because he expected to inherit; his mother's Close was at least security for borrowed money.

Edward Christian was six years older than Fletcher and is the best known of the trio, eventually becoming both the chief public defender of Fletcher and *bête noire* of Bligh, as well as advancing to Chief Justice of the Isle of Ely, Professor of the Laws of England at Cambridge's Downing College, Professor of Law at East India College, Commissioner for Bankrupts and an authority whose published opinions, notably in Blackstone, brought him considerable income. He is said to be better remembered in the Inns and Halls as a reliable bore and an eccentric of electrifying inventiveness. The Dictionary of National Biography says: *'He disappointed the high expectations of future distinction which had been formed from his university career and gradually sank so low as to become the subject of practical jokes* . . . The entry goes on to say that when he died in 1823, he was *'in the full vigour of his incapacity'*.

It's hard to know the basis for these comments for no evidence is given and we do not know if the author might have been pursuing some personal vendetta. It is perhaps best to judge Edward on his achievements outside the pages of the DNB, which were considerable.

Edward Christian was born in 1758 and from St Bees School went up to Peterhouse, Cambridge in 1775 but changed to St John's College in 1777, graduating MA as Third Wrangler in 1779, which means he scored third highest marks in his final year. At Cambridge he met and befriended William Wilberforce, who was to lead the move to abolish the slave trade.

The brother closest in age to Fletcher was Charles, born in 1762 and so only two years older. Of him nothing was known until my discovery of a document detailing 600 years of family history and of a copy made earlier in the 20th century of Charles's now lost holograph autobiography. This document makes him a vital new witness to the story of Fletcher Christian.

In 1779, when Edward was graduating in Cambridge and John XVII was fleeing the Isle of Man, Fletcher's social and financial security was whisked from beneath him. His long-widowed mother was bankrupt. By 1780 Moorland Close would be auctioned and it was largely the fault of those from whom Fletcher should have been learning how to manage money and his future.

In 1779 Fletcher's mother and older brothers had to admit finally there was something terminally wrong with their finances. Privately, Ann had been struggling for years and what seems to have happened was this: her elder sons had borrowed heavily against the estate in the interest of their careers and university educations with absolute disregard for reality. It was irresponsible, especially to their younger brothers and sister. Whatever happened, John and Edward would still have their smart Cambridge education and their Inns of Law and now Charles, Fletcher and Humphrey had no hope of such education and the connections this would give them. Their sister Mary would have no hope of marriage, for she would have no fortune.

Ann and her family had suffered vagaries of fortune even when Charles Senior was alive. He was either a bad attorney or a bad farmer or both. In 1752 he borrowed money from his eldest brother, John XVI and also leased two plots of land from him for £24/about £1850 a year. By 1755 Charles and Ann Christian had to discharge a debt to Thomas Brunsfield by giving him most of the extra pieces of land they had bought but by the time Charles died 13 years later, they had again been acquiring land from the Fletchers of Cockermouth.

The records of the family exist because first Charles and then his widow and children regarded the head of the Christian family as a pit full of money as deep as the coal mines he developed so well. Even when John Christian XVII was still a minor, he was lending Ann's eldest son John sums of £400/£30,000 today, at reduced rates of interest, with the approval of his guardian Henry Curwen. No reasons for the loans are ever given in the records and by 1772 Ann again owed Brunsfield a serious amount of money. This time she was bailed out by Roger Fleming, a mariner of Cockermouth, who paid £600/£52,000 today for the deeds of their plot Simonscales. I can only surmise that Brunsfield was some kind of regular and too-trusting supplier to the affairs of Moorland Close.

In December 1773 nine-year-old Fletcher Christian became a brother-in-law for the first time, when John married a woman identified only as Ann. The younger Ann's name appears on a document signed on December 17 together with that of her aunt, Mary Hardy of Wisbech, Cambridgeshire. Mary had given John £400/£30,000 on the condition that Ann junior be given an income of £30/just less than £2,500 per year. The capital was secured against the Close, but by 1775, John and his mother mortgaged the place for £1,500/£130,000 to James Christian. James was a Manx Christian, born at the family's ancestral home, Milntown. He had made a fortune in business in London, where he was well known as Master of Grigsby's Coffee House in Threadneedle Street, close to the Royal Exchange The respite for Moorland Close was short and troubled. Although James's fiscal spirit was healthy and willing, his physical state was not and in July 1778 he was dead, aged 73.

In 1779 Fletcher's mother and older brothers had to admit finally there was something terminally wrong with their finances.

There were distressing deaths for Fletcher to witness close to home, too. In August 1777, Charles the first child of his brother John and his wife Ann died. A few days after Christmas 1778 their daughter Maria was also buried in Brigham churchyard. After this they seemed to have given up for there are no further records of births or deaths. By 1779 Fletcher's grandfather Jacob Dixon was also dead but this had not made Fletcher's mother Ann absolute mistress of Moorland Close. The dead man had left his interest to his youngest sister, Frances, now married to a Christian family relative Patricius Senhouse of the Fitz, a grand Georgian eight-bay mansion just outside Cockermouth. She expected an income of £20/now £1750 a year. Presumably to ensure such incomes for them both, the two women, aunt and niece, rented troubled Moorland Close to John XVII, whose first wife had also died in 1779, supposedly from measles she caught from nursing her young son but the women were clutching at straws.

On March 9 1779 Ann Christian was *'carried to Carlisle for her son's debts'* (John's presumably) and shortly afterwards the Manx executors of James Christian wrote saying that they wanted their money to pay his bequests to charity. John put his grand house on Market Place Cockermouth up for sale. A large advertisement appeared on the front page of the Cumberland Pacquet on Tuesday October 19, describing its facilities and attractions and it continued each Tuesday until November 9.

Ann's nephew John Christian XVII had to step in and he offered James Christian's executors on Man a moiety of Moorland Close, retaining the other half as his own security for monies he had al-

The Fitz is just outside Cockermouth and here Fletcher's great-aunt Mary lived with her husband Patricius Senhouse and expected an annuity from his mother, which possibly contributed to the bankruptcy of Moorland Close. It's now a grand country-house retreat for up to 24 guests.

ready paid out. The executors agreed in principle as the putative value of the moiety of £750/today £65,000, suited their purpose, exactly half the mortgage having somehow been repaid. Months of niggling followed, for the Manxmen wanted Moorland Close's moiety proved by public auction.

The Letter Book of Charles Udale, painstaking and loyal agent to John XVII and John's letters from Brussels, show their patience with Ann and the extraordinary lengths they took to help her avoid such an auction, which would expose her plight to the public gaze and, apparently, antagonise an already spiky Frances Senhouse over at The Fitz.

Early in 1780 the family admitted defeat. Ann was in debt for the staggering amount of £6,490. 0s. 11d, well over £500,000 in today's values according to the tables of the National Archives. Ann owed money everywhere, so did her elder sons. As well as Moorland Close, everything else had to go and James's executors insisted on the auction.

The moiety of the Close that John XVII had publicly to encash was advertised in the Cumberland Pacquet on March 14 1780 and was to be sold at Cockermouth's Globe Hotel on Monday, April 10. No one wanted a moiety of a bankrupt farmstead, so and John Christian XVII cut further negotiations short by remitting the Manx executors a few shillings less than the money they wanted. He then took over Moorland Close's assets, including their sheep and wood, valued at just over £4,500, two-thirds of the debts he was paying.

For John XVII there was still no end to it. Edward Christian was writing to him, first hoping that some arrangement could be made so that one day *'some of us may be in such circumstances as to think it a desirable object to redeem the place of our nativity, which has been so long in my mother's family but if ever we should have such fortune, I hope at the same time we should have honesty enough to pay off as far as we were able the debts of the farm and restore your loss.'* Edward announces in the same letter that he had just been elected a Fellow of St John's, worth some 80 guineas/just over £7000 a year today, and from that he hoped generously to allow his *'poor mother twenty guineas* (£1800) *a year'*; the catch was that the annuity was actually to come from John and be repaid by Edward.

Ann had £20 a year from some other source (almost certainly John Christian XVII) and Edward goes on to say that with the total of £40 (about £3500 today) his mother would be able to live comfortably anywhere, sharp and urgent advice for she was still threatened with arrest for some of the debts. Edward suggested the family might remove to the Isle of Man *'or any other place, if they could be prevailed upon'.*

John agreed to finance the extra 20 guineas for his aunt, as seen by his account books. Only two years after John had bailed Ann and her family out of debt, she and her eldest son again owed him another £888. 6s. 4d., about £76,000.

With no home left in Cumberland, Ann Christian, now aged 39, went to live in Douglas, Isle of Man, probably in 1779. She took her daughter Mary and her two youngest sons, Fletcher and Humphrey. Charles joined the West Riding of Yorkshire Militia in Liverpool and by May 2 1780, he was marching to Leeds but planning to be a surgeon.

If Fletcher Christian had not known much of his Manx background before, or did not know the long ballad written about Illiam Dhone, one of his most famous ancestors, he would now have learned quickly. In the same year a broadsheet of *The Ballad of Illiam Dhone* was privately printed and widely circulated for political reasons. It was known and widely sung anyway and you would have had to be very sheltered indeed not to know the story, particularly if you were a Christian or related to them. Most Manx were one or the other, or said they were.

Among those on the Isle of Man during 1780 who would also have been reminded of the Christians' long history was Richard Betham, the Collector of Customs, as well as the naval officer who was shortly to marry his daughter, Betsy. His name was William Bligh and he had earlier been based in Douglas for several years. Now, after returning from his voyage to the South Pacific with Cook, he was again in Douglas.

The island was so small that Bligh must have known some of the Christians. This is when he either first met Fletcher Christian or at least heard of him. Among the places where he had stayed in Douglas was the Nunnery, where his hostess was Dorothy Christian Taubman, Fletcher's first cousin. It was her husband who was finally to bring these two men together.

For the moment Fletcher had still not decided to join the Royal Navy. Once, his privileged landed-gentry background guaranteed him a university education, a probable legal career and a future in society, perhaps even in Society. In 1779, banished to the Isle of Man when only 14 or 15, Fletcher Christian was assured of nothing, unless he re-created it himself.

CHAPTER 6

'RIOT... OBSCENITY... DRUNKENNESS... DEBAUCHERY'

In 1780 Douglas was chief port and township on the Isle of Man but mustered few more than 2,500 people yet life was busy with public assemblies for dances, with card parties, dinner parties and tea parties.

The island was in somewhat of a muddle. In 1765 the passing of the Revestment Act meant the British Crown owned the regalities and customs duties of Man, which it had purchased from the Duke of Atholl, the successor to the Derbys. The king of England was now the king of Man but the island's prosperity was smashed because until then the island had fattened itself on the profits of smuggling. When ports and towns were suddenly swarming with British Customs Officers, the Manx saw it as personal betrayal by Atholl.

By the time the Christians of Moorland Close arrived, things were simmering down and Douglas was a port from which the Royal Navy could strike out at smugglers from Ireland on their way to England.

The new family had as a co-citizen Thomas Scott, immediate younger brother of Sir Walter, by whom the author was supplied with the material for Peveril of the Peak. Thomas had planned his own literary work on the subject, but wearied of the task of transcription. The Heywoods had sold the Nunnery to the Taubmans and returned from Whitehaven in 1779 to live on The Parade, the newest and most fashionable part of town. The family head Peter Heywood had once been Second Deemster but now returned as Seneschal for the Duke of Atholl, who retained personal interests on the island. With Peter Heywood Senior were most of his ten living children, including six-year-old Peter, later to sail to the South Pacific with Fletcher in *Bounty*.

Herring fishing and the curing of the rock-hard red herrings (kippers were still to be invented) were the basis of any industry to which the lower classes, no longer smugglers, could apply themselves. During winter their boats were drawn up on the pebbly sweep of beach, for the herring is one of the few fish that is at its peak in midsummer. There was a small Old Pier, a small shipyard and a small brewery and St George's Church was just being completed. Most of the houses were on the flat land and there were then farms and gardens on the hills upon which boarding houses and bed-and-breakfast hotels today cluster.

There was a regular packet service between Whitehaven in Cumberland and Douglas that helped bind the ties with the island and the many families who had members on both sides of the water, the Christians particularly. Excitement for the richer young was confined to one another's company and the occasional great storm in which they might watch men and their boats being lost in the harbour. At the Pier, the Royal Navy's ships were constantly calling on their way to and from chasing recidivist smugglers. Privateers brought in their prizes, recruiting parties and press gangs were about, and riots by the *'uncivilised'* were not unknown.

A regular and important user of the packet from Whitehaven was Fletcher's 1st-cousin Jane Christian, sister both to John XVII and Mrs Taubman of the Nunnery. Jane's diaries show she was constantly at the opera in London, at Windsor, buying and commenting upon fashions, dining with the Duke of Norfolk in St James's Square, seeing the Prince of Wales and the Duke of Clarence at soirees and masquerades at St James's Palace. She regularly joined the knights, bishops, ladies and lords of her acquaintance and her family for teas and dinners of 25 people or more. When the Season in London was over, she travelled by *'easy stages'* from country house to inn to family dwellings all over England, several times eating each of three meals in a different house and sleeping in a fourth. Considering coach fares were equivalently far higher than today's railway prices, the cost must have been staggering but like all her sisters she had a fortune of some £3,000, over £250,000 today. This value is a third less than the same amount her sister Mary had in 1740 but Jane also had income from shares in the ships and mines the family continued to buy and for a long time she and her sisters shared a house in Whitehaven.

Fletcher Christian could not have avoided knowing what John, Jane and others of the Ewanrigg, Netherhall, Fitz and Milntown families were doing and, even if he was not taking part, he continued to share association with a privileged background and connections entirely different from anyone else who sailed in *Bounty.* What we don't know is if he was still at school and where this was on the Isle of Man; he was 15 in 1780 and there was little compunction or precedent for being at school at all.

In *The Life of Vice-Admiral William Bligh* by George Mackaness, Fletcher Christian is said to have been aboard HMS Cambridge in 1782-83, during which Bligh was the vessel's sixth lieutenant. Assiduous searching and diligent cross-checking did not reveal Fletcher's name on the muster sheet of *Cambridge* and so we currently do not know what Fletcher did between arriving in Douglas in 1779 and going to sea in 1783; perhaps he was continuing his education but we do not know.

Six months after Isabella Curwen secretly married John Christian XVII in 1782, 18-year-old Fletcher signed on as a midshipman, the entry-level rank of officer, aboard HMS *Eurydice*, Captain George Courtney. The date was April 25 1783, and the ship was moored at Spithead, the Royal Navy anchorage in the mouth of the Solent close to Portsmouth.

Joining the Royal Navy was the finest and only way Fletcher could acceptably recover and improve his social position and expected future in Georgian society. The Royal Navy was Britain's greatest defence, its fabled Wooden Wall against all comers. Here, ability was noted and rewarded unlike the Army, where you could buy yourself in, up, down or out whoever you were and the Duke of Wellington said at least three-quarters of army commissions were so filled. Naval officers were highly trained specialists with enormous responsibility for the king's ships and all who served on them. Unlike the Army, an officer of any background could be dismissed from the Navy for incompetence but he could also go all the way up to being a full admiral and be accepted socially.

Joining the Royal Navy was the finest and only way Fletcher could acceptably recover and improve his social position. . .

The anonymous writer of *An Inquiry into the Present State of the British Navy . . .* (1815) said: ''*It is no disgrace to the post captains of the English Navy who have many lords amongst them, that these are also the worthy offspring of tailors, shoemakers, farmers, ale-house keepers, sailors, pilots, haberdashers, milliners, in fact every calling under the sun.*'

Here was Fletcher's way to recover what had taken from him. If he discovered he had the ability required, some of which would have been proven in his education and helped his appointment, he had begun a certain climb towards security and position.

Eurydice was a 6th rater, with twenty-four 9-pound/4kg guns and although built at Portsmouth only two years before was of an almost archaic design. She alone of the many types of ship in the Navy could still be manoeuvred by rowing. She was 114 feet/35 metres long and 32 feet/9.75 metres wide, but had oars 113 feet/34.5 metres long, each manned by a team from among her 140 men, 25 of whom were marines to keep order and discipline.

When Fletcher joined, the ship had just returned from the West Indies, where in the previous October she had captured the French brig *Samea*. Refurbishing meant she was not going to sail until October 1783 and during the six months Fletcher was first aboard, he would have seen much of the worst aspects of naval life. Although officers and young gentlemen had some shore leave, most of this time would have been spent anchored off the port in the waters of Spithead. Able and ordinary seamen, guarded by armed marines, did not get shore leave, hence the custom of allowing women on board. Among the most enlightening pages of Ian Ball's *Pitcairn, Children of Bounty*, are those that illustrate this vicious aspect of naval life. Ball quotes a document privately printed in 1821 and submitted to the Lords Commissioners of the Admiralty. It was entitled *Statement Respecting the Prevalence of Certain Immoral Practices in His Majesty's Navy.* It said:

It has become an established practice in the British Navy to admit, even invite, on board our ships of war, immediately on their arrival in port, as many prostitutes as the men, and in many cases, the officers may choose to entertain to the number, in the larger ships, of several hundred at a time; all of whom remain on board, domesticated with the ship's company, men and boys, until they again put to sea.

The tendency of this practice is to render a ship of war, while in port, a continual scene of riot and disorder, of obscenity and blasphemy, or drunkenness, lewdness, and debauchery. During this time the married seamen are frequently joined by *their wives and families, (sometimes comprising daughters from ten to fifteen years of age) who are forced to submit to the alternative of mixing with these abandoned women, whose language and behaviour are usually of the most polluting description; or of foregoing altogether the society of their husbands and parents. These all inhabit the same deck, where whatever their age or sex or character, they are huddled promiscuously together, eating, drinking and sleeping without any adequate means of separation or privacy for the most part without even the slightest screen between their berths; and where in the sight and hearing of all around them, they live in the unrestrained indulgence of every licentious propensity which may be supposed to actuate inmates of this description.*

A Naval Officer of large experience asserts that from the time he entered the Navy, about twenty-eight years ago (1787?), he had served in no ship in which, while in port, the custom of permitting women of the very worst description to come and remain on board was not tolerated, and even encouraged, by the Commanding Officer. The Lieutenants and grown Midshipmen were obliged to sit at table and associate with them and to be witnesses to the debauchery and indecency which took place, not only there but among the men also. It was even common for the women to employ all their arts to debauch these youths who generally were caught in their snares, and became their prey.

The authors *'boldly and confidently'* attacked the normal defence of the toleration of prostitutes aboard His Majesty's ships, a cry that if women were not permitted the men would turn to one another, the penalty for which was death. The authors said, *'if there really exists a danger of this kind in the Navy, it arises more from the very practice which we have been reprobating . . . What can be more unnatural, more contrary to all the feelings of our common nature than the open, undisguised, unblushing promiscuous concubinage which now takes place. Is not the person . . . tutored in this school of impurity and licentiousness . . . less likely than others to shrink from any other abomination which may be suggested to his mind?'*

Many, if not most, of the ordinary sailors on *Eurydice* wore pigtails, and those who could not grow one would pin on a false tail of teased oakum, the fibre used for caulking, which was made by unravelling rope. Some of the prettier or more forward men affected curls or lovelocks over their ears, like women of high fashion, and many wore gold earrings. These were part of a sailor's superstition, supposed to improve eyesight as well as to ensure a burial if a body was mercifully washed up on a Christian shore. Some captains forbade them on the grounds that they were un-English.

Each man had two hammocks, one in use and one clean and they had to be scrubbed of the inevitable infestation once a week. Those not members of a watch had to be out of their hammocks by 7.30 am. If not, the hammock was cut down.

The horror of the food is worse than most imaginations permit. Ship's biscuit, euphemistically called bread, was the most liberal ration. They were cooked in the royal bakeries attached to the dockyards at the rate of 70 a minute. Each biscuit was about 5 inches / 12.5 centimetres in diameter, thick, brown, well cooked and stamped in the centre, which was therefore even harder than the rest. They should have been made of wheat flour and some pea flour but bone dust was thought often to have been added. It is likely. In an effort to please the lust for ever whiter bread, 18th-century bakers commonly added alum, which gave a harsh bitter flavour. They also mixed in chalk, lime, poisonous white lead, animal bones, and, it was charged, bones from the charnel houses. Those who had some sympathy for their customers included jalap, a purgative Mexican root, to counteract the costive effect of the alum.

. . . men methodically thumped their biscuits onto the table before each mouthful, hoping to dislodge what wildlife they could . . .

The pea-flour content of the biscuits never blended evenly but clumped together in threads or lumps of incredible hardness that could not be bitten through until age had softened the biscuits. When old, they developed a musty sour taste and became the perfect breeding ground for weevils and maggots, which in hot weather multiplied to an unspeakable degree. You could ask the cook to rebake the biscuits but it was more common to eat them in the dark, when the eye and the stomach were less consciously assaulted. It is said an enduring memory of naval service was the dull thumping rhythm of mealtimes that filled this darkness, as men methodically thumped their biscuits onto the table before each mouthful, hoping to dislodge what wildlife they could. If you bribed the cook, and it paid to, he might sometimes pound the biscuits with sugar and pork fat to make a sort of cake.

The salted beef and pork were worse and there was not even the brief pleasure of enjoying it relatively fresh at the outset of a voyage. The naval custom was that old meat had to be eaten first and the ship's company could expect to finish barrels of salted meat returned from other ships or which had been turned up in some forgotten corner of a warehouse. After several years in salt, even the best meat is unwholesome, shrunken and stone hard. There were wholly believable rumours that the mahogany-coloured stuff was horsemeat and Negroes who walked near victualling yards were said to disappear. Beef was worse than pork but sailors could carve curios out of either and both are said to have taken a good polish. Each four-pound/1.8 kilo piece of 'meat' was likely to be mainly bone, gristle and fat, and the provisioners packed the casks short anyway, so that a sailor's measure was said to weigh about three-quarters what it should. From time to time a man might ensure a better serving through trade and hoarding his ration over several days.

Breakfast was supposed to be burgoo, or porridge, but even a naval surgeon had said it was cruel to expect men to eat it. Most was given to the pigs and other animals on board, for the combination of bad oatmeal, filthy ship's water and the inattention of the cook made it wicked stuff. You could elect not to take this or any other ration and draw money or save credit for purchases in port. As an alternative, burnt ship's biscuit was boiled in water to make a thick, dark paste sweetened with sugar if available and called Scotch coffee.

Raisins and currants and flour were carried among the provisions and the messes would be given these in lieu of meat one day a week to make those boiled duffs and puddings that are still favoured by the English. Other substitutes might be rice instead of biscuit or equally foul cheese, oil instead of cheese, sugar instead of rancid butter. Each mess of six to eight men shared one table which was hooked to the beams when not in use and fixed between the guns when it was.

It is easy to see how the alcohol that was so liberally distributed was important to fill the stomach, to give some suggestion of nutrition and to dull the senses and any sensitivity that might be left. While it lasted, seamen were allowed a gallon/4.5 litres of beer a day, which was sometimes as little as 1% alcohol, then a pint/600mls of wine. It was just as well because the water carried quickly became slimy, foul with algae. Curiously most seamen preferred white wine to the more robust reds and service in the Mediterranean was unpopular because they got only red. They were also issued grog twice a day, which was one part of extra-proof rum to three of water, a UK pint/600mls of the mixture per man and at noon lemon or lime juice was added to help prevent scurvy. It was common to save the rum allowance for regular and painful binges, especially at Christmas time. Drunkenness was common and it was not thought unusual if one-third of a ship's company was less than capable of an evening.

For the new men and boys aboard *Eurydice,* the six months they spent afloat in Spithead would have avoided an early introduction to the worst of the food and drink. In port, captains were more or less obliged to supply fresh meat, vegetables and soft bread. Both in port and at sea the officers and young gentlemen fared only slightly better than the crew. Considering the waste products of the men and animals aboard each ship would go directly into the sea, with the hope tides would clear and clean, even rowing to and from a Royal Navy ship would not be a delightful interlude of fresh sea air.

Eurydice sailed for India on October 11 1783 to help mop up a battle with France. Fletcher Christian had turned 19 on September 25th and probably had enough sense and good advice to have laid in a store of treats for himself. Even then Fortnum and Mason at 181 Piccadilly, London, was recommended. Yet he was soon likely to be looking forward to eating biscuits and salt meat in much the way more experienced sailors did, for just as the free sex life in port brutalised libidos, naval food dulled the palate and its expectations.

Between September 1774 and September 1780, 175,990 men had been raised for the navy. Of these, 42,000 had 'run', leaving the number of men who served at 134,000, in round figures. Of these, 1,243 were killed in action but an appalling 18,541 were lost through sickness and disease, more than one in every seven, an average of 60 men a week for six years. In 1803, Lord Nelson said the life of an average naval man was finished at 45.

Able seamen were paid 6 shillings a week, about £25 today. The ordinary seaman had 4/9d, about £20, but deductions for clothing, extra food and support of a family at home easily halved this. Although the men often got a month or two's pay before a long trip, the balance was usually held in arrears, often for two years or longer. Even if the ship was paid off, they might not get all they were owed. The rates of pay and the inefficient system of payment had not changed since the days of the Commonwealth over 100 years before.

Fort St George in Madras was the East India Company's first fortress in India. These buildings are now incorporated into the headquarters of the State of Tamil Nadu.

CHAPTER 7

'A SUPERIOR PLEASANT MANNER'

Eurydice sailed for India via Madeira and Cape Town on Table Bay, at the southern extreme of Africa. On November 16 1783 Fletcher crossed the equator for the first time and David Laird, the master, noted, 'In order to celebrate the joyful event of crossing the Equator, the ship's company performed the ceremony of paying tribute to Neptune according to the usual custom of dunking into the sea those who had not crossed the line before'. This was always dangerous and it was forbidden by the time Fletcher Christian again crossed the equator southwards with William Bligh in Bounty.
The ship called at the spice-trading port of Tellicherry on India's western Malabar coast and by June had rounded the continent and was in Madras, now Chennai. There, Captain Courtney noted the arrival of a big East Indiaman called Middlesex. Fletcher's brother Charles was later to serve as surgeon in this ship and what happened when he did so could gravely have affected Fletcher's attitude to naval discipline.

Madras was the East India Company's first Indian fortress, started in 1664 on what had been deserted swampy land. By the 1780s it was a large town dominated by the magnificent, buildings of the company's offices, which were surrounded by walls and a ditch on the sea-shore. It was rather a jumble of irregular shape but any houses inside were only inhabited during siege. The countryside around the town was fragrant with the gardens of the white residents that were reached by a handsome brick bridge over the river south of the fort. The fort and its European environments were known as White Town and the native quarter, which was called Black Town, presented a spectacle of dirt and filth. None of the streets was anything but black earth, which mixed with the wasted fluids of men, animals, river and heaven to form stinking mud that engendered infection and allowed free passage only to the whimsical and awkward local carriages, drawn by two oxen. Perhaps it was the viscosity of the mud that required the palanquins of the rich to be carried by six slaves, rather than the four seen elsewhere in India.

Madras-cloth of cotton was hawked to visitors, including very fine handkerchiefs of large checked pattern and original colouring based on natural dyes, now very collectible. The Indian ships in the harbour were *parias*. They had masts of teak and rope of coconut hair and were extremely cumbersome, weighing up to 600 UK tons/610 tonnes even though only used for coastal sailing. An English captain of the time said, *'If their hull be defective, the manner of rigging them is not less so'.*

Fletcher's time in Madras was more than an exotic experience. On May 24 1784, after just seven months experience at sea and four months before his 20th birthday, he was promoted to acting lieutenant and subsequently entrusted with a watch for the return voyage. There can be no doubt, then or now, that he was not just a young gentleman in the right job but very good at it. Amongst more than 100 sailors on board *Eurydice* he had been recognised as superior.

In November the ship joined Sir Edward Hughes' squadron and sailed back to Bombay, now Mumbai, on the west coast. There, salt water was distilled into fresh water and the squadron headed back to Woolwich spending four weeks in Table Bay, where the only fresh water available was contaminated with whale oil. After three weeks at St Helena and a night at Ascension they were home by June 1785 after almost two years.

When Fletcher's brother Charles Christian was in India in *Middlesex* in 1786, he met Thomas William, who had been surgeon aboard *Eurydice*. After Fletcher's return from India Charles also met another unnamed officer who had sailed with Fletcher. Both these men corroborated Fletcher's own assertions about his abilities, saying that Fletcher was said to be strict, but *'ruled in a superior pleasant Manner'.*

Fletcher's other brother, Edward, also mentions *Eurydice* in *The Appendix* he wrote to Stephen Barney's *Minutes of the trial of the Bounty mutineers.* He says that when Fletcher returned to Woolwich he was met by a relative, whom he told he had been extremely happy under Captain Courtney's command, proud he had been entrusted with a watch all the way home. Edward, like Charles, gives a quotation supposedly from Fletcher's own mouth: *'It was very easy to make one's self beloved and respected aboard a ship; one had only to be always ready to obey one's superior officers, and to be kind to the common men, unless there was occasion for severity, and if you are when there is a just occasion they will not like you the worse for it.'*

This last observation is of keen importance in Fletcher Christian's future and is also agreed unreservedly by everyone who has sailed, whether or duty or for pleasure.

Fletcher's *'pleasant rule'* and the respect of the men aboard *Eurydice* was not much use to him, for peace had been concluded and in just a few days he had been paid off. Fletcher made up his mind to be appointed mate of a West Indiaman, feeling he had the necessary qualifications. It's fair to presume it would also be a young man's idea of adventure yet would not keep him away from home for years at a time or put him in danger of a sea battle. Fletcher turned to friends and family influence, because that's what you did. It was the Season in London and most useful people would have been there. Through his own endeavours or an introduction, Fletcher was treating with a merchant in the City when Captain Taubman arrived from the Isle of Man. His wife, Dorothy, Fletcher's first cousin, had died 18 months previously in January 1784. Taubman obviously felt an obligation and said he would write to William Bligh, a man whom he knew well, who owed him a favour or two and who I think already knew of Fletcher. Taubman said it would be very desirable for Fletcher to serve under so experienced a navigator as Bligh.

Bligh returned a polite answer, saying he was sorry he could not take Fletcher Christian, as he had his full complement of officers. Edward Christian later wrote that Fletcher Christian replied, prophetically observing to Bligh: *'wages were no object, he only wished to learn his profession and if Captain Bligh would permit him to mess with the gentlemen, he would readily enter the ship as a fore-mast man until there was a vacancy amongst the officers. . . we midshipmen are gentlemen, we never pull at a rope; I should even be glad to go one voyage in that situation, for there may be occasions when officers are called upon to do the duties of a common man.'*

Under the terms suggested, Fletcher did sail with Bligh but not for 15 months.

The Bligh family comes from Cornwall and in June 1967 I went there to make a television documentary about William Bligh, their most famed son, commander of the mutinous *Bounty*, later Governor of the restive colony of New South Wales. After a baffling search for the exact place of Bligh's origin it seems that his father, Francis, was born in St Tudy, and that William was born in Plymouth, at least he was baptised there. In the family Bible, his birth date is September 9 1754 and the church of St Andrew in Plymouth records his baptism on October 4.

William's father Francis had a developed taste for the wives of dead men, marrying a series of widows. The first of these was Jane Pearce, who was William's mother. She died before he was 16 and he was the single child of this marriage. The power of the movies is such that it is constantly necessary to reiterate that the picture of Bligh as an upstart red-neck who rose to position from before the mast is wrong. This assassination of character had its beginnings at the end of the 18th century and the Christian family cannot be said to be entirely free of involvement. Even so, many of the charges still brought against Bligh are difficult to disprove, for if his father attracted widows, William Bligh attracted dispute. As well as the mutiny aboard *Bounty*, he was court-martialled and convicted of using abusive language to a junior officer, was involved in the fleet-wide Nore mutiny, and was also ignominiously removed from his governorship of New South Wales, after another mutiny. He was, and is, controversial. His naval background was impeccable but, although he had every possible degree of education to suit him for the career, he had no advantage of patronage that was not dearly earned through personal endeavour or slavish devotion to duty.

William Bligh first appears in the Admiralty records on July 1 1762, aged seven years and nine months, as servant to the captain of HMS *Montrose*. Being registered at such an early age was a ploy to ensure rapid promotion later in life. A man had to serve at least six years at sea before he could qualify as a lieutenant. If his family had friends on the quarterdeck, he would be entered early on a ship's books and would be serving his time while actually labouring over copy books and eating cosy teas with mother in the nursery. It was much more unusual for Fletcher Christian to have waited until he was 18 to go to sea than for Bligh to begin putative service aged seven.

Promotion at the other end of a Royal Navy career promotion was equally illogical. Once a man had been promoted to post captain, exertion or appearance at sea became academic. Progression up the charts through the degrees of admiral was automatic, as those at the top died one after the other. Bligh's eventual rank of Vice-Admiral was not reward for skill at sea or any special achievement but because he had lived longer than others.

As to there being servants aboard, an Admiral of the Fleet on a large Royal Navy ship was entitled to an entourage of 50. A mere captain was allowed four servants per each 100 men of his ship's complement. Other officers, from lieutenants to cooks, surgeons, carpenters and boatswains were allowed one or two servants according to the size of the vessel in which they served. Often these servants were exactly that, hairdressers, tailors, footmen and artists. Many were also serious young men under that officer's direct patronage, boys who planned a naval career and who were starting the best way of all, under the special and noticeable protection of a single officer. It was a two-way affair. The patronage you gave helped your career and social life quite as much as the patronage you received. William Bligh was always acutely aware of this but it was years before he could use it to his advantage.

Young Bligh's supposed service aboard *Montrose* lasted six months but every little bit helped. It was 1770 before he was aboard *HMS Hunter*, a 10-gun sloop on which he was rated as able seaman but almost certainly served as a midshipman, an everyday bending of the rules. Fletcher Christian was offering to do it under Bligh's command years later.

Bligh was 15 when he joined *Hunter* after his mother was dead. Mackaness suggests that the legendary problems of stepmothers and the feeling of desertion by his father, so easily aroused in the breast of the lone adolescent, may have been the reasons for Bligh now going to sea. His nominal rating as AB lasted only six months and on February 5 1771 he was appointed midshipman. In September of the same year he was transferred to the 36-gun *HMS Crescent*, serving aboard her for three years until August 23 1774, and it is likely that this is the time during which his naval skills were honed

On September 2 1774, William Bligh signed on as an AB in HMS *Ranger*, there being no place for a 'middy'. *Ranger* was based at Douglas, Isle of Man, her duty to hunt and apprehend smugglers in the Irish Sea, under the eye of Richard Betham, who was HM Collector of Customs. It was an assignment offering neither spectacular career opportunities nor adventure. The ship spent most time being repaired, so Bligh sensibly capitalised on inactive service by improving his connections ashore, equally important for naval advancement.

As he approached his coming-of-age, William Bligh presented a striking, even beautiful appearance. For a man, he had skin of a surprisingly lambent, marble-whiteness that was commented upon most of his life. To that was added the startling combination of piercing blue eyes and raven hair. His face was quite broad, but in proportion to the length of his head and he had a nose as noble and fine as his skin demanded. The firm set of his chin was shadowed by a dark beard and contrasted strangely with lips that had never grown out of their childish cupid's bow. All in all, it was an appearance that demanded attention but it was to be no advantage.

By the end of September 1775, a few weeks after turning 21, Bligh was re-entered on the books of *Ranger* as a midshipman, a post he held until March the following year. His 18 months of duty on and about the Isle of Man were very useful. The Isle of Man was only a small fish in the seas of Bligh's ambition but he eagerly baited and secured his hook. His father was probably able to give an introduction to Richard Betham, being a Customs man himself, and Betham's daughter Elizabeth seems to have been a willing and enjoyable social partner. Their relationship was interrupted because Bligh was appointed master of *Resolution* and was to sail to the South Seas with James Cook, his idol and touchstone. The voyage sharpened Bligh into a great but flawed sailor, permanently affected by an inability to understand other men's views, making him a pusher rather than a leader of others.

It was not always a bad trait when on the open sea and in danger.

CHAPTER 8

THE MERCHANT SERVICE

Being appointed master of the flagship of James Cook's third voyage to the South Pacific is a clear signal of Bligh's professional superiority. The position of master aboard a ship was senior and responsible, requiring a man of great seamanship and above average ability in navigation. Both attributes were certainly true of William Bligh. It seems likely he was taken from his Manx obscurity because Cook saw journals or charts prepared by Bligh and chose him from this evidence of ability. At the time, Bligh had just sat and passed his lieutenancy exams. His certificate is dated July 1 1776, and states: *'He produces Journals kept by himself kept in the Crescent and Ranger, and Certificates from Captains Henshaw, Morgan, Thompson and Lieut. Samber of his Diligence, Sobriety and obedience to command. He can splice, Knott, Reef a Sail, work a Ship in Sailing, Shift his Tides, keep a reckoning of a Ship's way by plain sailing and Mercator, observe by Sun or Star, find the variation of the Compass, and is qualified to do the duty of an Able Seaman and midshipman.'*

From Bligh's many charts, journals and letters that survive, we know that he was gifted both in powers of observation and the ability to set what he saw on paper in words or pictures. The journals he kept as a midshipman, part of the training in observation required of all young gentlemen, were bound to have been exemplary.

Bligh's voyage with Cook, during which the Hawaiian Islands were visited and Cook was murdered, was both the final ingredient and the crystallisation of the complex man who was William Bligh. Cook's influence can be seen in so much of what was to come that Bligh's time with him must have been filled with hero worship and adulation, which led to a sense of severe personal loss on his death.

It was Cook who introduced new ideas of shipboard hygiene and food to the Royal Navy and in 1776 he was elected to the august Royal Society, awarded their Copley Gold Medal for his paper on the subject. Bligh was one of the few who adopted his ideas, perhaps because he had seen them employed to his own advantage when sailing under Cook. There was also an ugly side of Cook that Bligh seems to have copied. Cook was renowned for his vile temper and extraordinary cruelty. He flogged men more often and for less than Bligh ever did, yet inspired a devotion that bordered on veneration. Bligh was equally famed for his outbursts of invective, but his personal abuse was wounding in a way Cook's foul language never was.

As master, Bligh led the expedition home after Cook's death, helped immeasurably by having Kendall's K2 chronometer aboard, making it possible to be certain of longitude. Once back in the UK, Bligh said Cook's death might have been averted if men on the beach of Kealakekua had done their duty, or had been of the calibre expected of Royal Navy officers. He made no bones about the cover-ups he thought were employed to smudge the facts of Cook's murder as he saw them and he was bitter because charts he made were attributed to other hands.

Being the only officer to air such different opinions meant that in 1780 he was mortified to be the only man excluded from the general promotion of officers on the voyage. Passing the lieutenant's examinations did not automatically give a man this rank. There were many who continued to serve for years as midshipmen before being commissioned into full service and given the leg up towards being post captain. Some men who passed their examinations never became lieutenants.

Bligh thus returned from his first South Pacific voyage clearly demonstrating the basis of his many future downfalls. Confident always of a faultless personal morality, he thundered at the weakness of others and the weapon he invariably mounted was that of an acerbic and wounding tongue and was then surprised and hurt when he was not rewarded for his views.

Experienced, travelled, confident but snubbed, he returned to life on land, first in the West Country and then in the Orkneys, where a family called Stewart had welcomed him when *Resolution* had called. Eventually he returned to the Isle of Man, where he would quickly find himself at home again with the Taubmans, Christians and Bethams and with a wife

Elizabeth Betham Bligh has had nothing but good said about her. Indeed, it is hard to fault any naval wife of the time who remained loyal, domestic and fruitful. She was her father's second daughter, educated and vivacious, befitting the family's background of cultural connections with Scotland. By January all was arranged. Both aged 27 they were married on February 4 1781, at Kirk Onchan, just outside Douglas. Ten days after his marriage Bligh was called back to active service as master of *Belle Poule,* a captured French ship that had been returned to service against her former proprietors.

The battle in which he participated at Dogger Bank early in August was enough for him finally to be promoted to lieutenant. As such he was transferred briefly to *Berwick*, then as fifth lieutenant aboard *Princess Amelia*. In March 1782, he went to HMS *Cambridge* as sixth lieutenant and stayed aboard her until January of the following year. This ship took part in the relief of Gibraltar and Bligh was paid off in January 1783.

Once more Bligh returned to Betsy and the Isle of Man, where on half pay and expecting some overdue new appointment that might prevent them settling permanently, they moved into lodgings. By

May they were expecting their first child and moved into a house of their own, while Bligh was writing everywhere for an appointment, hoping his wife's family might help.

Elizabeth's mother was a Campbell and included among her relatives an Uncle Duncan. Duncan Campbell was an influential and a rich man, a merchant trader and a plantation owner in the West Indies as well as the proprietor of convict hulks. He was also involved in the slave trade but few as yet had conscience about such connections and William and Elizabeth Bligh were grateful for the interest of a man who seemed to offer more open doors to advancement than anyone else they knew.

Campbell advised Bligh that to hedge his bets he should obtain official permission from the Lords of the Admiralty for leave to quit the kingdom. That way he might be able to join the merchant service, for Duncan Campbell, ever the opportunist, saw in this nephew-in-law someone to whom he might entrust his ships and his business in the West Indies. Offered £500 a year, about £43,000 today, William Bligh was happy to comply. He moved his family to 100 Lambeth Road in London, close enough to the Admiralty so he could show his face as often as he wished.

Bligh successively commanded *Lynx, Cambria* and then *Britannia* for Campbell. Apart from his duties aboard, he was responsible for finding cargoes to be carried home, in competition with others. His letters to Campbell are a constant apology for bad business and lack of contacts, complaints about the weather and a retailing of misfortunes, as well as the usual affirmations of friendship and duty and a listing of compliments that had to be given. Even in the verbose and emotional context of 18th-century correspondence, the letters of Bligh always seem to slop over the boundaries of taste and convention and into self-pity and unctuousness, a style he never changed.

Fletcher sailed to the West Indies under Bligh in Campbell's ship *Britannia*. In April 1786, Bligh was still commanding *Lynx* and in Caribbean waters, so Fletcher was certainly not with him then. There are no letters I know of from *Cambria* that might fix the dates of that command but there is a letter dated November 30 1786 in which Bligh writes to Campbell about his great difficulties getting down the Channel at the start of another voyage. I think this is likely to be a letter from *Britannia*, at the beginning of the first voyage Bligh and Christian made together.

The sailing time to and from the West Indies was four to six weeks each way and allowing about the same time for trading once there, an average round trip can be reckoned to take three to four months. There would have been a turnaround time in England, too. Bligh and Christian sailed on two voyages in *Britannia* and at the end of the second returned to England in August 1787.

After Bligh had returned from his second breadfruit voyage to the South Pacific in 1793, he made a note that Fletcher Christian had been with him in the merchant service for three years. This is impossible. The maximum could have been only two years and on all other evidence, including Bligh's, the period was probably nine or ten months for the two voyages. Bligh's memory often played tricks to his own advantage. However long it took, Fletcher Christian was eventually aboard *Britannia* with Bligh, a man just ten years older than himself and with whom he could talk at length about so many things, the Isle of Man and his family in general, his naval experience in particular. He sailed on the first voyage as an ordinary seaman but ate with the midshipmen and officers. This was the spirit Bligh appreciated and it was no bad thing to have a member of the influential Christian family in his debt.

Fletcher sailed to the West Indies under Bligh in Campbell's ship Britannia.

Life in the merchant service was much easier than in the Royal Navy. Discipline was important for everyone's safety but flogging and food were all less severe than in the king's service. Edward Christian tells us that, when Fletcher Christian returned from his first voyage with Bligh, he said he had shared the labour of the common men but had also been helped by Bligh, who had shown him the use of charts and instruments, furthering Fletcher's knowledge of navigation. Fletcher added that although Bligh was very passionate, which presumably means temperamental, he prided himself on knowing how to humour him. So far, so good.

When we get to the scant knowledge we have of the second *Britannia* voyage, stories begin to diverge. Edward says Christian was taken out as second mate, that when he returned he did not have enough time to see any relatives but that in letters he had made no complaint about the captain.

Lawrence Lebogue, who sailed with the men both to the West Indies and to the South Pacific, was later asked if Bligh's treatment of Christian aboard *Bounty* had been the same as on *Britannia*. He answered, *'No, it would not long have been borne in the merchant service,'* suggesting Bligh's favouritism aboard *Britannia* swung in the other direction on *Bounty*.

Lawrence Lebogue was an illiterate sailmaker, possibly from Annapolis, USA, and at 40 was one of the oldest men who later signed on in *Bounty*. When Edward Christian and William Bligh were sniping at each other in pamphlets in 1794, Lebogue told Bligh he had never made the statement that Fletcher could not have borne Captain Bligh's conduct much longer, because, *'I knew Captain Bligh was always a friend to Christian when he sailed with him to the West Indies as well as afterwards . . . Captain Bligh was the best friend Christian ever had.'*

Lebogue had never been accused of saying that Christian could not bear Bligh's treatment aboard *Bounty*, he simply said he was treated differently aboard *Britannia* and *Bounty*. This confirms two points. The first, that men in the merchant service who were not serving under armed guards and fierce regulations, expected better treatment and had it in their power to demand it, or to take some kind of redress, without being hanged for mutiny. It also shows that Bligh behaved differently when employed by the king and perhaps Fletcher Christian did, too.

Edward Lamb, had been *'in the ship Brittanic [sic] when I was chief mate and eyewitness to everything that passed.'* He first refutes Edward Christian's assertion that Fletcher was second mate, saying he was put upon the articles as gunner but that Bligh had wished him to be thought of as an officer. This confusion may, of course, have been caused by Fletcher Christian, wishing his relatives to think his progress was rather more grandiose than it was. It is as likely that it was simply Edward Christian getting mixed up between Fletcher's actual position and the duties he was expected to do. Mr Lamb continues about Bligh: *'When we got to sea, and I saw your partiality for the young man, I gave him every advice and information in my power, though he went about every point of duty with a degree of indifference that to me was truly unpleasant; but you were blind to his faults, and had him to dine and sup every other day in the cabin, and treated him like a brother in giving him every information.'*

Edward Lamb goes on to give a tantalising extra picture of Fletcher Christian and his relationship with women as, *'then one of the most foolish young men I ever knew in regard to the sex'.*

Only one thing is certain about the *Britannia* voyages. Bligh and Christian were firm friends and well pleased with one another. In my view, because Bligh had sired only daughters and Fletcher's father died when he was less than four, it seems like an ideal mentor / student relationship, a substitute father / son relationship.

HMAV BOUNTY is believed to be the first Royal Navy ship to sail for neither warring, ruling nor exploring purposes but to profit from these. She carried an unprecedented crew only of voluntary sailors, which might have contributed to the later mutiny in the South Pacific

BY *BOUNTY* TO TAHITI

December 1787 to October 1788

The breadfruit of Tahiti was hoped to solve the problem of feeding the slaves on Britain's sugar plantations in the West Indies. When it finally arrived, the slaves largely refused to eat it

CHAPTER 9

BOUNTIFUL BREADFRUIT

Of all the breath-taking things you were expected to believe about the South Seas when the first explorers returned in the mid-1700s, two were spoken of particularly. The first was whispers of sexual licence and the second, more suited to the drawing-room, was breadfruit, the bread that grew upon trees. It must have seemed unfair that the British, unquestionably the most civilised of Christian nations, had to labour and pay for their daily bread, whereas indolent pagans of the South Pacific were given it *gratis*, just as the Lord's Prayer suggested it should.

The Standing Committee of West India Planters and Merchants was particularly interested in the virtues of breadfruit. While making their fortunes out of sugar plantations in the Caribbean, these men were also put to the unfortunate expense of feeding the slaves from Africa who did the work, the natives of the area having been exterminated for their hindrance of progress and white men's fortunes.

Bananas served the purpose much of the time but storms easily destroyed these delicate plants, which are herbs not trees as they have no true trunk but are supported by tightly rolled leaves. This meant the tiresome trouble and expense of buying and storing grain and other foodstuffs, particularly from the British colonies of the east coast of America.

In 1760 the Society of Arts offered a premium to encourage cinnamon production and then for the rest of the 18th century emoluments were extended to such other tropical products as cochineal, silk, indigo, cotton, cloves, camphor and coffee. Indeed, the whole world was being combed for botanical specimens that would benefit civilisation and specimen gardens were established in the West Indies to complement Kew, Chelsea and Edinburgh.

It did not take the Standing Committee long to realise the possibilities of the breadfruit tree. Breadfruit, *artocarpus altilis,* is one of the highest yielders of all food sources. Known as *'uru* throughout the Pacific, a mature tree will commonly provide a harvest of 200 kg of fruits a year. There are 100s of varieties, many fruiting twice a year, with ripe fruits varying from as little as 500g to 6kg. These can even be eaten unripe, when they are baked to give something that tastes a little like baked potatoes or bland bread. Nutritionally it is unbalanced although a good source of Vitamin C. Broadly, 70% is water, about 25% is carbohydrate (energy) but only 1% is protein, so it would give slaves energy but nothing to build or keep muscle strength unless supplemented with meat, cheese, milk or eggs, not that this would have been understood at the time.

It seemed as though a mature tree might feed a tractable native for a whole year, costing nothing. As well as being of a tropical disposition, breadfruit trees were strong and hardy and so could resist Caribbean hurricanes. The wood was useful and the sap exceptionally good at waterproofing and thus as a caulking on sailing ships. The idea was intoxicating, so much so that in 1775 the Committee soberly announced they were prepared to pay all reasonable costs of a captain willing to transport breadfruit plants from Tahiti to the West Indies and even suggested how it could be done. There were no takers and anyway, on July 4th the very next year, the American colonies took it into their heads to be independent. The breadfruit offer was swamped by martial news until it was noted that provisions from the ex-colonies for the slaves of the Caribbean's sugar plantations were becoming increasingly expensive.

The transportation of breadfruit became urgent and as soon as the Atlantic had resumed a semblance of normality on both shores, the Standing Committee got to work in London, starting what would now be called a PR campaign to convince king and country that their needs were of national importance and that the solution should be funded out of public monies. They, being nothing but poor farmers and traders, and having had time to think about their first offer of standing the costs, could look only to their government for salvation.

Being men of the world, the Committee knew the way to the king's purse was through his advisers and set about flattering Sir Joseph Banks Bt President of the Royal Society and the king's chief but unofficial scientific adviser on everything about the South Pacific.

As plain Mr Banks, he had been on Cook's first voyage to the South Pacific and in 1769 had gone so far as to eat breadfruit in Tahiti. Nothing then happened concerning the South Seas without the inclusion of Banks, created an hereditary baronet on March 24 1781.

In early 1787 Banks heard the news that breadfruit plants to the value of £2,000/about £155,000 today, had arrived in the French West Indies, adding fire to English West Indian complaints that their government was interested only in taxing them. Initially a ship that had first transported felons to Botany Bay was to collect plants from Tahiti and David Nelson, a botanist, was put in charge of plans. By March, Banks had changed his mind and thought an independent vessel, fitted under his supervision, would better suit the purpose. King George III was enchanted and quickly gave his blessings and orders, which he signed on May 15 1787. The Admiralty Lords were not pleased. They had war on their minds yet, as the king was more than a little involved, they supplied ample funds but little time or thought, dooming the expedition from the start.

Lord Sydney, one of the Principal Secretaries of State, wrote to the Lords Commissioners of the Admiralty on May 5 1787, detailing the exact nature of the voyage. Four days later the Navy Board was ordered to purchase a ship not exceeding 250 UK tons/255 tonnes. As nothing was found suitable among the 600 ships of the Royal Navy, advertisements let it be known that the Board would inspect vessels offered for sale the following Wednesday, also suggesting that someone acquainted with the expedition might attend. By May 16 the Navy Board had six ships to consider and quickly chose a snub-nosed coastal trader built in Hull, only two and a half years old, called *Bethia* and lying at Wapping Old Stairs on the Thames in London. The asking price was £2,600, say £200,000 today.

Sir Joseph arranged to meet Mr Mitchell, assistant to the Surveyor of the Navy, aboard the ship to give his approval. Meanwhile the Officers of Deptford Yard gave their assessment of the condition and value of *Bethia,* which in their opinion was at £1,820. 12s. 8d. plus £64. 15s. for anchors that had been offered with her, a total of £1,895. 7s. 8d. The sum eventually paid seems to have been £1,950, about £150,000 today.

Within days she was at Deptford Naval Yard and by June 6 had been rechristened *Bounty,* more a reflection of the expectations of personal gain she would bring to the West Indian planters than an indication of national benefit. She was a ship without precedent, marking a creative new twist to administrative thinking. *Bounty* was the Royal Navy's first ship bent neither on scientific exploration nor on colonial conquest but was simply to bear the fruits of such endeavours from one part of the world to another. Before all this she had to be adapted, both for the voyage itself and for her special task as a floating greenhouse. The Lords Commissioners were munificent, refusing nothing.

The priority was that the ship should survive such an extended expedition. Sailing's worst enemy was the shipworm *teredo navalis,* a voracious beast that effortlessly negated man's efforts to sail vast distances. Teredo worms could eat up a hull in a very short time. The simplest solution was to sheathe the hull in more wood, so that it could be replaced as it was eaten and *Bethia/Bounty* was so treated, having an outer hull of planks some 1 ¼" inches/about 32mm thick, secured by iron nails. This was still not suitable for a voyage to the South Pacific and back, so *Bounty*'s hull was then sheathed with English copper and all her fastenings were made of rustproof copper or bronze, an enormous but sensible expense, perhaps as much again as the purchase price. Then there were the masts and sails. Ships ordered to the South Seas always carried fewer sails than on other duties, so *Bounty's* masts were shortened, giving her less top weight and a lower centre of power.

There were very strict rules about the number of officers, men and guns that a ship might carry. The Admiralty wrote on June 6 to advise these things, as well as to announce the ship's new name and status as an armed vessel. She was to have four short-carriage 4-pound/1.8kg cannons, ten half-pound/225gram swivel guns. Her complement was to be 20 officers and 25 able seamen, 45 in all.

Bounty was to journey around the world under sail, yet at only 91 feet/ 27.75 metres long, she was hardly bigger than many a diesel-powered pleasure craft that today carefully hugs the coast. Her maximum width of 24 feet 4 inches/7.5 metres makes her sound squat, an impression not altered when you saw her almost flat nose and upright stern. Most surprising of all to the modern sailor, *Bounty* had no superstructure of day cabins, not even a galley cookhouse. Everything but steering and sail handling went on below the flush main deck, meaning cooking, eating, sleeping and the storage of supplies. Live chickens, sheep and pigs were kept both on the main deck and on the enclosed lower deck, together with their suitable feedstuffs.

Sailing ships smelled. There was black pitch used to plug holes in the decks and pungent tar that darkened and waterproofed the lines strung from masts and spars. Always there were unwashed men, incontinent animals, nauseating cooking smells and stale bilge water that slopped in the wooden hull, mixing all those and adding more unpleasantness of its own. Some of the sea water that washed onto the ship was bound also to dribble down into the bilge. It's said a big man o' war could be smelled downwind before it could be seen. The sickening stink of bilge water would soon permeate Fletcher's new ship and the closer they got to Tahiti the more the heat would concentrate the unpleasantness. There was little chance that sea breezes on the lower deck might help, where any light was likely to be from tallow candles that would add their own animal-fat smell.

Bounty would be a microcosm of self-sufficient life carried on in virtual darkness and impossible cramp. There were no portholes in the lower deck. Fresh air and light came only through the 3-foot/.9 metre hatches to the ladderways open only when conditions were suitable or when men changed watch, every four hours.

Officers and gentlemen abaft the mast had headroom of 7 feet/2.15 metres and for a few of them the doubtful advantage of tiny airless cabins, ventilated only by slits in the doors. The ABs before the mast had only 6 feet 3 inches/1.9 metres and shared one open space which directly abutted the galley and the pens in which goats, pigs and sheep were often kept. In this room they ate and slept. The average space a man could call his own was the 30"/76cms width he had to sleep on his canvas hammock, which was only 6"/15cms wider when opened flat.

Bounty's great cabin stretched across the whole of the stern and was the only part of the lower deck well-lit and ventilated. There were five tall, small-paned windows across the width and a large

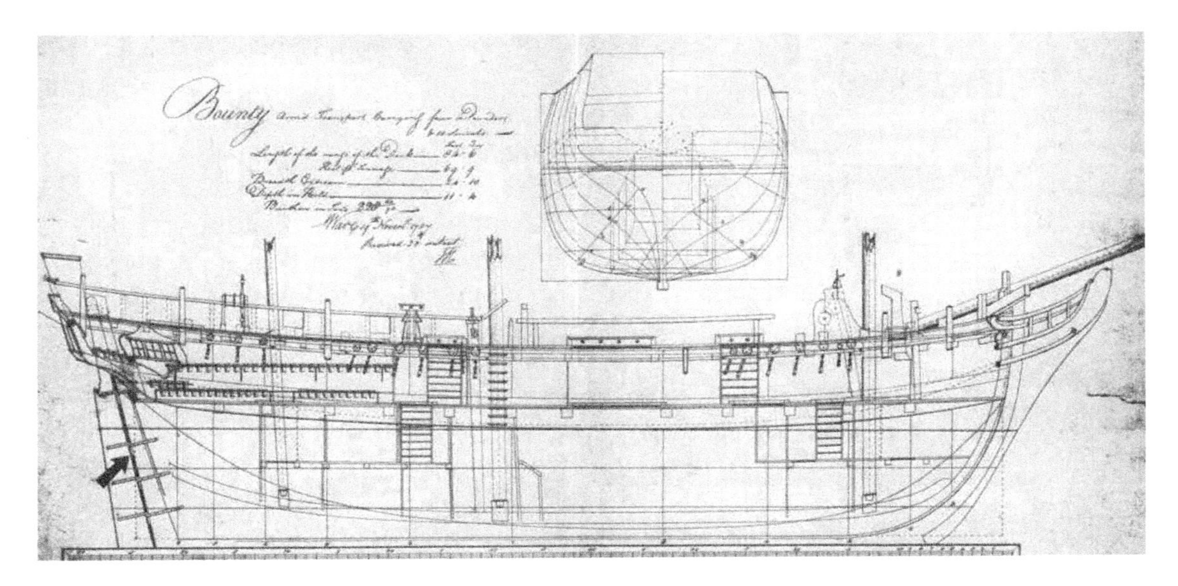

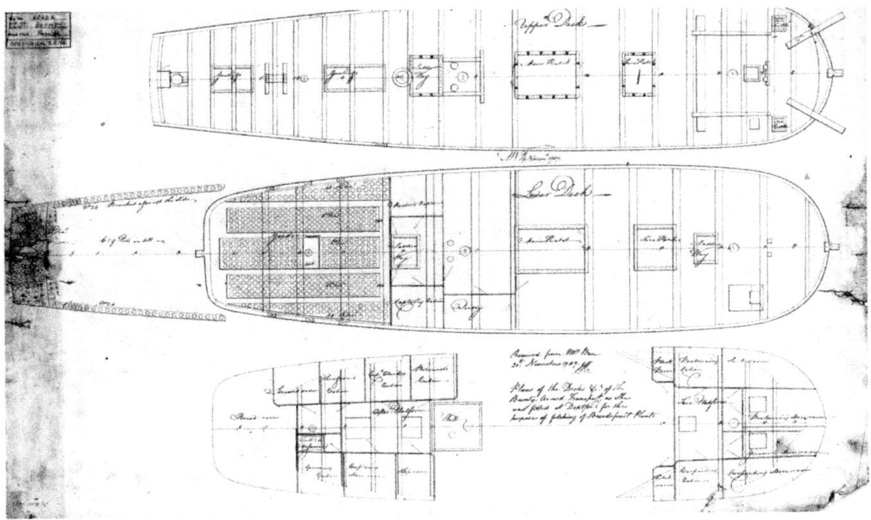

Although the rear cabin of Bounty was greatly extended, new cockpits below the lower deck meant greater space was created than was lost. With no portholes and with cooking, men and animals sharing the lower deck, Bounty's mission was a cramped and noisome challenge, even to seasoned sailors.

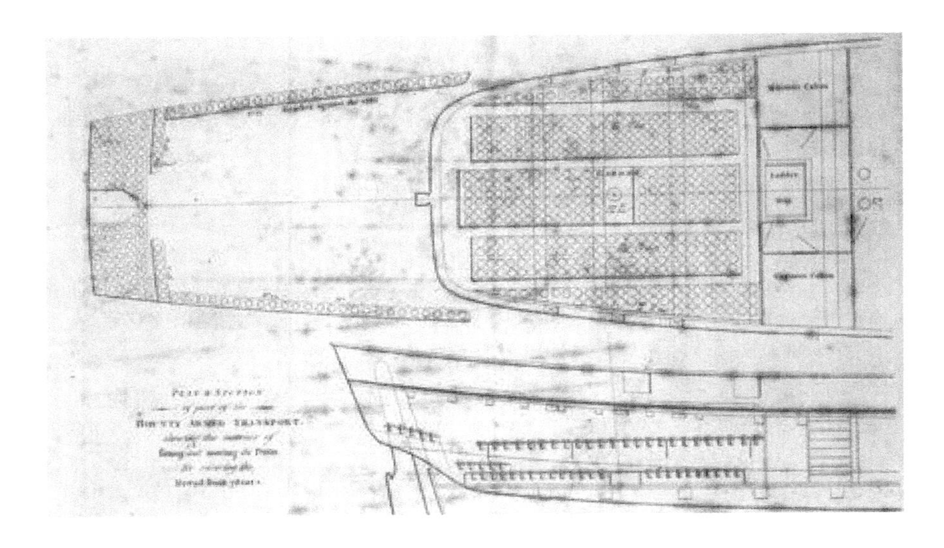

lantern window on both the port and starboard sides. That comfortable space should have been Bligh's headquarters but Banks instructions said this and more of the lower deck was to be a greenhouse for the breadfruit. A bulkhead was built from one side to the other of the ship, just to the rear of the aft hatchway, trebling the room's depth to 30 feet/9.2 metres. Bligh was reduced to a small unventilated and unlit cabin on the starboard side by these arrangements and 39 of the crew had to live for months on end with an average space of 30 square feet, about 2.75 square metres.

The others were down companionways in even smaller, darker smellier cells on newly built platform decks, where headroom was only 5 feet, just over 1.5 metres. These new decks, called cockpits, did relieve the congestion on the lower deck considerably. The fore cockpit accommodated the boatswain and the carpenter as well as the sail room and the storeroom of the boatswain, gunner and carpenters. The aft platform had cabins for the botanist, the surgeon, the captain's clerk and the gunner, as well as a steward room, captain's storeroom and a slop room for spare clothing. The added usefulness of the aft cockpit was that by boring holes through the floor of the greenhouse directly above, water which drained from the pots could be collected in barrels below. More deck space had been added than that taken by the new greenhouse but living was not any less cramped or in any way more gracious.

The arrangements for the breadfruit plants were wonderful. Having first had the greenhouse's deck made expensively waterproof by lining it with lead, the room was then furnished with three tiers of raised platforms that ran fore to aft with narrow gangways between them. The planks on the platforms had been pierced to hold earthenware pots. On the outer walls were suspended a line of similar potholders that were five and six deep across the rear windows so that, instead of the original plan for 500 pots, there were *'629 pots in all'* each carefully drawn and counted on the plans dated November 20 1787. New gratings on the decks and scoops in the sides provided added fresh air but not to the rest of the lower deck or cockpits. A stove was installed to keep the plants warm on the voyage around the Cape.

From the plans we see that the pantry, Bligh's responsibility as purser, was directly beside his somewhat smaller cabin. Bligh also had a place near the middle of the ship to eat in. This was against the bulkhead that divided the ship from the pantry on the starboard (right) side, precisely at the aft of the main hatch, forward of the mainmast. It's likely it was simply divided on either side by canvas screens from the berths of the mates and the midshipmen. Here also the arms chest was kept. It was exactly midway between the captain's cabin and that of the master, with only the ladderway between them.

As to the messy business of the men's more personal functions, most urinating would have been over the side and into the sea. For defecating there were perilous heads on either side of the bowsprit and Bligh probably used a chamber pot, attended to by his servant.

As *Bethia,* the ship had been designed to be sailed as a coastal collier by just 13 hands but that was not enough for a voyage to the other side of the world, when disease or accident might quickly reduce hands. There also had to be such as a sailmakers and carpenters and a gunner plus two gardeners.

A major criticism of the planning of the breadfruit expedition is that there would be no space for marines, the Royal Navy's policemen, whose tasks were to ensure discipline. Space was not the only reason for this omission. Marines were provided according to the category of ship and *Bounty* was simply not big enough to qualify. Bligh had no disciplinary backup of brute force in case of trouble or of any physical incapacity of the commander that might lead to insurrection. He didn't even have other commissioned officers to back him up. Still, *Bounty* wasn't going to war but on what must have seemed an enviable gardening leave in the South Pacific. What could go wrong?

The ship wouldn't even have the dignity of being HMS, His Majesty's Ship. She would be HMAV *Bounty,* His Majesty's Armed Vessel *Bounty.*

Fletcher Christian and William Bligh arrived back after their second voyage aboard *Britannia* on August 5 1787 and anchored at the Downs, a largely sheltered Royal Navy and Merchant Navy anchorage off the Kentish town of Deal, where ships waited for easterly winds to take them up to London. Records say there were sometimes as many as 800 sailing ships anchored here. Duncan Campbell found and boarded *Britannia* and revealed the breadfruit expedition plans. The very next day Bligh was writing fulsomely to Banks, with whom he had had no previous recorded contact:

August 6, 1787 Sir, I arrived yesterday from Jamaica and should have instantly paid my respects to you had not Mr. Campbell told me you were not to return from the country until Thursday. I have heard the flattering news of your great goodness to me, intending to honour me with the command of the vessel you propose should go to the South Seas, for which, after offering you my most grateful thanks I can only assure you I shall endeavour, and I hope succeed, in deserving such a trust. I await your commands, and am with the sincerest respect, Sir, your obliged and very Humble Servant.

Wm Bligh

Thanks to Duncan Campbell, Bligh at last had a patron of status and substance. Not having one in the past had hampered promotion terribly and he clung tenaciously to Banks for the rest of his life. It is likely Banks would have sniffed out Bligh anyway, for few men had the relatively low rank of lieutenant as well as the necessary experience of the South Seas. Bligh's time as master of Cook's *Resolution* showed him to be an excellent navigator and scientific man, capable of making charts, calculating, sounding and negotiating. Campbell was able also to recommend him as a businessman, a quality expected of the commander of *Bounty*.

Bligh was officially appointed on August 16 and took command at Deptford four days later. From a practical point of view, he was better off where he had been but ambition can be an imprudent master. His income was reduced by 4/5ths and this goes a long way towards explaining why Bligh was so pernickety about his money and interests. Yet, the voyage made promotion virtually certain and this was important to his pride.

When *Bounty* finally came out of dry dock at Deptford on September 3, the carpenters and joiners remained on board, well behind schedule and endlessly delaying the start of the voyage, a familiar situation with these trades. Four days later, six weeks after returning from Jamaica, Fletcher Christian signed on as Master's Mate. It's not known if Fletcher managed to get to the Isle of Man or tasted more of London Society to which he was previously introduced by his cousins Jane and Bridget; his older brother John was living in Pall Mall and Edward had chambers at Gray's Inn. It must be remembered that the great time taken to travel by coach, even to London, and the costs might have been beyond him. It would also have been prudent to stay close to Bligh and the prospect of an adventure in the South Seas.

What is true is that apart from the midshipmen, Fletcher Christian was one of the youngest on board and had the best education and most incontestable Establishment background of them all, albeit landed gentry rather than aristocratic. At 5 feet 9 inches/1.75 metres he was one of the tallest, too. Fletcher Christian was not just the perfect well-connected protégé for William Bligh but also one of the most talented and practical minded sailors, too. Today it would be called a win/win appointment and for Fletcher Christian, a few days before he turned 23, a huge boost to his confidence and his prospects of a secure future.

The frustrations connected with equipping *Bounty* for her voyage were shared by everyone. Sir Joseph Banks was hounded by Sir George Yonge, Secretary for War and a botanical enthusiast, who said he thought Bligh was almost totally unaware of his responsibilities in the South Seas. Banks replied, *'I am not a little alarmed at the rect. of yours: if Capt. Bligh is not sufficiently instructed and the purpose of the voyage consequently in danger of being lost, blame must lay somewhere and no one is so likely to support the burthen of it as your Hmble servant.'*

The full and largely unexplored correspondence between Yonge and Banks, now in London's Museum of Natural History, reveals many surprises and suggests that Bligh was playing the two men off against one another. It is impossible that Bligh, a stickler for procedure, would accept the command of a ship without first ensuring it was worthy of his talents and ambition. Yet he seems to have told Yonge that the botanist Nelson was responsible for provisioning the ship and for instructing Bligh as to his duties.

Naturally, Bligh wanted to make his mark on his ship and he found fault whenever he could. On September 15 he wrote to Banks regretting progress was slower than expected, *'owing to a great struggle to get the masts and yard shortened a circumstance I am happy I persisted in'*. He was subsequently proven wrong, as in the South Pacific he had to raise royals, small sails that fly high above the normal square rigging. The ballast of iron bars was also reduced, from 45 UK tons to 19/19.3 tonnes (the difference in tonnes is a multiplication of only .016), for he was certain too much deadweight in the bottom caused the misfortunes of many a ship in foul weather.

With fortuitous foresight, Bligh asked that the three boats be exchanged for bigger ones. The launch was biggest and became 23 feet/7 metres instead of 20 feet/6 metres and thus able to carry another four men. There was also a 20-foot/6 metre cutter and a 16-foot/4.9 metre jolly-boat. Once these were stowed and stacked on *Bounty*'s deck, there would hardly be room for the entire complement to be on deck at once, what with steerage, hatches, chicken coops and ladderways all taking up space as well. There was no escape from *Bounty's* gloom and cramp to private spaces for contemplation on broad and sunny decks.

On October 9 Bligh told Banks that the ship had been sent down to Long Reach in Dartford and there the guns and gunner's stores were loaded and he also thanked him for the present of a sextant of noticeably good workmanship.

The Navy Board ordered *Bounty* to be victualled with one year's supply of all provisions except beer. Only one month's allowance of the beverage was to be put aboard. Wine and spirits were taken in lieu of the rest, for they would last longer and better and beer could be made on board. Bligh disagreed, saying he should provision for 18 months and on September 18 the Admiralty instructed the Navy Board to provide additional supplies. Illustrating Bligh's advanced ideas of nutrition and avoidance of scurvy, the Commissioner of Victualling was directed to supply *'Sour Krout, malt [dried], wheat and Barley [instead of oatmeal], sugar [instead of oil], juice of wort [condensed brewing malt], and salt for salting fresh meat and fish'*. Bligh was given freedom to arrange the quantities of each and a week later there was further approval of five hundredweight/560 lbs/254 kgs of portable soup, something normally supplied in quantities sufficient only for the sick.

Chisels of many sizes, knives, nails, hatchets, looking glasses, beads of red, white and blue, stained-glass drops for earrings and coarse shirts were bought by the many gross, the dozen and by weight. By October 10 *Bounty* had £125. 6s. 4d worth of goods to exchange for breadfruit, about £22,500 value today. It can be safely assumed that Bligh's experience had been used to compile it and Banks' knowledge of the Pacific was doubtless called upon too.

HMAV *BOUNTY'S* MUSTER 1787

August 20, 1787

Lt William Bligh	33: commander with courtesy title of Captain
John Huggan	Surgeon
John Fryer	35: master
Thomas Hayward	20: AB to 1 Dec. 1787 then midshipman
David Nelson	Botanist
William Brown	About 25: assistant botanist

August 27

William Cole	Boatswain
William Purcell	Carpenter
William Peckover	Gunner
Lawrence Lebogue	40: sailmaker
Henry Hillbrant	24: able seaman; cooper
John Samuel	26: clerk
George Stewart	21: able seaman serving as midshipman to 30 Nov. 1787, then acting master's mate to 2 March 1788
Peter Heywood	15: AB to 23 Oct. 1790 then midshipman

August 29

William Elphinstone	36: master's mate
Peter Linkletter	30: quarter master

August 31

Isaac Martin	30: able seaman
Joseph Coleman	36: armourer

September 7

Charles Churchill	28: master-at-arms
Alexander Smith	20: able seaman; his real name was John Adams
Fletcher Christian	21: master's mate to 2 March 1788 then acting lieutenant
Thomas Burkitt	25: able seaman
John Millward	21: able seaman
Thomas McIntosh	25: carpenter's mate to 31 Dec. 1787 then carpenter's crew

John Mills	29: gunner's mate
James Morrison	27: boatswain's mate; responsible for floggings
John Williams	26: able seaman
John Sumner	22: able seaman
John Hallet	15: midshipman
Robert Tinkler	17: able seaman
October 7	
James Valentine	28: able seaman
October 8	
George Simpson	27: able seaman to 14 Oct. then quartermaster's mate
Robert Lamb	21: able seaman; butcher
Thomas Ellison	19: able seaman
Thomas Hall	36: able seaman; ship's cook
October 13	
John Norton	34: quartermaster
October 16	
John Smith	36: able seaman: Bligh's servant
October 23	
William Muspratt	27: able seaman: cook's assistant, tailor
Matthew Thompson	37: able seaman
Edward Young	21: hired as able seaman but sailed as midshipman
November 16	
Michael Byrne Able seaman, blind, ship's fiddler	
November 21	
Charles Norman	Carpenter's crew to 1 Jan. 1788 then carpenter's mate
November 28	
William McCoy	Able seaman: also known as McKoy
Matthew Quintal	Able seaman
December 6	
Thomas Ledward	AB to 11 Dec. 1788: surgeon's assistant then acting surgeon

CHAPTER 10

A LAUDABLE RECOVERY

Fletcher Christian's sail-mates on *Bounty* are thought to have made up the first British naval ship ever to have sailed with no pressed, that is forced, men aboard. Perhaps this heady mixture of willing appointees and adventurous volunteers was yet another reason that her eventual capture by mutineers was so simple it seemed inevitable?

In his biography of Bligh, Gavin Kennedy reckoned that by attending to the fancies of others, as well as to a few of his own, Bligh diluted the strength of his crew. He believes that *Bounty* should have had 25 able seamen to stand watch and to manhandle the sails, yet the ship had only 13 (for which she had been designed as a coastal trader), one of whom was Michael Byrne, an almost sightless Irish fiddler. The berths that might have gone to other ABs had been whittled by such arrangements as the employment of five midshipmen instead of the two properly allowed. As well, there had to be an armourer to oversee the guns, a commander's cook, a steward, and at the last moment an assistant surgeon, all signed on as able seamen. The latter appointment was a most sensible precaution, for the surgeon was a sot, who had been foisted on Bligh with a degree of insensitivity to the voyage that borders on criminality. He seemed useless until his death on Tahiti. Yet, to be fair to the man, who was liked by the Tahitians, we only have Bligh's word of his disabilities.

The two botanists were also men whose berths had to come out of those allotted to ABs, although the instructions of assistant-gardener William Brown make it clear he was to stand watch. As to the sailing duties of the rest of those thought extras, we have no exact knowledge but the midshipmen were certainly not expected to haul ropes and climb masts and I doubt if a commander's cook or an assistant surgeon or stumbling violin player would.

It looks as though the ABs were in for an even tougher time than usual, but this was not the case. I think Bligh's belief in his methods of diet, exercise and hygiene at sea may have allowed him to play deliberately with his crewing, certain that there would always be enough men to sail the ship. He was probably justified. Scurvy and other diseases caused far more thinning of the seamen's ranks than favouritism ever did. It was not unusual for a ship to return from a long journey with fewer than half its original complement, all lost through disease. Or perhaps 25 ABs really were too many and it was, like so many aspects of Georgian life, simply a form that had to appear to be gone through? It is unlike Bligh to do himself down and I think the latter is most likely. The real problem aboard *Bounty* was not of manpower but of having no marines or other commissioned officers aboard, simply because the ship was too small to qualify for them.

19 men assigned to *Bounty* did not sail with her. Three were transferred to other ships and all 16 pressed men deserted. Gavin Kennedy offers interesting observations about their replacements, who had plenty of opportunity to change their minds and futures before sailing, thanks to the slowness of the artisans. Of the volunteer replacements, six became mutineers of whom Young, McCoy and Quintal were hardliners. Others who sailed off to oblivion after the mutiny were Brown, appointed to the ship before Bligh, and the American, Isaac Martin, who signed up on August 31, a week before Fletcher Christian was appointed.

Although nothing official would be announced until she was well into the Atlantic, rumours and assiduous questioning made the destination of this extraordinary vessel common knowledge. Volunteers for such a journey must have been inquisitive, adventurous and footloose by nature and thus were intrinsically more likely to want to stay in the Pacific than men who had been forced aboard by a press gang.

Bounty was also competing for men with a general muster, for there were new problems in the Netherlands and the Baltic. To career-minded naval officers the approaching battles were a boon because war meant faster promotion. The drawback was obvious. War also meant the possibility of death and sensible ambition includes some time for the enjoyment of one's achievements. I am persuaded that many a volunteer who sailed aboard *Bounty*, careerist or not, had the voyage in mind as an adventurous and possibly profitable way to avoid war's permanent interruption to his plans.

Bligh would have liked another change aboard *Bounty*, one just as important than ballast and boats. He wanted promotion to captain, both as a right and because he then could have one or two commissioned lieutenants under him, which in his later opinion would have helped ensure there would have been no disciplinary problem. He continued to write about promotion from *Bounty* with a decided whine to his quill as long as he could send letters. Although moral right was certainly on his side, there were rules and regulations about what rank could command what type of ship. It was impossible to bend them and Banks was either incapable of forcing the issue, or felt he had done quite enough for his begging protégé. Added to his burden of difficulties with the Admiralty and the delaying of his ship's departure, the snub to Bligh's sense of justice must have made him feel well justified to sulk. Having no other commissioned officer aboard forced a solitariness of command upon him even though, from a practical point of view, there was nowhere to berth them suitably.

His father-in-law, Richard Betham, sympathised with Bligh and agreed he should be promoted but pointed out what was common knowledge. In peacetime virtually no promotions were made other than at sea. Betham told Bligh he should expect to be given sealed orders that were not to be opened until he had crossed a certain latitude. Promises were not enough for Bligh. He could not get lieutenants or new insignia out of salt air.

In the meantime, Bligh must have been very glad to see, perhaps had even encouraged, the slow growth of a coterie of young men aboard who were in some way associated with his past or his family and who could thus be expected to show respect over and above that normally commanded by a mere lieutenant, even if custom permitted him to be addressed as Captain Bligh. In his capacity as commander of a naval vessel Bligh was for the first time in a position to help others towards preferment, a heady and treacherous thing. It was easy to be blinded by a man's social position and friends, dangerous to hire a man for these assets rather than for his qualities at sea. If others did not know their duty to him, *Bounty's* captain was certainly showing that he understood the duties of patronage and Fletcher Christian was soon surrounded by others who owed their appointment to Bligh.

Fletcher Christian's status aboard Bounty *was as a midshipman, the most junior position on a Royal Naval vessel that had officer status.*

Although being master's mate is of great importance, Fletcher Christian's status aboard *Bounty* was as a midshipman, the most junior position on a Royal Naval vessel that had officer status. It was an apprenticeship, where training began to be a commissioned officer. A warrant officer ranks beneath a commissioned lieutenant and is usually a technical expert of some sort, such as a carpenter or armourer. As a midshipman you might be appointed assistant or mate to a warrant officer but this did not change your status; a warrant officer could not be flogged but his midshipman-mate could be.

George Stewart arrived as a midshipman on August 27 but was rerated down to AB on November 30, when young men more deserving of the title had been employed. He continued to serve as a young gentleman, not as an AB, and the demotion was a matter of paper administration. Stewart was well educated from a very good family on the Orkneys and Bligh must have moved fast to get him on *Bounty* only two weeks after his own appointment. When *Resolution* had called at Stromness in the Orkneys in 1780 on her way home from the South Pacific, Bligh had been well looked after by the Stewart family and promised interest to George, aged 13 at the time. Now 21, he must have been at sea sometime because Bligh later said that his seamanship would have recommended him, quite apart from his undertaking to help the young man.

On the same day Bligh also welcomed 15-year-old Peter Heywood, whose case had been pleaded by Bligh's father-in-law, who had long known of the plans through Duncan Campbell. Peter's father in Douglas, Isle of Man, had suddenly fallen from the security and profit of being Seneschal to the Duke of Atholl, the Lord of Man. Betham wrote saying he thought young Peter *`ingenious . . . a favourite of mine and indeed everybody . . . I hope he will be of some service to you, as far as he is able in writing or looking after any necessary matters under your charge.'* He underlined the duty he felt towards the Heywoods and how great would be their disappointment if he had not applied to Bligh successfully, his emotional blackmail disguised only by the thinnest turn of phrase. Bligh acquiesced, so yet another family connection - and his father-in-law - were obliged to him.

Then it was Betsy Bligh's turn. She recommended 15-year-old John Hallet, the brother of a friend from London, whose father had written to her. Mackaness tells us the fourth midshipman, 20-year-old Thomas Hayward from the agricultural village of Hackney also came through her influence. Thus, four of the five men who were to sail as midshipmen came directly through Bligh connections. The fifth was Robert Tinkler, not of the same social class but who was recommended by John Fryer, *Bounty's* master. The lad was Fryer's brother-in-law, so Bligh could hardly refuse. By September 7 the four young gentlemen were embarked, Peter Heywood having stayed with Elizabeth Bligh at 100 Lambeth Rd, today a B&B called Captain Bligh House. If Bligh had later cause to regret having any of these young men on board, he had only himself to blame, which might explain some of the vitriol he directed towards them.

Two other faces were familiar to Bligh, Lebogue the sailmaker and young Tom Ellison, who was to serve as an AB. Both had been on *Britannia*. Ellison, small and thin, had the special interest of Duncan Campbell, so also generated kudos.

Six weeks later yet another young man was signed on as an AB but who would sail as a midshipman. Edward Young was 21. Born on St Kitts in the West Indies, he was perhaps a mulatto and described as having *'a bad look'*. Bligh says, in the journal he later wrote during his epic open-boat voyage, that he was recommended by Sir George Young, *'a Captain in the Navy'*. It has long been claimed Edward was the nephew of Sir George Young, Bart, later an Admiral. A note I found made by the Admiral's son says: *'It was unpardonable to say Young was my father's nephew or mine! There was an Edward Young, son of George, the son of the Rev. James Young and a first cousin once removed of the Admiral, who may have been this man'*. Bligh gives no further amplification, except to note Edward Young *'proved a worthless watch'*. Young is constantly referred to in Bligh/*Bounty* books as an officer but this rather exaggerates his status as a midshipman, a trainee officer.

Fletcher Christian's age is given on the muster sheet as 21 but this is not true. He was 22 and would be 23 just weeks later, on September 25. In absolute contrast to Bligh's pallor, Fletcher Christian was brown skinned and dark haired. He was 5 feet 9 inches/1.78 metres and his body was notably muscular, marred only by a slight outward bending of his knees, which in breeches would emphasise the appearance of being positively bowlegged.

Christian and Bligh had more in common than sailing together and their connection with the Isle of Man. They also had similar expectations of the future and a belief in their inherent right to rule other men. Fletcher Christian's was based on the birthright of centuries; Bligh's was the authority donned with his uniform.

Overall, *Bounty* had a remarkably young crew, with most under 30. Bligh at 33 was one of the oldest, perhaps by design and, apart from the midshipmen, Fletcher Christian was one of the youngest, so to have his position as master's mate is startling. Fryer the master was 34 and newly married and William Elphinstone, the other master's mate was 38. Christian must have been indebted to Bligh for an excellent recommendation and must also have been outstanding at his job. When he returned, he was virtually certain of becoming a lieutenant, years before Bligh had reached that rank.

Fletcher Christian had already sailed to Africa, India and the West Indies. He had never been in trouble; indeed, he had been singled out for special favour by Captain Courtney on *Eurydice* and by Bligh on *Britannia*. For a young man whose world had been ripped from under his feet only eight years before, he had made a laudable recovery. If he continued the same way, Fletcher Christian would easily replace the status, security and future that he had lost through his immediate family's financial ineptitude.

Yet, Fletcher had a particular problem with his ambition, curiously similar to that which exercised Captain Bligh. Both were isolated. With no marines or other commissioned officers on board, Bligh commanded alone, whereas Fletcher Christian had no peer aboard *Bounty*, no-one who had equal family background, social experiences and friends. No-one else aboard would have shared the same conversations as he did with his smart cousins Jane and Bridget, or would understand what he was missing or envying in the properties and income and revolutionary ideas of cousin John Christian XVII. Who else would understand why he once thought marriage to Isabella Curwen might be possible?

Fletcher's background and a need of reminders of its values and manners perhaps explain why he took a Chinese porcelain bowl with him – and who now knows what else that has been lost?

Marlon Brando's clipped accent and foppish clothes in his 1962 portrait of Fletcher Christian are exaggerated yet properly showed quite how different Fletcher was from any other man on board

Bounty, a difference that brought with it an expectation of behaviour from others that would have been met aboard *Eurydice* and that could be borne on short journeys to and from the West Indies. Once aboard *Bounty* Fletcher Christian was never again to have the support of companions of his intellect and social class.

Fletcher's cousins Jane and Bridget Christian were frequent guests of the Duke and Duchess at Norfolk House, 31 St James's Square, London. First opened for viewing by Society in February 1756, it featured one of London's earliest dedicated Music Rooms, the first three windows on the left of the nine on the first floor, the piano nobile. The restored Music Room is now in the Victoria and Albert Museum and I like to imagine young Fletcher here, too. His brother John lived steps away in Pall Mall.

CHAPTER 11

'ANIMALS, FIERCE AND RESENTFUL'

The twin aggravations of a dilatory Admiralty and obstructive weather delayed Fletcher Christian's departure for the South Seas far more than the fiddling craftsmen. On November 4 1787 *Bounty* had sailed as far as Spithead, the sheltered anchorage off Portsmouth that was actually a sea-water cesspit, as so many ships tipped every type of human and animal waste overboard. Bligh had written to Banks on the way, to say it would have given much pleasure to have shown him the ship, as *'I flatter myself she is the completest that has sailed on any expedition'*. A flattering contribution to Bligh's pleasure was that *Bounty* had been furnished with Thomas Kendall's K2 chronometer, one of the very first timekeepers capable of constant time. At sea, this meant being able precisely to calculate how far west or east a ship was from Greenwich, giving an accurate fix on longitude that had been impossible and responsible for countless tragic losses of life.

Until then, sailing ships had little idea of their true position, something of great advantage to Fletcher Christian when searching for a South Pacific refuge after his mutiny.

Bligh had used Kendall's K2 on his voyage with Cook and this must have inclined the Board of Longitude to supply him, for the instrument was valued at £200, over £17,000 today, more than 10% of *Bounty's* purchase price.

The weather was then perfect for Bligh to sail down the Channel and continued to be so for three weeks. Yet he did not have his sailing orders and without those he could only fret and write fine phrases, inspired by a not unreasonable exasperation.

For Fletcher this delay was life-changing. He discovered the East Indiaman *Middlesex* returning to Spithead from Madras and he knew his brother Charles was on board as surgeon. My unearthing of this unexpected meeting of the brothers showed how it affected the future of Fletcher Christian and of everyone else in *Bounty*. The repercussions still continue.

Fletcher hired a small boat and boarded *Middlesex*. He had much to tell Charles. Their only surviving sister Mary had died on the Isle of Man that year, and so had Uncle Edmund Law, Bishop of Carlisle, who had married their father's sister Mary.

Charles had been away for almost 20 months. The two brothers and one of *Middlesex*'s officers, who had once been in the Royal

Navy, were rowed ashore and spent the evening and night together, something I discovered in a copy of Charles's unpublished autobiography, now in the Manx National Archives. Although writing nearly 25 years later, Charles remembers the vivid impression Fletcher made.

'. . . he was then full of professional Ambition and of Hope. He bared his Arm, and I was amazed at its Brawniness. 'This,' says he, 'has been acquired by hard labour.' He said, 'I delight to set the Men an Example, I not only can do every part of a common Sailor's Duty, but am upon a par with a principal part of the Officers.

I [Charles] met with a Surgeon [Thomas William] in East India who had gone out with him in the Eurydice *Frigate, commanded by the Honble Captain Courtney, and also with an Officer on my Return to London who had sailed in the same ship with Fletcher. They corroborated to me his Assertions Captain Courtney had appointed him to act as Lieutenant [by giving him charge of a watch]. They said he was strict, yet as it were, played while he wrought with the Men he made a Toil a pleasure and ruled over them in a superior, pleasant Manner to any young Officer they had seen.'*

If this were all Charles tells us it would still be an important addition to the sparse first-hand material we have about Fletcher Christian. In fact, there was much more to say but, like all men writing about themselves, Charles had left something vital out of the document, not in any way to obscure facts, which were well known at the time, but because the details were largely irrelevant when he wrote his autobiography. He writes:

'I am persuaded that few men [i.e. himself] *had a stronger propensity to Beneficance or possessed a greater share of Benevolence, or a more anxious Disposition to be pleasing and serviceable to all Classes of the Community until a chagrined Turn of Thought arose, the offspring of the Middlesex East Indiaman Voyage. In that Ship, as well as in every other Ship or in every House . . . it might be a peace-preserving Mean to have the following Mottoes always placed conspicuously . . .*

`Beware of whom you speak, to whom, of what and where.' `Give every Man thine Ear, but few thy Voice.'

`Take each Man's censure, but reserve Judgement.'

. . . The Precepts and Doctrine of our Religious inculcate the Forgiveness of Injuries, but when Men are cooped up for a long time in the Interior of a Ship, there oft prevails such a jarring Discordancy of Tempers and Conduct that it is enough on many Occasions by repeated Acts of Irritation and Offence to change the Disposition of a Lamb into that of an Animal fierce and resentful What can not the Power of Provocation bring to pass on Land, where there is a free Range of Separation?'

I knew I had to look further into Charles's experience on *Middlesex* and what else he told Fletcher. My researcher Michael Brook found the Captain's Log in the archives of the India Office in Whitehall. In the entry for September 5 1787 is the origin of the canker that distressed Charles for the rest of his life. Incredible as it may seem, there was a mutiny aboard *Middlesex* and Charles Christian is cited as one of the officers in the conspiracy, protesting against a captain who was cruel and mean with supplies.

Charles Christian was in his 50th year in 1811 when he wrote about his life, in his mother's house in Fort Street, Douglas, Isle of Man. He wrote in letter form, as a defence against rumours that he had forged banknotes on the island. He relates fluently and picturesquely in non-consecutive episodes a life packed with adventure and emotion, which included horrid details of serving as medical officer on a slaver in West Africa that was later captured by the Spanish. He retails graphic first-hand experiences of the slave trade I have not read elsewhere and it seems he owned some of the slaves aboard, bought as an investment.

A superficial look at his autobiography might easily lead a browser to dismiss Charles Christian's tales of misfortune and persecution as the delusions of a paranoiac, indeed, this is exactly the impression of Vio Christian and her daughter Rita, who found and read the document early in the 20th century. If true this would clearly cast suspicion on the veracity of the evidence and careful consideration had to be given to the possibility that the journal was the work of a deranged man. Even if so, it doesn't change the fact that he had taken part in a mutiny.

After much thought and research, I cannot see that Charles was paranoiac as he both prefaces and punctuates his autobiography with an open willingness to believe he might be in some way mistaken or responsible for his bad fortune and that is certainly not the manner of paranoia. Charles was an eccentric and he possibly had a personality that made him a natural butt of malicious humour, yet throughout his troubled life he never lost his extraordinary optimism or religious faith.

After his mother fled with him, Fletcher, Humphrey and Mary to the Isle of Man in 1779, Charles joined the West Yorkshire Militia in 1780, where he remained for three and a half years and then he spent two years at the Medical School in Edinburgh. Although only in his 20s, Charles was an avowed celibate. Like many over-informed medical men, he was convinced he would never live out the following year and used this reason to excuse his constant desertion of the women to whom he was attracted, some of whom he says were very rich and very interested. In February 1786 Charles signed on as surgeon on board *Middlesex*, an East Indiaman, Captain John Rogers. She was bound first for India, then Macao on the China coast. When she left Madras for England, she must already have been an unhappy ship.

September 5 1787 was a day of many a lamb turned into *'an Animal, fierce and resentful'*, and a time of great fear. At 6 pm Captain

Rogers reports that he confined W. Greace in irons for presenting a loaded pistol to his captain's breast, and for threatening he would put the first man to death who touched him. The first officer, Mr G. Aitken, was dismissed for aiding and assisting the mutiny's commission. Three hours later Mr D. Fell, the second officer, was also dismissed for *'drunkenness, Insolent Language and striking at his captain on the Quarter Deck . . . At this time I dismissed the Surgeon, also in the conspiracy.'*

Aitken and Fell were locked away, for on September 7 Captain Rogers reported they broke out of their cabins but were reconfined *'to prevent further evils'*. Surgeon Charles Christian must have been locked up too.

As participants in mutiny on a Royal Navy vessel Greace, Aitken, Fell and Christian would have been court-martialled and if found guilty could have been hanged from a yardarm. This wasn't true on an East India Company's ship. Their action was neither against the king nor was it a civil offence, having occurred at sea, which is why Charles was able to leave the ship when it arrived in Spithead and on November14 he was paid £52.12s.2d back pay, the equivalent of about £4000. Nonetheless, there were actions at the disposal of the Captain within the framework of the East India Company and its Court of Directors but from the minutes of the EIC's Court of Directors for October 19 we learn that Captain Rogers had not made an official report of the mutiny. The Directors resolved that his actions were reprehensible in that he had not entered in the Company's Log, quite separate from that of the Captain, *'the several transactions relating the behaviour of his officers in September'* and he had not advised immediately on arrival his dismissal of the chief and second officer. The action of the guilty men was considered *'extremely censurable'* and they were each suspended from service with the East India Company, Greace permanently, Aitken for three years, Fell and Charles Christian for two years. These men would have been given an opportunity to explain themselves and the Company obviously felt Rogers was more than a little to blame. He was fined £500, almost £40,000 today, for use of the East India Company's Poplar Hospital, and suspended from trading and profiting for one year.

So, Fletcher's last conversations in England were about mutiny, an astonishing discovery. More than that, they must have been about the duty of officers towards their men and the actions that could be taken against captains who were cruel or thoughtless. No wonder Fletcher remained ashore all night. The conversation with his brother must have made a deep impression as Charles explained and defended himself.

On November 10 1787 Bligh wrote it would be hard if instead of some good reason for being detained he were to discover it was simple oversight by the Lords of the Admiralty. He later said they all deserved to be punished for precisely that sinful attitude towards his expedition. Fletcher Christian and the crew were subjected to a further fortnight of tedium and complaint during favourable winds before *Bounty's* sailing orders finally appeared, just as the winds changed. On the 28th the crew was given two months' pay in advance and *Bounty* tried to sneak off during a brief change of wind but she was forced into St Helens anchorage off the Isle of Wight, across from Spithead.

Bligh was well aware of the hardships that lurked at the tip of South America. Every day of delay made them more frightening. At this rate he would be attempting to round the Horn when every force of the southern elements would be against him. He was confident of the ship but already practising that ill-advised habit of belittling the character of his complement. On one hand he noted his *'men and officers all good and . . . happy under my directions'*; on the other, he was doubting their ability to stand the challenges of the Horn.

So, Fletcher's last conversations in England were about mutiny, an astonishing discovery.

By December 3 *Bounty* was ignominiously back in the mouth of the Solent in Spithead anchorage. Three days later, still wheedling for promotion, Bligh attempted again to clear the Channel. This time *Bounty* was blown almost to the French coast and again scuttled back to St Helens anchorage for shelter, with many of her crew suffering severe colds.

Bligh was not prepared to risk further *'damaging the ship or hurting the health of my men and officers'* and waited until events augured better for his departure. The added time in port gave him a welcome chance to consider further his orders to reach Tahiti by the shortest route, that is via the Horn. He knew failure to carry out orders would reflect badly on his character rather than on the Admiralty and he was in no mood to be fodder for them, writing to Campbell, *`. . . they took me from a state of affluence from your employ with an income of five hundred a year to that of Lieuts pay of 4/- a day to perform a voyage which few were acquainted with sufficiently to insure it any degree of success'.*

The difference in income in today's money was from the equivalent of about £770 a week to £108. Although a current difficulty. this was also a spur to Bligh's determination that future success was as important for his family as to his reputation. Bligh expected that once this voyage was over, all his *'concerns over money would have been solved'* and his *'poor little family'* would be financially secure. The bait of future solvency was certainly a stimulus to success and might explain the extra edge Bligh gave to his carping about exertion and excellence.

Deciding that the prospect of the Horn out of season was too big an obstacle to risk without the backing of alternative orders, he wrote to Banks asking if he could coax discretional orders from the Admiralty. There was no hesitation. Bligh had his new orders within two days, allowing him to sail via the Cape of Good Hope if he arrived too late to breach Cape Horn.

Bligh once more said goodbye to Betsy, where they had managed some secret and unexpected time together, presumably on shore in Portsmouth. On December 23 the winds finally moved to the east and *Bounty*, William Bligh and Fletcher Christian were finally headed for the South Pacific. After a day of gales Christmas Day was altogether calmer and the festival, which Fletcher Christian must have once hoped to celebrate in sub-tropical waters, was marked by the issue of extra rum to accompany beef and plum pudding. Next day unremittingly cruel, steel-grey Atlantic rollers smashed over the struggling ship. On the 27th *Bounty's* stern windows collapsed under the weight of gale-driven salt water. The icy flood that raced through the ship broke an azimuth compass and Bligh managed to rescue the precious Kendall chronometer only with great difficulty. The crew's fear was further tormented because even though chilled, wet and frightened they had no hot food, for the stove could not be lit. Instead, rations of grog were added to their beer and they had to fill up with biscuits.

When the storm abated, anxious stock was taken of the damage. Extra spars had been washed away and the ship's three vital boats had been damaged. Their repair had to take priority in case there was another such storm, or worse. Seven full hogsheads of beer lashed together on the deck had disappeared and two casks of rum had split, their contents dribbling into the bilge. Most seriously, the sea water that crumpled the stern had contaminated *Bounty*'s vital supply of biscuit, stored in a cockpit directly below the greenhouse. No-one could have foreseen this sort of flooding. The proposed breadfruit nursery had drainage in its floor so overflows from the irrigation system could be collected. The stores of biscuit would have laboriously to be checked and repacked.

Bligh concentrated on drying the men's gear, their bedding and the ship's interior. When the stove was finally lit, two men from each watch were detailed to wash and dry clothes. The hatches were opened to air the ship and the lower decks were rinsed with vinegar to dissuade the growth of mildew.

On New Year's Eve men were still sorting and repacking the spoiled ship's biscuit and hoisting planks from the hold to repair the boats. When they opened some barrels of meat, four 4 lb / 1.8kg. pieces were missing from the pork and three 8 lb / 3.6kg. pieces from the beef, but this was expected by men in the king's navy.

On New Year's Day 1788 the sorting of the biscuits was finished but a whole cask of cheese was found already to be rotten. Still, the weather was fine enough to complete repairs to the rigging and to clean. Men were put to drawing yarn to make replacements for sodden mats and all sails were set. They were on their way at last. The superstitious on board may have been forgiven for wondering if the storm had been an ill portent.

Twelve days after leaving England and without the opportunity for her officers and crew to settle into the routine that ensured shipboard safety, *Bounty* was almost shipshape and sighted Tenerife, dominated by the snow-capped peak of Mt Tiede, a volcano that still smoked ominously. This was *Bounty*'s final civilised opportunity to repair, replenish and replace before heading for Cape Horn.

Bounty anchored in the roads (sea lanes) of Santa Cruz, the island's capital, on the morning of January 6. Bligh immediately sent Fletcher Christian to pay the captain's respects to the Governor, the Spanish grandee Marques de Branchforte. This is positive demonstration of the faith Bligh placed in Christian's superior social skills rather than a simple favouring. It was not unusual for Bligh not to make first contact and socially confident Fletcher Christian would not have protested about being given such a flattering duty. Without it he might not have gone ashore on this subtropical gem of the Fortunate Isles. As representative both of Bligh and King George, Fletcher Christian was directed to tell His Excellency that he would be saluted by *Bounty*'s guns, provided an equal number were fired in return. The reply he carried back was thought insulting. Branchforte said he returned an equal number of guns only to persons equal in rank to himself. Bligh had met his match in sensitivity of public image and omitted the ceremony with no further comment.

None of this hindered Bligh's tasks there but the lack of supplies in mid-winter did. Their delayed sailing now meant that Indian corn, potatoes, pumpkins, and onions were scant and twice summer's price. Beef was scarce and inferior. A shilling (12 old pence) was then worth the equivalent of almost £4 today. Ship's biscuit cost 25 shillings for 100 pounds/45 kgs and chickens were an outrageous three shillings each. Even the sub-tropical fruits, normally so prolific, were in pitiful supply except for a few dried figs and bad oranges. Water was bought at five shillings/£20 a ton, delivered.

Ferrying goods out to *Bounty* was risky, so Bligh treated with local boatmen, arranging to have everything handled by them. Wine was the only thing that gave Bligh some cheer. It was good and it was plentiful. Thinking it would be better for his men in the tropics than spirits he bought 863 1/2 gallons, nearly 4000 litres. He also took aboard two hogsheads, each about 240 litres, of the very finest quality Canary wine for Joseph Banks, judging it *'not much inferior than the finest London Madeira'*. It would be considerably improved as it crossed the equator and returned. After five days at Tenerife, *Bounty* sailed toward the coast of Brazil with Bligh recording the ship's company all in good health and spirits, except himself. As well as commander, Bligh was the purser, in charge of the ship's accounts and thus of the food stores. It was not unusual for the posts to be combined on a small ship but such autocratic control of supplies was open to enormous abuse, a fact both known and expected by the others aboard.

Everyone who has been at sea will know that food is an explosive subject. Meals are one of the few pleasures hard-working men expect or get. Food is their reward, a touchstone for the start and finish of wearying repetitive duties. Sailors are more sensitive about food quality and quantity than men and women ashore can possibly imagine. This is just as true today as it was in the time of sailing ships.

Bligh had to balance not only his books but also the conflict between what the seamen thought were their rights and what he thought practical to give them, allowing for delays and damaged stores. It would be obvious to Bligh from the start that he would be suspected of cooking the books rather than enough food and indeed he was.

Criticism hurled at Bligh about meagre portions, stealing or substitutions is largely ill-founded. Certainly, he expected to make profit from the pursery and we know how important he felt this to be. Even if he were slightly dishonest, it's clear he considered his men's health at least as much as his own pocket. That *Bounty* had always such a record of good health is proof enough that his men never suffered from any profiteering Bligh might have enjoyed. Yet there were problems and he mentions them in a curiously frank letter to Duncan Campbell, who could naturally be expected to be sympathetic and knowledgeable about such matters. Bligh continues his cry that he has no assistance worth mentioning, the transparent ploy of a man begging for sympathy and blaming others as a precaution against later failure, but hardly the way to engender loyalty or comradeship on the ship: . . . *Having much to think of and little or no assistance, my mind is pretty well employed, and as my pursing depends on much circumspection and being ignorant in it with a worthless clerk, I have some embarrassment, but as I trust to nothing or anyone and keep my accounts clear, if I fail in the rules of office I do not doubt of getting the better of it . . . Tom Ellison improves and will make a very good seaman . . .*

CHAPTER 12

A BOY AT THE HORN

Once they were away from the Canary Isles, Bligh could tell the men aboard officially where they were bound. Explaining that he had no way of knowing the length of the voyage - not even to the nearest six months - he announced he was cutting the allowance of ship's biscuit by one-third. There was nothing sinister in this. In his journal Bligh explains that he had noticed men usually only ate two-thirds of what was given them and by saving the rest he could ensure there was something for later. Clear though the logic is and even though the men were assured that any such shortage would be made up to them in cash at the journey's end, there were bound to be grumbles. As though compensation, they were told all their water was to be filtered through a drop stone purchased in Tenerife, which would greatly improve its palatability and safety. Finally, Bligh said he had been assured that promotion would come to all on their return to England.

In another enlightened decision that benefited everyone on board, Bligh changed the established routine to stand three watches instead of the usual two. In the classic system no man who was part of a watch was ever able to sleep more than four hours and so at times of danger when all hands were called, he might be on duty for 12 or more hours without rest. With three watches the men were on four hours and off eight, so even if there was an emergency, they had a good chance of rest and were more likely to react with alacrity to the orders of their superiors, who were equally well rested.

The arrangement also meant there was time for relaxation and Bligh insisted some of this took the form of dancing rather than indolence. He must have planned to switch to three watches before he sailed from England, having gone to much trouble to find Byrne, the almost blind fiddler who was expected to play for several hours each evening from 4 pm, a time of day when even most men on watch could join in. Dancing as exercise was continued on board Royal Navy ships into the 20th century, as photographs testify.

Each watch needs an officer to command it and Bligh gave charge of the new third watch to Fletcher Christian, a great compliment, even though such responsibility had been given him before, when he was even younger and in *Eurydice*, a considerably bigger ship.

On Thursday February 7 *Bounty* crossed the equator and Bligh allowed all the usual customs and pranks performed upon those who had never done so before, except the ducking, which he considered brutal and had anyway been banned by now. The officers had to pay forfeits of rum to the men, which Bligh agreed to reimburse and there was a special spirit of comradeship during the evening's dancing.

The south-east trade wind was now fresh and steady and the weather dry. More of the ship's bread was put into casks to protect it from vermin, another innovative experiment of Bligh's that proved most successful. For almost a month Bligh wrote mainly of marine matters and how he felt ships should be maintained in tropic waters. By the end of February, they were only 350 miles / 560 kms from Brazil and the jungle-clad wilderness of South America. A westerly wind carried to *Bounty* some of the continent's jewel-like butterflies and insects that looked like the English horsefly. Only a week later the weather had turned cool enough for the men to lay aside their tropical clothing and to wear heavier ones Bligh had ordered in England.

A month after being put in charge of a watch, Fletcher Christian was further honoured. On Sunday March 2 1787, after inspecting the cleanliness of everyone aboard (he cut their grog rations if they were dirty), Bligh conducted Divine Service, at which the Articles of War were always read, and announced he had given Fletcher Christian, barely 23-years old, a written order to act as a lieutenant and thus as second-in-command.

Countless comments claim this was unfair to Fryer, an insult, but this is not true. It would have been flying in the face of all naval precedent to promote a master at sea. The position was extremely important and responsible. The master was chosen with more care than any other officer and had the status of a lieutenant, although ranked below any who would also be on board. On shore, the master was in charge of ensuring all the sailing supplies necessary for the voyage and then for stowing the hold to safeguard the ship's balance and to protect it against overloading. At sea the master checked the ship's position at least daily and then set the sails according to the ship's projected passage and the weather. The master supervised the lowering and hoisting of the anchor when a ship had to be secured off-shore and when docking and undocking where there was a quay. He was the man to whom any issues with a ship's chandlery, sails, armaments or supplies were reported and, as well as other navigation-based Log entries, noted all financial expenditure. The master was in charge of the entry of parts of the official log such as weather, position, and expenditures.

If such a man was promoted to other duties, who would have the skills to take over? The master's mates were certainly accomplished but they were assistants, not replacements. Fryer would not have expected the promotion and never once said he did. The only man who might have been miffed was Elphinstone, who was first mate and already 38. Not having been promoted to lieutenant by that age, he must sensibly have resigned himself to permanent midshipmanship even though he must have been a very competent sailor to be appointed to that role.

The wide skills required of a ship's master neatly focus what an exceptional sailor Fletcher Christian must have been to be appointed master's mate on an expedition to the far side of the world and also explain how he was subsequently able to sail *Bounty* so far and so successfully in search of a home. He had served aboard *Eurydice, Britannia* and *Bounty* satisfactorily and his promotions were an expected reward. George Stewart, reputedly also a gifted mariner, was promoted to Christian's old post and henceforth was acting master's mate.

Bligh obviously thought he needed support of a special kind and at this stage trusted Christian more than anyone else. Having Christian as acting lieutenant should have made Fryer's job much easier, as he would have had his contact with Bligh lessened. Still, it would be unwise to overlook the possibility that in spite of the regulations Fryer might have been upset by Christian's promotion. The evidence can certainly be interpreted that he was and even Bligh says Fryer behaved spitefully thereafter. As there is absolutely no precedent that I can find for the promotion of a master while at sea we must dismiss this as the major reason for any disenchantment real or imagined shown by Fryer subsequently.

Some other motive must be forwarded for the tension between Fryer and Bligh, perhaps a simple jealousy of Fletcher Christian. If Bligh was publicly voicing the doubts and dislikes of his officers that he was writing privately in his cabin and if he included the master among those he belittled it is no wonder respect for the captain was quickly cancelled for such men as Fryer who, unlike Fletcher Christian, had not learned to humour Bligh's moods.

Bligh now had someone to share command, as he always wanted. For Fletcher Christian the promotion must have been gratifying but it changed nothing. He had no peers aboard and no naval rank or privilege would change that. Being acting-lieutenant might have isolated him more, especially with Fryer being piqued.

Bligh had hoped to complete the voyage without punishing a single man but he might better have tried to walk to Tahiti. His care of the men's health and comfort and his attention to the hygiene and conditions aboard were considered reason to expect exemplary behaviour but hardened seamen are not so easily bought. A week after Christian was made acting lieutenant, Bligh ordered Matthew Quintal flogged for insolence and contempt of Fryer. According to the Articles of War, which Quintal had heard read every Sunday, Bligh had no choice. He was sentenced to 24 lashes, inflicted by the scholarly and articulate boatswain's mate, James Morrison.

Although MGM and Dino de Laurentis amongst others would

have us think so, the voyage south aboard *Bounty* did not consist entirely of suspicions of theft and flogging. That whatever unpleasantries had happened were commonplace and unnecessary to report, is perhaps best seen in a previously unknown account of *Bounty*'s voyage from Tenerife to the Cape of Good Hope, which I discovered in the Cumberland Pacquet of November 26 1788 where, in the adjoining column, Fletcher's 1st cousin John Christian XVII is thanked for his gift of £2 (about £150 today) by the convicts of Carlisle jail.

Although the author is anonymous, he is almost certainly the very young midshipman Peter Heywood. He was distantly related to Fletcher Christian and the two families must have known one another socially on the Isle of Man. Age difference notwithstanding, the men struck up a firm friendship aboard *Bounty*, with a degree of hero worship by Heywood that he acknowledged even later in life.

Extract of a letter from a midshipman (aged sixteen) on board his Majesty's ship 'Bounty', commanded by Capt. Bligh, now on her voyage to Otaheite under the immediate patronage of His Majesty . . . for the purpose of conveying . . . that valuable production of the vegetable kingdom, the BREAD FRUIT TREE BOUNTY, Simon's Bay in False Bay (Cape of Good Hope) June 17, 1788.

'I shall give you a short account of our passage since leaving Teneriffe [sic], and of the exceeding bad weather we experienced off Cape Horn. I do assure you, the account which Lord Anson gave of it is very true, and not in the least exaggerated, as has been generally supposed; and the report which Captain Bligh will give (as most likely his voyage will be published) will (I dare venture to affirm) correspond with mine in every particular; and perhaps deter future navigators from attempting to double that Cape at so improper a season of the year.

We left Teneriffe on Thursday the 10th of January, after staying there four days. I wrote to you from thence by a Spanish packet, which was to have sailed in two days for Cadiz. After we left Santa Cruz, we shaped our course westerly towards the coast of Brazil, and from that road, till we got into the latitudes of the 30s we had the most pleasant weather imaginable and always plenty of fish. I have drawn one of every sort we caught, and also such birds as I could get a good sight of: so that I hope, by the time I come home, I shall have a tolerable collection.

On Saturday morning, the 16th of February we saw a sail, which next morning we came up with, and found her to be a South Sea Whale-Fisherman, bound for the Cape of Good Hope. In a few days afterwards, we got out of the north-east trade, entered the variables and now and then met with a gale that used us rather roughly, and which went very much against the grain, being so uncommon in the delightful climate we left behind.

The number of large whales which we daily saw, in running down the South American Coast is wonderful; and two or three of them at a time frequently came alongside to windward of the ship and blew the water all over us; and were thereby so troublesome, that, to make them set off, we were obliged to fire at them with muskets charged with ball. They frequently bore three shots before they offered to stir. On Saturday morning the 23rd March at 2 o'clock, we made the land Tierra del Fuego bearing south-east and by this bearing found ourselves in sight of land above Cape St Diego, and of course too far to the windward of Staten Land, to attempt going through Straits le Maire, as the wind was south-west, so we immediately halted off east.

At noon the east part of Staten Land made its appearance. This land is exceedingly high; the summits of the mountains are chiefly rocks, most of them entirely covered with snow and have altogether a very wild and desolate appearance. The only natives belonging to it, and which we saw in vast numbers, are seals, porpoises, and whales; and the birds are wild ducks, albatrosses, quebrantaussoes, petrells, and many other sea birds. Cape St John which is the east point, is in the latitude of 54.47 S. And 63.47 W. We had pretty good weather for a day or two after we left the land; but as soon as we were clear of it we began to feel the effects of the Cape Horn climate! from the 25th of March till the 18th of April, was one continued gale as it seldom ceased for four hours together. During the 29 days we were beating off the Cape, we had to encounter the most violent storms that I suppose were ever experienced; and I can safely say, the wind was not 12 hours easterly during that time, and we never had more canvas spread than close reefed top sails; but most chiefly, when not lying to, reefed courses. The porpoises we caught off the Cape were thought delicious morcels; and a sea pie, made of albatross, (which, you may judge, must be very fishy, when caught above 100 leagues from the land) went down very well. After beating above three weeks, to no purpose, and the ship at last beginning to be leaky, so as to oblige us to pump every hour; and many of the people being ill, by the severity of the weather, and want of rest, (there being seldom a night but all hands were called three or four times) the captain, on the 18th of April, in the forenoon, thinking it dangerous, and very improper to lose so much time, bore down for the Cape of Good Hope, to the great joy of everyone on board. [Bligh says this happened on the 22nd.]

This was Fletcher Christian's first great challenge as second-in-command.

Peter Heywood's description of Bligh's attempt to round the Horn suggests only a small part of the terrifying ordeal the ship of wood endured. Bligh wrote on April 2 that, *'the storm exceeded what I had ever met before'*. There were another three weeks to go.

This was Fletcher Christian's first great challenge as second-in-command. He would have been everywhere, not just relying on his great physical strength for his own survival but also setting examples of duty without fail and using his documented sympathy for others to sympathise and encourage those who coped less well. If he had failed his responsibilities Bligh or Heywood or many others would have said so but none did.

On April 3 Bligh reported the snow was no longer lying on the decks but came in large flakes, the hail was sharp and severe and the entire ship was battened down. The only access to the lower deck was through the aft hatchway and Bligh's own mess place. By the 12th Bligh had given up his cabin at night, which allowed those who had no dry hammock due to leaks to sleep in comfort. He did everything he could to keep the ship warm and dry and ensured no man went on watch in wet clothes.

When at last he knew his men could take no more and as the danger of their position increased, with more and more men injured or ill with rheumatics and other complaints, Bligh ordered the ship to bear away toward the Cape of Good Hope, for which he was loudly cheered.

It is Bligh's combination of determination and sailing skill at treacherous Cape Horn that first led me to call him Foul-Weather Bligh. When the sea and weather was at its worst, Bligh was at his best and I'd want him to be in command if I were in peril on the sea. It was only in fair weather or in port, when he had to deal with men who might not agree with him, that his lack of empathy caused mis-communication, the effect of disinterest and his famed use of insulting language. For the moment, Bligh was magnificent because he was utterly in charge in ways few other captains could be.

The ship had carried poultry, sheep and hogs in great numbers but almost all had been victims of the weather. Fresh hog meat was roasted from the few that remained, the aroma of which would have made a cheering comfort. Bligh exerted himself to improve his men's *'jaded'* physical and mental health. He made each man have a breakfast of boiled wheat with sugar and a pint of sweet wort (a malt extract) each day, plus fine old rum, sauerkraut, mustard and vinegar. At first some of the men were worse after abandoning the Horn, reaction to the horrific mental and physical challenges and dangers that affirmed *Bounty* had turned away just in time.

Fletcher's discomforts were not solved by now having a following wind and better weather. It was only with the greatest difficulty the cook could light a fire in the galley or then get anyone to bear the smoke this caused on the lower deck. The foresail is generally supposed to be the cause of a ship's stove smoking when going before the wind but hauling it in made no difference and Bligh regretted not bringing a small caboose for cooking on deck. Four men had to take turns getting the dinner boiled, each in the utmost discomfort. Three officers and two men were seized with violent sickness, vomiting and severe headaches which, added to those already sick, meant *Bounty* had 12 men out of service.

Now let Peter Heywood complete his narrative and take his ship to the tip of Africa: *From that day, till we made this land, we had the wind constantly from the westward so that we had only been a month and three days making the run between the two Capes, which was I dare say, as great a run in the time, as ever was performed; and have the happiness of telling you that the Bounty is as fine a sea boat as ever swam. She does not sail very fast; her greatest rate is 8 or 9 knots; but once she went ten, quartering, which is quite sufficient. We made the Table Land on the 23rd of May and anchored in this bay on Saturday night following.*

We shall leave this place in about a fortnight and proceed for Van Dieman's Land, to wood and water; afterwards to New Zealand; and then to Otaheite.

I suppose there never were seas, in any part of the known world, to compare with those we met off Cape Horn, for height, and length of swell; the oldest seamen on board never saw anything equal to them, yet Mr Peckover (our gunner) was all the three voyages with Captain Cook.

Bounty's arrival at the Cape was marked by a salute to the fort that was satisfyingly returned by an equal number of guns. The distance from Cape Town to their berthing at Simon's Town in False Bay possibly helped the men concentrate on making the needed repairs because the entire ship had to be recaulked and every piece of the stores and provisions checked. Fresh meat and vegetables with soft bread were served daily and Bligh was rightly proud to note that, although other ships arrived direct from Europe with sick crews, tales of dead men and outbreaks of scurvy, he had braved a longer and more horrifying voyage yet had delivered his ship with every man in excellent health.

It took 38 days before they were refitted and provisioned and *Bounty* sailed at four o'clock on the afternoon on Tuesday July 1. She saluted with 13 guns, which were again returned. Those who were superstitious about their departure from England should have been shivering at this 'unlucky' number. It was the last time Fletcher Christian was ever to see a European settlement and, if one of his shipmates can be believed, he had not enjoyed it much.

It is revealing nothing to remind even a newcomer to this story that Fletcher Christian was to found the remote community of Pitcairn Island and that there is a mystery about when and where he died. One of the seamen who sailed with him signed on as Alexander Smith but later reverted to his real name of John Adams, which is how I refer to him. He was a co-mutineer and the only one alive when Fletcher Christian's hideout of Pitcairn was found almost 20 years after the mutiny. It is from Adams that we have most of the versions of Christian's death and later I will apply myself to the reasons he gave so many conflicting stories. Naturally Adams gave his opinions about the mutiny. In the last and most reliable of these accounts he mentions something about the Cape of Good Hope that always stuck in my mind. Adams told Captain Beechey in 1823 that Christian was under some obligation to Bligh and that their *'original quarrel'* happened there and *'was kept up until the mutiny occurred'* in a greater or less degree:

'Mr Christian . . . was under some obligations to him [Bligh] of a pecuniary nature, of which Bligh frequently reminded him when any difference arose. Christian, excessively annoyed at the share of blame which repeatedly fell his lot, in common with the rest of the officers, could ill endure the additional taunt of private obligations: and in a moment of excitation told his commander that sooner or later a day of reckoning would arrive'.

So, Fletcher Christian was obliged to Bligh for his posting to *Bounty* and for his promotion to acting lieutenant but if the relationship was complicated by financial dependence, nothing would be more certain to sour the friendship, especially if Bligh were continually carping about it with his famously wounding tongue. The idea of money as the reason for the mutiny was again current late in the 19th century but it was largely dismissed because it came from the Pitcairn Island family of George Nobbs, an 'outsider' who married Christian's granddaughter Sarah. That source is precisely what should have made historians listen.

Cape Town was certainly a place where Fletcher might have needed money—to spend on entertainment, to buy private stores of food or souvenirs. Letters I discovered in Carlisle showed he had previously brought shells from the West Indies to his cousin Jane, the diarist and socialite. Perhaps he, too, wanted to help his mother financially? Once I knew, as other writers have not, that Fletcher's immediate family had lost their fortune, it became obvious that if he asked Bligh for money in Cape Town, Bligh would have enjoyed making Fletcher squirm, reminding him of everything he already owed.

During my research in the Mitchell Library Reading Room of the NSW State Library in Sydney I discovered a draft letter by Bligh to Banks that proves all the above, saying: *. . . and Mr (Edward) Christian knows from his Brother's Note of Hand (which he received) that he was supplied by him with what money he wanted.'*

This gives credibility to stories that Bligh and Christian had argued about money at Cape Town. Bligh's exaggerated comment that he *'for three years kept [Christian] employed'* and Christian's need to borrow money from him both explain why Bligh considered Christian so obliged to him.

As Fletcher Christian sailed away from Africa, any change in his relationship with Bligh must have been very unsettling. *Bounty* was isolated in a way that we can barely understand today. Fletcher had no means of communication with the outside world and ahead neither the southern Indian Ocean nor the vast South Pacific offered him hope or escape in a friendly or useful port of call.

Fletcher Christian was acting second-in-command of a small piece of Great Britain, remote and afloat in the immensity of the

Indian Ocean. Ahead lay only vaguely known coastlines and the even bigger Pacific Ocean. Fletcher must have felt specially isolated because it seems the quarrel was not patched up. What could be worse when Bligh typically considered himself alone and unaided, labelling his officers incompetent and assuming unnecessarily full responsibility for the ship and the success of the expedition?

The ship's passage from the Cape of Good Hope to Adventure Bay on Bruny Island off Tasmania's south-east coast took several weeks. Peter Heywood said that Fletcher Christian enjoyed demonstrating his strength and could make a standing jump from inside one barrel directly into another. He would hold a heavy musket at arm's length and ask that it be measured as absolutely straight. At other times he helped Heywood complete the education that had been interrupted, with lessons in mathematics and classical languages. It is no wonder Heywood sympathised with Fletcher Christian all his life.

Bounty had enjoyed westerly winds during her passage but they were boisterous, sometimes bringing hail and snow, sometimes running seas so high that once the helmsman was thrown over the wheel and badly bruised. They stayed for a couple of weeks in Adventure Bay, contacting the Tasmanian aboriginals, seeing eagles, blue herons and parrots and catching fresh fish for the galley. Mr Nelson measured a tree of more than 33 feet/10 metres in girth and picked up an opossum measuring 12 inches/30.6cms from its ears to its tail, which was the same length again.

Fletcher Christian and the gunner William Peckover were in charge of the important wooding and watering parties and it's not fanciful to imagine how much Fletcher would have enjoyed being involved, not just being on land but actively laying out and planting gardens. Not many on board would have that childhood background to fall back on. Thinking of future ships, a garden of fruit trees was planted. Near the watering place they planted onions, cabbage roots and potatoes, presumably also for the use of the locals. They were described as being *'a dull black'*, for they painted themselves with soot, laid so thickly over their faces it was difficult to say what they looked like. The *ti*-tree was abundant and the leaves were gathered and dried for use as a sharp, slightly citric tea and because the thin branches made excellent brooms.

What Bligh omitted from his *Narrative of the Mutiny*, although it was fully recorded in his Log, was the problem he had with the carpenter William Purcell. On August 26 Purcell directly refused to assist with such general duties as hoisting water into the hold. This was probably a sulk, for Purcell had already earlier felt Bligh's disapprobation of his conduct with the wooding party, when Purcell had answered back *'in a most insolent and reprehensible manner'*. Bligh should have confined him until he was tried, but that could not be until the ship returned to England. Loathe to lose the assistance of an able-bodied man, he sentenced him instead to labourer's duties aboard the ship. Like an errant schoolboy, Purcell was given a *'chance by his future conduct to make up in some degree for his behaviour . . .'* Purcell did not labour on board, somehow convincing Fryer, who was in charge of these duties, that he was exempt because he was a warrant officer, or because he had carpentering to do. Refusing orders was mutiny, or near enough, and if *Bounty* had not been so far away from civilisation Purcell would have been in the deepest trouble. It was a vital moment for Bligh and he could not ignore the event. He gathered evidence from all those involved useful for a trial in the future and then ordered that, until Purcell worked as commanded, he should have no provisions. Severe punishment was promised to any man who dared assist him. Purcell was *'immediately brought . . . to his senses'* but he had not been punished as he should have been.

Captain Bligh's Log says: *'It was for the good of the voyage that I should not make him or any man a prisoner. The few I have even in the good State of health I keep them are but barely sufficient to carry the duty of the ship, it could answer no good purpose to lose the use of a healthy strong Young Man in my situation. I therefore laid aside my power in the particular for the good of the Service I am on, altho' it continued in force with equal effect.'*

Bligh was deluding himself. His power can no longer have had equal effect if he had not used it when challenged so severely. No single man is vital to a ship. The carpenter could just as well have been lost overboard or have died of scurvy and the expedition would have gone on. Considering how healthy they were so far from a home port, Bligh had more men to sail *Bounty* than would normally have been expected. For the good of the voyage Bligh should have demonstrated his power and confined Purcell, instead of writing about it. He was not allowed to flog Purcell because of his rank, so he was saved that decision. Although the cinema has created the impression that Bligh was prone to punish too easily and too severely, the opposite was true, again and again. My impression is that he hated inflicting any sort of punishment but not on any humanitarian grounds. Bligh recoiled from it because of his own insecurity and needed desperately to be admired and this was thought more likely if he forgave rather than punished. What complicated this behaviour was that when he did punish, he was inconsistent, a menace to naval discipline. By not confining Purcell, Bligh had side-stepped a personal crisis but laid a firm foundation for a bigger one. It is a course of avoidance he took again and again and each time he did he lost further respect.

As Fletcher Christian had said to his brother, sailors expected discipline and punishment when deserved and relished it for the sense of order and security it gave. Picking and carping may have made Bligh feel a better man but severely reduced any chance of him being respected the way Cook was, who punished severely but was loved unreservedly.

Boatswain's mate James Morrison, who left two accounts of the voyage, also writes of events in Adventure Bay saying that here were sown seeds of eternal discord. Bligh made officers and able seamen alike collect wood and water. When Bligh joined them *'only to criticise'*, according to Purcell, he left the ship in the hands of seamen rather than officers and came on shore with a guard of armed able seamen. Both on land and sea, responsibility for duties and status were blurred and this broke the rules these men lived by, undermining both officers and seamen.

Morrison wrote that Bligh *'found fault with the inattention of the rest, to their duty, which produced continual disputes everyone endeavouring to thwart the others in their duty and in this way they found their account and rejoiced in private at their good success.'*

Bligh was deluding himself. No single man is vital to a ship.

It seems a remarkable situation, with no man doing his duty except to win points off one another and to avoid Bligh's criticism. Bligh's was the rule of a bad-tempered schoolmaster, petulant and niggling. I cannot believe that Purcell would have behaved so unless Bligh had helped create the situation. Why would a warrant officer deliberately sabotage himself in the early days of a long voyage if there were not both provocation and, more to the point, some hope of getting away with it.

On October 9 Bligh was to sign the expense books of the boatswain and carpenter, thus formally approving them as correct and ensuring the officers were paid properly at the end of the voyage. The master Fryer also had to sign the books, something done every month or so. This time, Fryer refused unless Bligh first signed a certificate saying that his behaviour had been blameless throughout the voyage, which was not even half completed. It was gross impertinence and Bligh had to be authoritative. Telling Fryer that he did not approve of any man doing his duty conditionally, Bligh ordered all hands aft and read the relevant Articles of War. Fryer's bluff was called and *'this troublesome man saw his error and before the ship's company signed the Books'*. Fryer had capitulated but like Purcell he had not been punished. Neither did he take up Bligh's offer to allow him to write his reasons for not wishing to sign the books at the page where he subsequently put his name. Morrison is no help in explaining the circumstances, writing that Fryer did not wish to sign *'for reasons best known to himself'*.

What might those reasons be? Why should the master of a vessel on an outward voyage put himself in great danger of present imprisonment or later court martial unless he felt he had grounds for feeling righteous? There is the possibility Bligh was being too generous to himself in some of his pursing *'circumspections'* and that Fryer was offering to swap oversight of this for the certificate of good behaviour but a mutually beneficial arrangement could have been made privately, with no mention of a certificate. Why the need for a certificate?

Through long study of Bligh's behaviour, I can only presume Fryer did not expect Bligh to be fair to him on their return, based on his demonstrated inconsistency. It is easy to see how a man could reach this decision. Fryer, working so close to the captain, would be a regular target of Bligh and would certainly know of his habit of belittling and insulting everybody, regardless of their true worth and ability. Perhaps encouraged by Bligh's demonstrated weak attitude towards punishment, Fryer became an opportunist, seeing a chance to protect his career from Bligh's unfair criticism and chronic inconsistency.

Bligh might have thought himself a humanitarian by not punishing Fryer and could easily have rationalised it as being for the good of the expedition, far more believable in the case of a master than that of Purcell the carpenter.

In reality, the captain had shown weakness and was ripe for exploitation, especially by those to whom he had been most unpleasant.

CHAPTER 13

TEN MONTHS TO TAHITI

Fryer was not the only man whom Christian saw made Bligh angry on the last weeks of crossing the Pacific. The surgeon Huggan was suspected of sabotage or at least of insubordination. Three days before the Fryer incident, Bligh was incensed to find able-seaman James Valentine from Montrose, reportedly one of the strongest and healthiest of men, near death because of infection after blood-letting, treatment for an unspecified ailment. Bligh thought it monstrous it had happened, infamous that he had not been told. Once more, Christian and all the officers were to blame for not having let Bligh know what was happening and in this he was probably correct. Yet remember, Morrison wrote that every man was simply trying to do his job with as little trouble as possible. In other words, they were avoiding Bligh. There was little or no knowledge of antiseptic practice and so the blood poisoning was not necessarily the fault of the surgeon but could have happened independently or been the fault of Valentine himself.

Under normal circumstances Huggan, his assistant Ledward or another officer should have alerted Bligh. Valentine died of sepsis on October 9, the same day Fryer refused to sign the books. It's tempting to think his intransigence was symptomatic of the ship's carousel of blame and avoidance rather than something personal.

A few days later Huggan declared that three men were suffering from scurvy. Bligh had said the appearance of scurvy on a ship was a disgrace yet here were three cases. Or were there? Huggan had been severely castigated by Bligh for Valentine's death and might have been attacking in return. It is possible that the men did have scurvy in its early stages because none of the 'new foods' that Bligh had purchased contained much vitamin C, the vital preventative. They certainly gave a broader spectrum of vitamins and minerals than the usual ship's diet and dramatically improved general health aboard. Still, if a man were lax or silly about what he ate, scurvy was a distinct possibility, especially if he did not take his lunchtime grog, which included citrus juice.

To Bligh, the sick men had rheumatics or something else. He fed them essence of malt and increased the vigour of his routine for health and hygiene. On Sunday October 19, Bligh describes the muster of all hands thus: *'I think I never saw a more healthy set of Men and so decent looking in my life.'* Even today, anyone who has sailed at sea for many months would find that hard to believe.

That evening Bligh supervised the dancing but two men refused to dance. When assistant-gardener Brown said he could not dance because of pains in his legs, Huggan backed him by diagnosing scurvy. Bligh could find no symptoms and said so in his Log. The continuous tussle over diagnosis, perhaps aided by the mischief of some men and encouraged by knowledge of Bligh's unwillingness to punish, continued for weeks. The atmosphere was one of suspicion and discontent but of a silly schoolboy nature rather than anything threatening or malevolent. With every day, Bligh's authority and dignity were further undermined.

It is sad to see this happen to any man, even when he is the victim of his own attacks. It can be sadder for some of those around such a man. William Bligh had been a mentor and guide to Fletcher Christian for some years now, perhaps even a father figure, so the sight of Bligh being humiliated and tormented must have been excruciating. Fletcher Christian was as likely as any to have been lashed by Bligh's tongue and had the dreadful double pull of being faithful to a friend and loyal to his suffering brother officers. No man likes to be associated with an object of fun and Fletcher Christian must have observed Bligh's situation with great personal discomfort.

Bligh could have avoided the nonsense about diagnoses and treatment simply by insisting Huggan stop drinking. Huggan's continued inebriety and incapacity was just as insolent as the actions of Purcell and Fryer. Again, Bligh seems to have been reluctant to confront the reality head on. On October 21 he asked Huggan to stop drinking *'in a most friendly manner'* but Huggan was insensible to the suggestion. In his private Log Bligh clearly bears out my belief that he shrank from punishing anybody, hoping that problems would disappear simply because of his superior caring. He wrote: *'The Surgeon kept his Bed all this day and always drunk without eating an ounce of food. If it is ever necessary this should be publickly known, I may be blamed for not Searching his Cabbin and taking all liquor from him; but my motive is that . . . hoping every day will produce a change in him, I forbear making a public matter of my disapprobation of his conduct, in expectation as he has done many times this voyage, he may turn sober again.'*

How can a captain consider himself responsible if, once he has identified a problem and knows the solution to it, he simply sits back and hopes the man will independently come to his senses? It is inviting every other man aboard to behave as he will. For much of the time *Bounty* was sailing towards Tahiti, discipline aboard seems only to have been what the crew wished.

Two weeks after Valentine had been buried at sea, *Bounty* made the final manoeuvres that would bring her to the sanctuary of Tahiti's lagoons. She swept east of the island and then turned to sail westward. On Saturday October 25 they sighted Mehetia, an abrupt single volcanic cone, 60 miles/100kms east of Tahiti. At 6 o'clock the same evening Fletcher Christian first saw the sharp mountain tops of Tahiti, illuminated by the last of the sun.

Conscious of his duty to the expedition's success, mindful of his reputation for scrupulous attention to detail and ever solicitous of his men's health, Bligh ordered the ship's complement to be examined by Surgeon Huggan, for *'it was not expected that the intercourse of my people with the natives would be of a very reserved nature'*. Huggan pronounced a clean bill of health, suggesting no one aboard *Bounty* was venereally infected.

It is puzzling to me why, in view of the importance of not introducing more of the venereal disease brought by earlier European sailors, Bligh did not resort to the opinion of his more sober assistant-surgeon, Thomas Ledward. It is perhaps another example of the captain's belief that inflexibly following rules was the greatest service he could perform for his masters.

Quite who had which venereal disease aboard *Bounty* and from whom is fascinating but pointless to conjecture. Detective work by other authors indicates that some had been treated on the outward voyage. Medicine at the time knew little about the difference between syphilis, gonorrhoea or any kind of non-specific urethritis, common enough among men with rarely washed foreskins. To say, as some incessantly do, that Fletcher had what we would today call an infectious venereal disease, is transparently malicious. If it were true, he would have passed it on, taking it to Pitcairn Island and of this there is not a whiff.

Of course, Bligh was right to be so careful. He knew sexually transmitted disease now existed on Tahiti and did not want the responsibility of worsening it. Any man found infected would have been prohibited from free alliance with Tahitian women, so bribery may have encouraged the surgeon's clean bill, possibly as deliberate sabotage. Still, even if Huggan's opinion of universally sound genitals was as questionable as his insistence upon the presence of scurvy, Bligh would have felt he had done his strict duty and he had written it all down. He also gave strict instructions about general behaviour and bartering and the crew was told they must not under any circumstances reveal that Cook was dead.

At 4 am on Sunday morning *Bounty* hove to, waiting to get a final fix at sunrise. Point Venus and Matavai Bay were four leagues/22 kms away and, in these last hours of isolation, *Bounty* was vexed by slow, variable winds. As the sun rose behind Tahiti, the ink-green shadow of the startling land of crumpled lava they were skirting gradually focussed into three dimensions. At last, Fletcher Christian saw Tahiti's morning mist lift to form sun-gilt clouds and land that gathers itself into deep folds and sweeps suddenly into two, sharp raw peaks that rise 7000 feet/over 2 kms from valleys and ravines of deep, twisted black-green. Stupefying rock pinnacles and isolated steeples hurl themselves into the sky in slopes of such acute gradient that man may never set foot on most of them. Neither the

Immediately in front of the site of the breadfruit camp, this view seems little different from what Fletcher would have seen as he stepped on to the beach from the breadfruit camp

mountains of Cumberland nor any of his past landfalls would have prepared Fletcher Christian for his first sight of Tahiti.

Today's visitors agree it's hard to believe Tahiti is real

By the time *Bounty* carefully managed the passage through the hissing reef it was 9am and she was then overwhelmed by a welcome of gift-laden canoes. The number of men and women soon aboard *Bounty* grew so that Bligh found it hard to see his officers and crew. It was a journey's end as intoxicating as the physical surroundings.

The black beach that edged the crowded lagoon of Matavai Bay is unremarkable, a gentle, shallow curve, just over a mile/1.6 kms long, of volcanic sand notable for its fineness and intrusiveness. As Fletcher Christian looked to his right, he would see the naked and rufous outpost of cliff called One Tree Hill. To his left, the extremity of the arc was formed by the low flat peninsula named Point Venus by Cook, which would be the expedition's land base. Behind the beach the land is flat for some distance and watered by a river that divided Point Venus and poured its limpid waters from high in the interior into the salt-blue Pacific.

Much of the land ahead was thickly blanketed from Fletcher Christian's view by gardens owing little to the artifice of man, a wonderful perfumed complication of tropical creeper and vine, a muddle of tall coconut, great breadfruit, orange and vee-apple, mango, hibiscus and gardenia, plantain, banana and sweet potato. There, he would discover each man lived with his family independent of others, contented and commanding in his corner of paradise.

For now, Fletcher Christian could only guess at these pleasures. After sailing 27,086 miles, almost 44,000 kms, further than circulating Earth and being at sea for the best part of ten months, they were finally in Tahiti. The press of golden bodies, the garlands of scented flowers, the gifts of fresh fruit and meat and vegetables proved it was real.

It was Sunday but there was to be no rest. The gifts, the visitors, the insistent questions and constant assurances of goodwill increased as the tropical shadows lengthened. Bligh noted sourly that only chiefs of lesser rank came to greet him on the first day and that most seemed more anxious to know the fate of Cook. Bligh wanted to move *Bounty* to a safer and more permanent anchorage but dared not do this while so hindered with visitors. He deferred the operation until the earliest hours of Monday, finally anchoring in seven fathoms/13 metres of water close the shore of Point Venus. This dawn start may not have been widely appreciated, for although Bligh ensured that all Tahitian men left the ship overnight, any woman who wished was welcome to stay, probably on deck because Tahitian priests forbade anyone to have another stand above a person's head.

Extraordinarily, Tahitian priests had once foretold pale-skinned men with clothes that covered their bodies would come in a big canoe with no outriggers. This had come decades earlier and was happening again. There was little change in the original belief on Tahiti that these hairy, pink-skinned men were Gods of a sort.

It was only a matter of days before Bligh was satisfied with formal receptions by those he thought the most important chiefs. He was mistaken in the relative ranks of some but managed quickly to be invited to take as many breadfruit plants as he wanted, in return for the gifts that were being so freely distributed. The Tahitians must have thought King George very strange, giving away such real treasures as metal tools and glass beads for mere plants, which grew abundantly. By Sunday November 2 Bligh had paid his respects to the child Tu, who was carried out for a mutual inspection across a river. He was too sacred to be approached closely, even by Tahitians. Bligh thought him Tahiti's supreme chief but he was only chief of one of the districts that divided Matavai Bay.

Henceforth nothing was thought likely to hinder the collection of breadfruit shoots and that same day Fletcher Christian was sent with a party of eight men to erect a tent on Point Venus. This was to be the breadfruit nursery and with the help of officials from the local districts Bligh fixed a boundary over which Tahitian men and women were not to cross without permission. Matavai's priests and lesser chiefs were happy to oblige, even helping with the policing of the site because this was a surprisingly easy way to pay for the gifts *Bounty* brought.

This view from high inland across Matavai Bay shows buildings on Point Venus where Fletcher Christian's breadfruit camp had once been

An aerial view over Point Venus today
(Tahiti Tourisme)

TITREANO IN TAHITI

October 1788 to April 1789

It's easy to think the top print is inaccurate and romanticised, but men and women did drape tapa *cloth like togas and dancers did use coconut shells like that, as we still found on Bora Bora.*

CHAPTER 14

FREE LOVE AND DEATH

Fletcher Christian's shore party included Peter Heywood and gunner William Peckover, who had been here on all Cook's voyages. He spoke excellent Tahitian and had a perfect understanding of the Tahitian mind, so was placed in charge of the trading and purchasing of provisions, which neatly avoided inflation caused by competitive buying. The botanists Nelson and Brown were naturally members of the permanent shore party but the other four may have been rotated from among the entire complement. Otherwise, Bligh allowed only two men shore leave each day. So, most of the time, the majority lived on the ship, hotter and more airless than ever but constant visitors gave *Bounty*'s men ample opportunity to make friends of both sexes. Soon few were without a *taio*, a blood brother whose family immediately became theirs.

With his second-in-command permanently on shore, Bligh, or *Parai* as his name was pronounced locally, was completely responsible for the ship, entertaining and protocol. He entertained chiefs and their wives to gift giving and huge meals, often having to put the food or the wine for which they quickly developed a taste into the mouths of the rulers. On shore he was royally entertained with ceremonies and presentations. He was danced and sung for, carried over rivers, dragged up streams in canoes. Women dropped their clothes, leaving them as presents. Bligh wrote everything down for he was as assiduous a questioner as his hosts. Throughout he wore his full-dress uniform with its lace and skirted coat. At many ceremonies he was draped in bark cloth *tapa*, a great honour that added both to his dignity and the heat engendered by his heavy coat of finest English wool. It was too much, even for an Englishman and a naval lieutenant. Several times he felt faint, a situation aggravated by the crowds who stood close. They had to be waved away.

Bligh was writing about life in Tahiti, not living it. If certain customs did not suit his temperament or inclination, he adapted them to his way of thinking. Every Tahitian and *Bounty* man was expected to bare their shoulders in the presence of royal personages or at certain places belonging to them but Bligh removed only his hat, saying that was the way he would salute his king.

Bligh lived as much like an Englishman as he could while Fletcher Christian was daily living more and more like a Tahitian. Apart from the nudgings and winkings of a prurient press, few people know what 18th-century Tahiti was really like. There was no cannibalism and none of the excesses of nudity, poverty, slavery, squalor, riches and disease that Fletcher encountered in India, Africa and the Caribbean. Instead, he found himself confronted by everything Rousseau had said was possible and preferable - a tall, noble race of golden men and women with few, if any, old, maimed and unwanted. Food grew abundantly with little attention and great variety and there were fish in the rivers, the lagoon and in the open sea.

Most disconcerting of all for Fletcher Christian, these charmed creatures were draped in yards and yards of snow-white cloth, like fancy-dress togas. It was as though the sirens and heroes and farmers and sailors of his classical Greek and Roman education had been resurrected, or had never died.

The 2018 *Oceania* exhibition at London's Royal Academy, which should have known better, allowed criticism of early artist-visitors for depicting Tahitians like Classical Greeks and Roman but those critics were ill-informed. Tahitians did dress like this and the higher their rank the more they piled on the draperies of white cloth, just as the Tahitian Mai did for his portrait by Joshua Reynolds. Tahitians looked and dressed like golden gods and goddesses and like them were willing to share life's physical pleasures, provided strict rules were understood. If they were broken, the miscreant could be bludgeoned to a post-*flagrante delicto* death.

We know a lot about ancient Tahiti, thanks to Bligh's zealous observations for the book he hoped would make him rich, and the anthropological interest of Cook, Banks and James Morrison, among others, who excused their detailed observation of sex and paganism by reference to Christian reforming zeal. There are also the three scholarly volumes of Douglas Oliver's *Ancient Tahitian Society* and thinner, more accessible ones, like Bengt Danielsson's *Love in the South Seas*, all of which repay with understandings of a life that, although cruel to those who were not considered suitable, was seen by those who never experienced it as the nearest to Eden anywhere on Earth. Bligh even tells us that when *Bounty* arrived there were no mosquitoes and that would indeed be paradise for most visitors to the South Seas, even today.

Tahiti in the 1780s had no single king. It was a series of small competitive clandoms, each managed by a ruling class of a few families, numerous gods and presumably, goddesses. This was the *ari'i* class and the top people of most districts were related in one way or another, so there was essentially a royal clan on the island but no overall sovereign. Each district's chief, who could be a man or a woman, was so sacred that their hair and nail clippings, faeces and urine had to be hidden lest mortals use them to cast spells upon his family. All that a chief's feet touched belonged to them, so when one was about on business, they were carried shoulder high. When you saw them, you were expected to bare yourself to the waist, even if Royal Navy officers.

The Tahitians were eminently practical and did not let such things as living gods interfere with day to day convenience. Thus, *ari'i* chiefs and chieftainesses were not carried about all their lives, for this was a class bred strictly for maximum height and weight accompanied by minimal colour, all marks of the regality and divinity they claimed. Men and women were regularly over 6 feet/1.8 metres tall, towering over the average pale Englishman and even the lesser and darker Tahitian classes might have done so, too. Once an heir had been born to an *ari'i* chief, and pronounced worthy of life by virtue of their size and colour, the new-born was proclaimed sovereign, allowing the parents then to walk freely and unacquisitively while their child became the darker man's new burden. Thus, the sacred child Tu was a ruling chief but had a living father and grandfather.

To ensure the continuation of privilege and the purity of their noble lines, marriage with the darker and shorter lower classes was forbidden, although the *ari'i* men might use their women for pleasure. Incestuous relationships, otherwise anathema to the Tahitian, were arranged for hierarchical reasons within the ruling *ari'i* class. Marriages between half brothers and sisters were considered ideal. This relied on an element of hybrid vigour of strong genes to create a child ever bigger and paler but if that failed, the new-born was killed at birth.

There was enormous prestige in being a chief but power struggles were usually controlled within each clan by delegating most duties and privileges to other families, so that although one man had the position, others wielded much of the power. It worked extremely well until the introduction of European firearms by *Bounty* and idea of a sole monarchy.

The self-indulgence and leisurely life possible on Tahiti for *ari'i* is most clearly showed in ritual fattening. At intervals, a group or an individual would retire to a special thatched building in a pleasant place and do nothing but gorge on starches and rich pork and keep out of the sun, emerging fat to the point of grossness, pale to the point of divinity.

It was not just the *ari'i* who were careful about the colour and number of their children. Abortion and infanticide were as common as eating and far more children were conceived than were ever allowed to live. Bligh recorded speaking to women who had lost six and eight children respectively without qualm or regret but it seems unlikely he would have understood any woman's misery. Certainly, the majority of girl children would be smothered or have their heads beaten in before they gasped their first lungful of Tahitian air, something explained a little later. We may shudder but birth control was vital to the community's balance and well-being, ensuring those who survived would contribute to the community rather than be a burden. Tahitians never risked a population too big for the island's resources, inexhaustible though these appeared.

Mechanical birth control was unknown and *coitus interruptus* unthinkable but abortion was readily induced by herbs, direct interference with the foetus and by a method of deep massaging, which must have dislodged the foetus and its placenta from the wall of the womb. Women gave birth squatting, often held from behind by a man who was there to decide in an instant if the infant was to live or die.

Once children survived initial assessment, they did not remain as they were born. Girls almost immediately had their faces massaged, while bones were still very soft. This went on a long time as the objective was to flatten the face, particularly the nose. If this was not done with care, the septum would be broken and breathing affected all their lives. It was an aspect of Tahitian femininity and appearance that early European visitors must simply have accepted without ever knowing it was artificial. Boys faces were left as they were. Instead, their new-born heads were bound to boards, so the occiput, the back of their head was flattened and forced up to a peak, so they looked bigger and fiercer as warriors.

Flat-faced or pointy-headed, absolute freedom of self-expression and lightly supervised independence within an indulgent, extended family were the promises given to those children allowed to live. The undivided rooms of the low, cool houses made concealment of intercourse impossible and birth was also public entertainment. Although children knew their real mother and father, they would each have many more surrogates, all called by the same name. Within the immediate family and relatives, children were virtually public property from birth. The only drawback from the child's point of view was that all six of the men or women called parents might choose to chastise them at once. It was actually very rare for children to be severely reprimanded and Bengt Danielsson says it would have possibly resulted in a pit full of food being specially cooked for the child, as an apology for having his dignity, or bottom, affronted. Morrison says that if a mother hit a girl for screaming while she was being tattooed, the mother would have been put to death but this is unsubstantiated.

For all the admired pallor of the parts of their skin that had not been tanned, Fletcher Christian and his fellow officers had little chance of physical entanglement with the daughters of the *ari'i*. Even though so little visited, the Tahitians had already decided on conventions for the comforting of visiting ships, based on the convenient breakdown of both the crew and their society into three strata. The captain was allowed, indeed expected, to bed only an *ari'i* woman. Bligh never seems to have brought himself to conquer these mountains of corpulence. On the other hand, he was such a stickler for protocol and so prissy it is just as likely that he did but was simply too nauseated to record this for posterity and the public in his Log and journal. What would Betsy and the girls have thought? He had the precedent of Cook, who avoided or concealed the truth of his royal duties, too.

The officers of a ship were given considerable freedom among the women of the aristocratic *ra'atira*, the important class of land-owners and minor chieftains and this is how we know the probable background of Mauatua, who later went with Fletcher Christian to Tubuai and Pitcairn Island. Ordinary seamen had to make do with the women of the *manahune*, the darker-skinned labouring class, a generally happy mutual arrangement, for there were more of them, they were freer to give affection and wanted only copper nails in return for their favours.

Sensuality was not the sin Christianity taught it to be, not on Tahiti. It was free and easy and constant. Women were expected to pursue and to enjoy sexual activity every bit as much as men but stories of uninhibited public couplings and universal concupiscence are largely exaggerated. Any that can be verified are applicable only to the *manahune* women, who naturally took every chance to gain status by having a white, and thus, in their eyes, an upper-class lover. To make themselves more attractive, some would bleach their skins. Sex was indeed on offer, in spite of Tahitians having to overcome the horrors of the *Bounty* crew's unwashed foreskins, their pubic and other body hair, rotten teeth, execrable manners and a universal mantling of stale body odour and foul halitosis, all abhorrent to Tahitians, who constantly plucked their pubic hair, bathed twice daily and had magnificent teeth.

. . . this is how we know the probable background of Mauatua, who later went with Fletcher Christian to Tubuai and Pitcairn Island.

I would like to see Fletcher Christian as combining the best features of his re-creators Wilton Power (1916), Errol Flynn (1932), Clark Gable (1936) Marlon Brando (1963) and Mel Gibson (1980) on film and David Essex (1985) on stage. Yet he and Bligh and the rest undoubtedly stank. Washing was never a primary interest of the English sailor and rinsing clothes in salt water is hardly enough to banish months of sweat and grime. Incursions into their health, teeth, gums and hair by the malnutrition and vitamin deficiency of earlier voyages would not have been cured by Bligh's more sensible provisions. A 21st-century observer of the welcome afforded these men might well feel the hospitality of the Tahitian women exceeded that of even the politest hostesses.

Tahitians chose with whom they lived at an early age and because nothing social or sexual was hidden, and there were no enormous time-consuming responsibilities to the family or community and no rules about what the age of maturity was, they could grow up at their own speed, entering adult life and activity when it suited them. Parents would as easily suggest that a fractious child should masturbate as today's might suggest Play Station, X-Box or U-Tube. Children emulated copulation and played at being mothers and fathers to a realistic extreme. Girls used soft unripe coconuts as infants, which they would drop from between their thighs with appropriate grunts to the cries of approval of their playmates.

What we call teenagers formed gangs and went off to live at one another's houses, or co-habited wildly and communally in one they had constructed themselves. For as many years as it pleased them, they devoted each day to sport, music, dancing and love, sowing and reaping their wild oats until such time as one of their partners seemed more attractive than the others and marriage was considered. In many ways this self-policing stage was an extended initiation rite. There were few others except for tattooing for both sexes and a type of circumcision for the boys.

Male circumcision on Tahiti was based on extremely ancient but largely forgotten religious grounds. It had been continued both as a tradition and because health considerations had been attached to the operation, as they still are today in many countries. On Tahiti, the penis was supercised rather than circumcised. When the boy was judged old enough, perhaps by the appearance of pubic hair, he was attended to by a man who specialised in the operation, probably on the family marae, the equivalent of using a local church for Christian christening. The upper foreskin was stretched over a shell or piece of smooth curved wood, then quickly slit with a shark's-tooth knife and ashes applied to the wound to stop the blood flow. It was not terribly painful but the boy would limp and wince for a day or two to tell everyone of his new status. If he was royal it was more of an ordeal for others. When a noble's prepuce was being slit on his ancestral family marae, one or more human sacrifices were suspended by sennit strings through their ears from a sacred *toa* tree under the direction of the high priest of the holiest marae, the equivalent of Anglican England's Westminster Abbey

From this time on, a boy was expected to cover his penis in public. To reach full maturity and not be supercised was considered perfectly disgusting and you were shunned by women and laughed at by men.

Fletcher Christian and his compatriots were among the first Englishmen to submit to the agony of extensive tattooing. The word tattoo is an anglicisation of the Tahitian *ta'tau* and the *Bounty* men who returned to England, including Peter Heywood, introduced the custom to sailors at large. Tattooing on Tahiti was then a painful and dangerous process, infection from which could be fatal. Both boys and girls were tattooed and the latter were not considered suited to take their place among women until they had been. There were meanings to some of the marks, others were purely decorative but without them any claim to beauty was hollow. Sadly, missionaries destroyed Tahitian culture more profoundly than on any other South Pacific island and there is little known about traditional Tahitian tattoo design.

Sir Joseph Banks gives the earliest description of what was done, and how: *The colour they use is lamp black which they prepare from the smoak [i.e. ash] of a kind of oily nut used by them instead of candles; this is kept in cocoa nut shells and mixt with water occasionally for use. Their instruments for pricking this under the skin are made of Bone or shell, flat, the lower part is cut into sharp teeth from 3 to 20 according to the purposes it is to be used for and the upper fastned to a handle. These teeth are dipped into the black liquor and then drove by quick sharp blows struck upon the handle with a stick for that purpose into the skin so deep that every stroke is followed by a small quantity of Blood, or serum at least, and the part so markd remains sore for many days before it heals.*

I saw the operation . . . performed upon a girl of about 12 years old, it provd . . . most painful . . . every stroke . . . hundreds of which were made in a minute, drew blood. The patient bore this for about 1/4 of an hour with most stoical resolution, by that time the pain began to operate too strongly to be peaceably endured she began to complain and burst into loud lamentations . . . she was held down by two women who sometimes scolded, sometimes beat and at others coaxed her.

This particular tattooing session lasted over an hour and was to decorate one of the girl's buttocks with the solid black thought so attractive in both men and women, one of the few facts we know. When she eventually submitted to the additions that curved up from the base of the spine across the small of the back the pain would be worse but this was the decoration most highly prized and commented upon.

Bligh mentions briefly *'tattows'* of the men who later took his ship. Fletcher Christian had a Garter star tattooed on his breast and was tattooed on the *'backsides'* Bligh said. It is likely he entered so fully into Tahitian life that he submitted to the blackening of his entire buttocks, the sign of a mature Tahitian male, indicative of the time he had to himself and the closeness he felt to Tahiti and Tahitians.

CHAPTER 15

'THE THIRD PERSON WAS MORE HORRIBLE'

Unfettered passion was tempered by practicality for young Tahitians, who easily absorbed the tenets of survival during childhood. The three necessities of life in Tahiti were food, clothing and shelter and as there was none of the artifice and complication required by Europeans, instruction was consequently very simple, too. Proficiency was expected from both boys and girls in the best methods of cultivating breadfruit, bananas, plantain, yams and other roots. Both learned the best baits, the seasons and places to lure each type of fish, what could be eaten from the sea and what could not. They would know how to rear dogs, swine and fowls and women were instructed in a social polish of sweetness and grace that attracted men to them and that they passed on, to the improvement of the manners of their swains. Young girls were taught early on how to beat the bark of the paper mulberry tree, *broussonetia papyrifera*, into the cloth, known today as *tapa*, that everyone so gracefully wore and that was used to furnish their houses and beds. Boys were expected to know each plant and tree and the use of its leaves or timber for making habitations, boats, paddles and sails but women were the ones who knew and prescribed cures made from them. Tahitian women taught their daughters that the way to avoid pregnancy was to change partners constantly.

Fletcher Christian would have recognised the universal dual morality of males, who were allowed to spread their amours far and wide while inhibiting those of their wives. Even so, as well as her husband, a Tahitian woman was expected to have sex with all her husband's brothers and his *taio* or blood brothers as well as with honoured guests. It is through a misunderstanding of this carefully policed intra-familial freedom that some stories of universal licence have originated. Europeans were appalled at the ease with which a man offered his wife to guests, even though it was represented as good manners. One missionary, who refused and then foolishly stayed to sleep, wakened to find his penis being handled and discussed by women who thought his refusal indicated he did not have the usual necessities. He was so anguished by this rape of his righteousness that he was never mentally the same again.

As in any society there were males who were attracted more or as much to their own sex and to sodomy but no-one seemed to notice or care that much. Both boys and girls were thought to have begun sexual activity as early as seven or eight, something hard to prevent when they regularly saw sexual intercourse in the undivided interiors of their houses. Of course boys would play with other boys and when they were older some would choose discreetly to continue such relationships, even though husbands and fathers. Privileged older men bribed much younger boys, as also seems universal. Tu, later King Pomare I, is said to have demanded that a stream of young men attend to such needs.

There was a much more obvious kind of gender fluidity, the *mahu*. They were males who had been brought up as females, who walked, talked and dressed as women and performed most of the domestic duties of the sex. They were limited in number and *mahu* did not usually choose their way of life themselves but were selected at an early age. This was sometimes done because there was a shortage of feminine help in a family group and a boy child who was perhaps more girlish was then sacrificed to the hearth. Sometimes the boy was chosen because he had a small penis and the status of being a *mahu* was considered some compensation.

Bligh wrote that he had closely examined a *mahu* or two and said that their genitals were somehow pulled back between their legs and seemed to have shrunk. Other men, he said, obtained pleasure between the *mahu's* thighs and he also made oblique reference to further aberrant sexual practices that he believed were rife, possibly bestiality with dogs.

There was no shame in being a *mahu*, indeed there was rank and advantage. Just like the palace eunuchs of the Ottomans or in Beijing's Forbidden City, many *mahu* were senior servants to chiefs, who were forbidden women attendants. Most were available for sexual purposes but are thought performed only fellatio and perhaps inter-crural gratification.

Daily life for the Tahitians had little pattern to it, other than bathing at least twice in a cool stream and preparing the one meal of the day. Besides being clean and fragrant, food was probably uppermost in your mind. You might fish, you might garden or pick a little fruit, change a flower in your hair or stroke oil into it. You plucked out or shaved your pubic and all other body hair with bamboo shards or sharpened shells, cleaned your teeth, then possibly enjoyed a massage with scented oil. Sometime during the year women pounded for days at the bark of the paper mulberry tree to make tapa cloth.

Women could not touch much of men's food and were forbidden to eat most good things anyway. It was men who prepared the great earth ovens, men who placed the food on the heated stones and covered it with banana leaves and more earth until it was cooked. Men and women ate separately, for if a woman were to touch a man's cooked food it was thrown away. If a woman touched his eating implements, they had to be reconsecrated or replaced and it was forbidden for men to wear cloth over which women had walked.

Perhaps Tahiti's most startling customs were associated with the *arioi*. These were bands of men and women sworn into a secret society devoted solely to pleasuring themselves and others. All three classes of Tahitian men and women were permitted to join, starting at a position in the ranks equal to their origins. *Arioi* status at any time was identified by the type and positioning of cabbalistic tattooing and wherever the *arioi* went they had to be welcomed and fed without stint, easily depleting the reserves of a small community or island within days but to deny them could mean death. In return, the host was royally entertained.

Penis manipulators were real stars. William Bligh was horrified by them and insisted their actions be terminated but still recorded what he had seen: *. . . they suddenly took off what clothing they had about their hips and appeared quite naked . . . the whole business now became the power and capability of distorting the penis and testicles . . . The person who was ready to begin had his penis swelled and distorted out into an erection by having a severe twine ligature close up to the os pubis applied so tight that the penis was apparently almost cut through.*

The second brought his stones to the head of his penis and with a small cloth bandage he wrapped them round and round up towards his belly, stretching them at the same time very violently until they were near a foot in length . . . the two stones and the head of the penis being like three small balls at the extremity.

The third person was more horrible than the other two, for with both hands seizing the extremity of the scrotum he pulled it out with such force the penis went in totally out of sight and the scrotum became shockingly distended (as far as the knees) . . .

In this manner they danced about the ring for a few minutes, when I desired them to desist . . .

The lasciviousness of *arioi* dances under great flares stimulated great orgies but consummation was normally discreetly veiled in the bushes. There could be absolute abandon and crossing of every social, age or other barrier, for the *arioi* were not permitted children and any produced were aborted or killed at birth. The *arioi* were the planned and practical pressure valves of a society that relished personal freedom but clearly understood that freedom only works if policed by the strictest of rules. People always look to what is forbidden and the *arioi* pandered to that but controlled any unwanted outcomes. In time of war, *arioi*, trained by their martial games, were prohibited sexual adventure so their energies were directed solely to battle.

Not only the *arioi* stimulated orgies, neither were they the only ones who danced. A high point of life was the *heiva*, a formalised festival where hundreds of men and women would sing and dance to thrilling drum rhythms, each in perfect unison as they went through actions both acrobatic and sexual that flabbergasted first visitors and that have the same effect on modern ones.

The enjoyment of Tahitian life would be simple for Fletcher Christian and his shore party. Within the rules, sexual liaisons would be trouble free and at night they could abandon themselves to the Tahitian ethos, as long as one eye was on the collected breadfruit, so comfort and duty might be pursued concurrently.

CHAPTER 16
TITREANO

It is commonly assumed that *Bounty* stayed at Tahiti for six months because when she arrived the season for collecting breadfruit plants was over. Although possible, breadfruit do not usually reproduce by seeds and seedlings but by shoots or suckers which spring from the roots of established trees, in the same way as bananas.

Less than two weeks after arrival Nelson pronounced the collection of breadfruit plants could begin. Work began on November 7, when 110 shoots were collected. By the 15th they had 774, so in just a week Fletcher Christian's shore party had done all that *Bounty* had come to do. After waiting another month to complete repairs and to ensure the shoots had taken successfully *Bounty* could have sailed home but instead she stayed a further 20 weeks.

There has been much speculation about this, including criticism of Captain Bligh for what seems self-indulgence.

Bounty was the first British ship to be at Tahiti during the rainy season, or summer, which lasts approximately from November to April. This is also the cyclone season and even today sailing ships prefer not to be on the Pacific in the worst of these months. As well, Bligh was expected to return via the Endeavour Straits between Australia and New Guinea. If he sailed at the end of December, crossing the Pacific during the unpredictable rainy season, he might be prevented from continuing west through Endeavour Straits by head-on prevailing winds, just as at Cape Horn. From every point of view the success of the project was probable only if he delayed at Tahiti until weather conditions improved and the wind directions changed.

Bounty was not detained at Tahiti through over-indulgence in Tahitian life but because she sailed so late, a victim of the sloppy, mission-delaying inattention the expedition was given in England in 1787. If the delay is thus explained, the use Bligh made of it is not. It is perplexing that he did not keep his men more attuned to life at sea by embarking on short surveying and map-making expeditions to close island groups. Even allowing for the treacherous changeability of the weather it would have seemed that Bligh's thirst for discovery would have sent him exploring. Instead he contented himself with discoveries on shore, daily hob-nobbing with those he thought the island's sovereign families. He was writing a detailed journal, which because he would have stayed in Tahiti longer than others, was bound to be filled with novelty and become a very big source of income when published. The delay was annoying but had distinct advantages for Bligh's determination to improve his family's financial situation. Bligh's royal progress was not without tedium and startling reminders that rulers here had highly individual attitudes to life and love. He was especially exercised to hear whispered that the servant who normally put the food into the mouth of Teina, the father of Tu, was also the lover of his wife, the masculine Itia, who scorned gifts of beads and mirrors, demanding nails and iron. Teina was 6'3"/1.9 metres tall and very fat and was a vacillating, lethargic whiner. Itia was quite as big but had the opposite attitude, so determined to prevail over visitors that she first arranged for the gifts Bligh gave to be kept on board *Bounty* in a specially constructed chest. Then, in case Bligh stopped giving she daily transferred some of the contents, so the chest always had room for more. Teina and Itia even supervised Bligh's munificence to other chiefs and he noted they were not as generous as he would have been.

. . . in just a week Fletcher Christian's shore party had done all that Bounty *had come to do.*

Early in December a startling thing happened. The wind changed and Matavai Bay became as rough as the open sea. *Bounty*'s hatches were lashed and the crew remained on board as the ship rolled and pitched furiously. Storm-water swelled the river that ran along Point Venus, threatening to flood the potted breadfruit shoots. By midday on December 6, the wind had abated just enough for Teina to risk the waves and tearfully clamber aboard *Bounty,* not to see if Bligh were well but regally to farewell everybody on board before they were dashed to the shore or the reef and perished. Nelson struggled out to report that Fletcher's team had diverted the stream by digging an emergency channel and that the breadfruit plants were safe.

No serious damage was found on board or ashore, apart from a sharp deterioration in the precarious health of surgeon Huggan. On the 9th he was wanted ashore but when an officer went to fetch him, he found him helpless in his dark, fetid cockpit-cabin, from which he had not moved for days. Sensing the stupor was something more than drunkenness, the officer ordered Huggan to be moved into better air and light but this was more than his ravaged constitution could stand and he quickly expired. He was buried on Point Venus in a grave that Tahitians dug facing east - west as they had already been taught by Catholic Spanish visitors. Ledward was appointed surgeon.

With no precedent for staying at Tahiti in these months, Bligh did much questioning and found the storm *Bounty* had just ridden out was by no means a freak. Matavai Bay became positively dangerous between November and April, the entire rainy season. He resolved to move to one of the sheltered bays of Moorea, the high-peaked island seen from Matavai and less than a day's sail away, an idea greeted with horror.

The chiefs of Tahiti's competitive districts were still jockeying for status, certain there was much more treasure from King George aboard *Bounty*. They barely wished to share with one another, let alone with another island. Teina was the most anxious because he had been half promised some muskets and Moorea was home to his arch enemies. If Moorea should get the muskets and the friendship of the Englishmen, his ambition of overall sovereignty for his family and his son Tu would be dashed. In an act of histrionic genius and injured pride, such as only a manipulative Tahitian could muster, he accused Bligh of ingratitude and of treachery. Wailing and wheedling, Teina said Bligh was risking the life of his crew, now all his dear friends, if he sailed to Moorea. Why not move around the corner to Toaroah in his more sheltered district of Pare?

Bligh was not convinced but walked over One Tree Hill to inspect the suggested harbour and found it was rather a good idea. Quickly the ship was prepared and Fletcher Christian was ordered to transport the breadfruit pots to *Bounty*, which seems an unnecessary labour. The breadfruit might have remained where they were, for the new anchorage was only a few miles away. If the move really was necessary, they might simply have been carried. There were plenty of Tahitians whom Teina could have commanded to carry them.

As it was, the short move to Toaroah harbour almost dashed the entire endeavour, ship, men and breadfruit. Just as *Bounty* slowly approached the entrance to her new anchorage a combination of bad observation from the crow's nest by Fryer and the tardiness of the launch crew in throwing a line aboard *Bounty* meant she went aground on a coral shoal inside the reef. Her copper sheathing and snail's pace meant was not holed but then her crew bungled the relatively simple matter of getting her off and the change of anchorage took a whole day of tedium and effort.

Gavin Kennedy blames Christian, saying he must have been in charge of the slow launch (must he?) but the grounding could also have been entirely Fryer's fault. Those who point at Fryer or Christian might equally wonder if Bligh had given orders as concise as they might have been. Those who have sailed in these waters know it is extremely dangerous to move through reefs and shoals unless the sun is directly above, giving a clear view of underwater hazards. *Bounty* was sailing in mid-morning, not an optimum time for best vision and, as commander with South Pacific experience, Captain Bligh should have known that.

I cannot help thinking this was because *Bounty* was no longer efficiently crewed, something that would not have happened if the ship had sailed on exploratory charting expeditions rather than wallowing in Matavai Bay. The teamwork that seems so effortless at sea rapidly dissipates in port and only slowly rekindles into efficiency. Perhaps personal fault is why Bligh's reports of the incident mention only Fryer and no other officer and why they resound with rather too much protestation of his own innocence? Was he protecting himself, rather than Fletcher Christian as has been suggested?

When *Bounty* was finally anchored and buoyed at Toaroah there were celebrations all around. Teina and Itia were jubilant. Not only was Bligh still in Tahiti but he and his ship were now in the waters of their own Pare district, giving increased status and even greater opportunities of obtaining muskets or anything else. Teina ordered his holy men to perform a sort of thanksgiving ceremony. Bligh in turn ordered Christmas celebrations for all on December 28, which included a demonstration of the power of the ship's cannons and swivel guns. Teina, the timid giant, was perfectly terrified by their noise and thus resolved even more firmly to own some.

Fletcher Christian's new breadfruit camp was not much different from that on Point Venus, though the surroundings were somewhat less salubrious. On a small point of land, just a few minutes' walk from the camp, was the main marae of Teina's family and district. It horrified Europeans because it was covered with the bones of sacrificial victims and was the site of ceremonies that required interminable, chilling chants and the blare of conch-shell trumpets. They could not ignore the bodies of the sacrificed simply by averting their eyes because there was a smell, too.

It was forbidden to kill on the marae itself, so there was understandable unease throughout the community when there was to be a ceremony requiring a cadaver or two. Often these were collected over an unnerving number of days or weeks and the mere knowledge that one tribute had been killed was never the signal for release from constant vigilance. Women who suspected their husband or son might be chosen kept an especially close watch on their men. There was no ritual foreplay to the killing. The first the victims or their family knew about it was when the man was felled with a sudden blow to the head. If a woman were quick enough to touch the body before it could be whisked away to sacred places, it was desecrated, so she could then take it for private mourning and burial. The body was also abandoned if it had lacerations from fingernails or teeth from avid lovemaking. If the victim had been especially choice, a mother who saved her son's body might find spitefully herself conjoined in his state of oblivion but even dead women were not allowed onto a marae.

Fletcher Christian was known as *Titreano,* for this was the closest the Tahitian tongue could come to his family name. For four more months, he and his precocious friend Peter Heywood were surrounded by the exotic life of Tahiti, day and night, sacred and profane. They had little to do but enjoy themselves. There were watches to be kept over the obediently flourishing plants but

Nelson and Brown were the experts and these were primarily their concern.

Off duty, there was a substitute or addition to whatever alcohol Fletcher had been allowed to bring ashore, the mildly narcotic drink *'ava*, called *yava* by Morrison and better known today by its Tongan name *kava*. This drink is made by mashing and sometimes masticating the root of *piper methysticum* with a small amount of water. It has an anaesthetic and calming effect, promoting well-being and contentment and causes subtle changes of perception of sight and sound without inducing hallucinations. Unlike alcohol, it is never associated with loss of temper, belligerence or confusion of thought and speech. It's no exaggeration to say that if Pitcairn Island had the *'ava* plant as a substitute when *Bounty's* alcohol ran out, history would have been very different indeed.

Fletcher Christian and Peter Heywood soon became part of the community. Heywood was proficient enough at the language to commence a dictionary and both men submitted to tattooing. John Adams too was highly tattooed, and among those who found themselves enamoured of one woman rather than several were George Stewart and Charles Churchill.

After the melancholy of their extended voyage, the men from the northern hemisphere appreciated food as much as women. The European has no difficulty in taking to the bland sweet diet of the South Pacific, high in starch and flesh. There was plenty of pork and fresh fish and fruit, and vegetables cooked in the same pit were sometimes plain, sometimes made into a sort of pudding wrapped in banana leaves. Bligh even wrote the recipe of one, a mixture of grated starchy root and coconut milk that nowadays you are more likely to eat it on Pitcairn Island than in Tahiti. And then there was the breadfruit usually baked in pit ovens, then peeled and eaten warm, when it is not unlike a mild freshly baked bread or sweet potatoes. In times of plenty a pulp of the flesh would be stored, when acetic-acid fermentation of its sugars meant its keeping qualities were greatly improved, as with sauerkraut. That sharp taste was rarely liked by Europeans.

Fletcher Christian must have looked as though he had been born on the island. With his height, his easy manner and willingness to enjoy the physicality of Ma'ohi life rather than observe it, Titreano must have been a welcome and popular guest, even if more naturally dark-skinned than the Tahitian ideal of Europeans.

CHAPTER 17

AN INDIVIDUAL PROBLEM

If the journals of Peter Heywood had survived, we might know what Fletcher did in more detail but would they change our views that much? The ease of Tahitian life was not the cause of what became the most celebrated mutiny of all time. It simply served to remind men like Fletcher Christian that there was an alternative to a life of insult and humiliation from an unhappy, vain and petulant captain, even if he had once been a friend.

I do not know what Fletcher Christian and his plant-nursery team did from day to day, so it would be dishonest to paint a detailed picture of their life. I shall tell you more of what those contemporary accounts and later research say they probably saw and heard and what influenced the people with whom these European men were in daily contact. These are the traditional morals and mores of the Ma'ohi men and women who later sailed with Fletcher on Bounty, escaping before the malignant influence of French and British missionaries destroyed Tahitian culture and society. As he walked, he would hear his new friends call constantly, inviting him to eat, to drink, to talk, to join the acrobatic games of children and men that he enjoyed so much, like wrestling, javelin throwing, stilt walking and kite flying. Fletcher would drink coconut milk, fish under blue skies or by torchlight from canoes, wear garlands of perfumed flowers as fragrant oils were massaged into every muscle, rub noses in greeting, and sit cross-legged on piles of soft tapa at gargantuan feasts while dancing girls graphically and energetically indicated where, and precisely how, he should next direct his attentions. There is no basis for saying Fletcher Christian had nothing to do with Tahitian women but at this stage it seems he had no special attachment.

Douglas Oliver's *Ancient Tahitian Society* tells that because growing, harvesting and fishing took such little time, Tahitian men seem always to be playing. Archery and javelin throwing doubled as preparation for battle but there was sport for absolute enjoyment, too. Tahitians surfed on long boards outside the reef and I can't imagine Fletcher would not have been fascinated to try this. Astonishingly, there were ball games we would recognise. In one, a small ball was hit with long stick with the aim of getting it through two sticks at either end of a field. In another, a much bigger ball was kicked between two teams, with the same objective but the use of hands was forbidden. Tahitian styles of wrestling and boxing were very physical and as a fit, muscular man with time on his hands, Fletcher would have competed. Compared to most of *Bounty's* men, largely confined on the ship, life in the breadfruit camp under coconut palms on black-sanded Point Venus was filled with activity and joy for months on end.

To women of middle-class *ra'atira,* an officer like Fletcher would have been irresistible, not just physically but because of everything else that surrounded him, especially of metal. Scissors cut hair better than bamboo shards and metal tools lasted longer than anything fashioned from stone. Like men, Tahitian women plucked their body and pubic hair and for good reason. On other islands Tahitians were known as nit-eaters, because they were infested with nits, which they constantly reaped and then crushed in their perfect teeth, something ensured because they used soft sticks to avoid food being stuck there. Nits are why Tahitian women were made to have short hair (sorry MGM and others), whereas men had long hair that was constantly combed, oiled and styled with flowers and shells by servants.

In contrast to the stinking bilges of *Bounty* and to the middens, latrines, chamber pots and street-sewers endured back home, the air was sweet in Matavai's village of individual dwellings, each neatly fenced to keep free-running pigs at bay. When Tahitians bathed in the rivers running into the sea, they also emptied their bowels, so their wastes were naturally flushed out into the sea. Gauguin illustrates this graphically in his painting *Te Poipoi* but the reality of what the woman featured is doing seems not to have been explained when it sold for almost $US40 million in 2007.

On Tahiti, Fletcher Christian and *Bounty's* crew were given time to become individuals once more, a state a responsible sea captain would have obviated by regular exercise, discipline, sailing expeditions and the employment of good sense. Individuals cannot sail a ship or be safe in one. Fletcher's shipmates were volunteers, more adventurers than sailors and long, golden sojourn on Tahiti was enough for them to slough off the narrow, simple acceptance of discomfort and community that was the lot of the 18th-century seafarer

While *Bounty* swayed in the lagoon, Fletcher Christian dallied ashore and Bligh observed and noted on land and entertained aboard. As long as the breadfruit grew and noble visitors kept coming, Bligh evinced little interest in the breadfruit camp, although he did order the construction of more tubs and pots so more specimens could be collected.

Nothing his crew did there was really so astonishing, considering the attractions of Tahiti, the length of their stay and the slackness of discipline. The petty tasks found on board for the men each day were no substitute for sailing. His officers and men could not be the paragons Bligh expected and it is remarkable they did not cause him a great deal more trouble.

The most serious exasperation was the discovery at 4 am on January 5 1789 that some men had deserted. The watch relief found that the launch, the ship's biggest boat, had gone. Duty officer Midshipman Hayward was asleep on duty, which was not unusual for him even though the penalty was death. He did it so often Bligh must have known, or should have.

Bligh roused the ship for an immediate roll call and found three men missing, able- seaman John Millward, Bligh's personal cook William Muspratt and Charles Churchill, master-at-arms with overall responsibility for discipline in lieu of marines; Muspratt and Churchill are special surprising because both worked closely with Bligh and would be thought to be specially loyal, an indication of how far from normal life had become.

Hayward was put into irons but Bligh blamed the desertions on his entire officer complement, calling them neglectful and worthless. The missing men had managed to escape with eight muskets, probably for use as barter goods to ensure the secrecy of their hideout. Fryer should have kept the keys to the arms chest but it was more convenient to give them to Coleman, the armourer. What their part was in how the escapees were armed has never been explained.

Although hot for action, Bligh was prevented from setting off in pursuit. Informants from shore, anxious to help in case they were rewarded, told him the launch had been sailed from Toaroah around One-Tree Hill into Matavai Bay but then abandoned. The runaways then stole a local sail-canoe and headed for the atoll of Tetiaroa, some 30 miles/38 kms north. Ariipaea, who was Teina's braver and more trustworthy brother, agreed to lead a search party but was prevented from leaving for a whole week by bad weather. Meanwhile, Bligh discovered some of the stowed unused sails were rotten. Neglect of a ship's sails was probably more serious than desertion.

Rotting sails may have been the particular fault of the master Fryer and the boatswain William Cole but this was ultimately Bligh's responsibility. The situation was an affront to Bligh's vanity, a reminder perhaps that he had been paying too much attention to his noble guests and journal but not enough to the well-being of his ship. Bligh's deserved reputation as an extraordinary sailor was based on handling serious threats to a ship's safety at sea but in port and in calm weather he could be a perfectly ordinary, rather lax, commander.

When the runaways were eventually found and arrested bravely by Bligh himself, they were treated leniently, even though Bligh said he found in their possessions a list of men who planned to mutiny and stay in Tahiti by damaging the ship in some way. Bligh later wrote to his step-nephew Lieutenant Francis Bond that Christian's name was on the list. Christian laughed when challenged and Bligh believed his rebuttal at the time, although he came back to it when defending himself in later years. I think there is rather more imagination than fact in the story. Those who could write or read, which would be few, would know that a list of would-be mutineers is not the thing to make but if there were one it is more likely to have been of those thought ripe rather than of certainties.

At other times tools and equipment went missing from *Bounty* and Bligh ordered a Tahitian to be flogged far more severely than he had punished any of his own miscreants, an inconsistency everyone, English or Tahitian would identify. There was a mystery about how one of the ship's lines was cut. Was it the result of friction of coral or was it part of the plot to hole *Bounty* so the stay would be extended? Maybe it was a Tahitian trying to revenge his *taio* Hayward, whom he did not think should be in irons?

Bligh did have to flog some of his men in Tahiti but these events further illustrate his leniency of punishment, tempting admirers of Bligh to cite this as proof of his humanity. I agree with Kennedy in suggesting it is likely that harsher punishment, more correct punishment, would have meant better discipline. The fact that many of the hard-line mutineers had been flogged by Morrison on Bligh's orders is grossly overplayed. Flogging was as much part of naval life as weevils in the biscuits. If there is any conclusion to be made from the later action of those who had been flogged, it is simply that they were the most hot-headed aboard.

By the end of February, the tropical idyll was ending and Bligh was preparing his ship and the breadfruit plants for the return voyage. He made an all-out attack on cockroaches and insects and let cats have free run of the ship. Accommodation for the extra tubs of specimens was made out of part of the chicken coop behind the ship's wheel. Furious rain interrupted preparations for almost two weeks until, on March 25, Bligh sent the cats ashore, rather harshly told the sailors they could take only as many souvenirs as could be stowed in their private chests and ordered a thorough search for stowaways. Over 1000 flourishing breadfruit plants were ferried to their floating home and pens were filled with 25 pigs and 17 goats. It was not until April 4 that a wind blew suitable for *Bounty*'s departure. Bligh was loaded with gifts for King George and Teina finally got his muskets, some rounds of ammunition, pistols and the ship's two dogs, Venus and Bacchus. His tears were for his future safety and fortune rather than at the departure of Parai and Titreano. With masterly succinctness Bligh recorded: *at five o'clock . . . we bade farewell to Otaheite, where for twenty-three weeks we were treated with the greatest kindness and fed with the best meat and finest fruit in the world.*

The first few days at sea, even after as little as a week on shore, can be agony as men attempt to smother individuality. The inevitable talk of home and hearth was discomforting but helped insulate them from the awful and immediate dangers of sailing so far in a wind-driven wooden ship. Fear wasn't lessened by memories of Bligh's threats of what would happen to his officers in the Endeavour Straits

The contrast between the safety of the land and the danger of the sea has probably never been greater than for Fletcher Christian and his shipmates, for few crews were so ill-prepared as was *Bounty's* for the long journey home, more than half-way around the world.

In this famous image painted and then engraved by Robert Dodd, Bligh stands in the launch and a mutineer is throwing one of the four cutlasses Fletcher Christian eventually allowed. Fletcher stands centrally on Bounty's stern, unable to undo actions that would have profound repercussions for as long as both men lived and that continue today.

THE MUTINY OF FLETCHER CHRISTIAN AGAINST WILLIAM BLIGH

April 28 1789

CHAPTER 18
THREATS OF ERUPTION

Captain Bligh and acting-lieutenant Christian's joint responsibility was now to see *Bounty's* officers and men settled down to a reliable, safe routine. To sail home, they would pass through dangerous waters, so, taking advantage of the relative safety of the Pacific, the crew was set to practise handling sails and lines, preparing to cope swiftly with sudden squalls amid the sunken dangers of the Endeavour Straits.

On April 13 *Bounty* discovered an island they found to be called Aitutaki but it was two days before the weather was calm enough to allow three of its inhabitants to board. They fell on their knees to kiss Bligh's feet, giving him pearl-shell breast plates, which had hung from their necks by braided human hair. In return they were given knives and beads plus a boar and a sow, as they had no previous knowledge of pork. That evening, a whirlwind ripped past the ship, almost turning her back to face the way she had come.

It's possible William Bligh saw and approved Dodd's original painting of the launch astern of Bounty. So are these images his memory of what Fletcher Christian looked like and how he thinks he appeared?

Fresh breezes alternated with calms and opposing currents, so it was not until April 23 that they reached Nomuka (now Annamooka), a low island on the eastern limits of the Friendly Islands, today the Kingdom of Tonga. First contact had been by the Dutchman Abel Janszoon Tasman, whom Morrison says reduced the islanders to

good behaviour. That goodness had not lasted, for now they were fractious. They tried to take casks from men collecting fresh water, they grabbed at the axes of men chopping wood. They were so insolent and heedless of firearms that they countered any threat from *Bounty's* men with a raised club or spear.

The men and women of Nomuka were of average size and well made. The women were judged handsome and were not tattooed but had circles of weals burned into their shoulders by hot sticks of bamboo. The men were tattooed from knee to waist, so it looked as if they wore tight breeches. They all wore just one piece of cloth tied about the waist. Both sexes dressed their hair with lime or burnt shells, which although originally black, soon turned to red, purple or white. Morrison tried to be fair about the overall impression given by these Friendly Islanders, saying the men's countenances were open *'yet they have something in it that gives an unfavourable idea to strangers: perhaps this might have been heightened in our eyes by their actions which did not correspond with their name'.*

Bligh knew Nomuka from a visit with Cook but he failed to see the social chitchat he enjoyed with Nomukans on board was different from encountering them in ugly mood beneath the tropical canopy of their island. The constant prodding and teasing of the ship's watering party by the islanders was more than exasperating it was frightening, because this was the first time any of *Bounty's* crew had been threatened by armed warriors.

Fletcher Christian was in command of the watering party and found the attentions of the martial islanders such that he and his men could not carry on with their duty and informed Bligh accordingly. Bligh publicly damned him for a cowardly rascal, asking if he were afraid of a *'set of Naked Savages while he had Arms'*. Fletcher Christian replied, *'The Arms are no use while your orders prevent them from being used.'*

This further inconsistency by Bligh and the cross words the two men exchanged was not the first clash since they left Tahiti. Fryer said that while they were working the ship in case of later difficulty Mr Bligh and Mr Christian *'had some words when Mr Christian told Mr Bligh: 'Sir your abuse is so bad that I cannot do my duty with any pleasure. I have been in hell for weeks with you. . . several other disagreeable words passed which had been frequently the case in the course of the voyage.'*

There is no reason why even close friends on board a small ship should not disagree. Disagreements are often part of friendship and nothing much should be made of them. Being in *'hell for weeks'* is different. If there had not been the Tahitian interlude Christian might have said *'in hell for months'* if the clashes did begin in Cape Town. On April 26, the day after being called a cowardly rascal in front of others, Christian was again collecting water with a party of men. Bligh ordered Fryer to go and hurry the party along.

The watering hole, previously used by Cook, was a quarter of a mile/about 400 metres inland from the beach, which was crowded with islanders. Fryer had to ask directions from the two men who were guarding the party's boat, for there seemed to be several paths through the plantations in front of him. A good-looking Nomukan man and woman, whom Fryer understood to be nobles, took his arm to lead him to the pool. Fryer first gave orders that the crew who had rowed him ashore were to take the boat far enough off shore to be out of harm's way and he then went inland. Matthew Quintal was rolling a cask of water to the boat, surrounded by islanders. Fryer accompanied him back to the shore, saw the cask loaded and then returned with Quintal, when they were met by the same man and woman.

This time they indicated Fryer should join them to eat. He excused himself but gave the woman a gift of a Jew's harp and a few small nails. Almost immediately Quintal cried out, *'Mr Fryer, there is a man going to knock you down with his club.'* Turning, he was surprised to find the young noble brandishing a club above his head. The man escaped, somehow dissuaded from his murderous intentions, although Fryer was not armed even with a stick.

Returning to the watering hole, Fryer noted Christian was getting the casks filled as quickly as he could but there were islanders all about who frequently heaved stones. One chief repeatedly pointed a very long spear at Fletcher, who was armed with a musket and bayonet. In this tense situation Fryer told Christian to get the casks down to the boats empty or full and using bribes of nails, employed some of the troublemakers to help. On the beach they found even more aggravation. The sailors in Fryer's boat had ignored his orders and, instead of standing off with oars, had anchored with a grapnel and gone ashore. While they were playing tricks with local boys and girls, someone had stolen the grapnel.

Fryer asked Nomukans what had happened and was told the anchor had been taken by men from another island and that they had already paddled away with it. When Bligh was informed *'he was very warm about the loss of the grapnel'* and said he would detain some of the chiefs on board until it was returned. Fryer said it was unfair to trouble the men aboard or to hold them, as they could know nothing of the anchor or its loss. He added they had plenty of grapnels on board and plenty of iron to make another. Fryer said he did not feel the loss was great and by reasonable standards it was not. Bligh was not being reasonable. *'The loss not very great Sir, by God! Sir if it is not great to you it is great to me.'* Fryer's reply to this outburst was that he was sorry about the loss but that being sorry was no use in righting the situation.

Bligh purposefully pursued vengeance. The boats were cleared and hauled in and the anchor heaved. As Fryer was supervising the unfurling of sails, Bligh unexpectedly ordered the ship's crew to arms. Assuming there was insurrection of some kind, for there were still islanders aboard and canoes at the side, Fryer dashed from his duty to Bligh, to learn his commander had taken prisoner the chiefs aboard. Two of the men were of an extremely high-ranking family and Bligh had bragged about their visit to his ship. Keeping four, he sent a relieved fifth back to shore. Once he was in a canoe, he made signs to *Bounty,* confirming that the anchor had been carried off to another island. Bligh, conscious that many of his men were awkward and uncommitted about bearing arms in such an unnecessary circumstance, threatened them, calling them a *'Parcel of good for nothing Rascals'*, adding he would *'trim them all'*. When the chiefs protested against their arrest, they were sent down to peel coconuts for Bligh's dinner, a terrible insult but Bligh was in an insulting mood. He claimed that with just four other men and stout sticks he could easily disarm the ship's entire complement and aimed a pistol at William McCoy, threatening to shoot him for not paying attention.

By four in the afternoon the canoes following *Bounty* had fallen astern except for one double-hulled vessel. This was filled with women and men, including Nomuka's oldest chief, all of whom were weeping and injuring themselves in a terrible and bloody manner. One struck himself with a paddle blade several times with a sound that could be heard at a considerable distance. This was Nomukans way of showing grief and most people, including children, had lost much hair and a finger or two through self-injury.

At last Bligh freed the chiefs, gave them presents and ordered the canoe alongside. Their resentment was plain to see but there was little they could do to revenge the insult. The general opinion aboard *Bounty* was that if a weakly manned ship were subsequently to call at the island it would pay handsomely for Bligh's treatment of the chiefs. It was amazing, quite irrational treatment of the Nomukans, of Fletcher Christian and of *Bounty's* men. Bloodshed was only just avoided and Fryer had no doubt that if Bligh had gone ashore his fate would have been that of Cook and he would have been murdered.

The contradictory orders on Nomuka, where once again Christian and other officers were made to look publicly foolish, continued what Bligh began long before in Adventure Bay. If this was typical of Bligh's behaviour on *Bounty,* if he constantly gave silly orders and then used them to insult and aggravate his crew, there were few men of dignity who could honestly feel respect for their captain. For Fletcher Christian, it must have been horrifying and he was more than an observer. He was right in the firing line and worse was to come, because Captain Bligh's irrationality had become typical.

Only conjecture can suggest the root of Bligh's behaviour but there was a clear need to get his ship disciplined and unquestioning again, a difficult thing when the crew was much healthier than would be normal. They were at sea after so many months on Tahiti of all places and not one of them was a sullen, pressed man anxious only to get home.

Bounty sailed overnight towards Tofua with difficulty, for the wind dropped almost to a calm. As they approached the island, they could see its volcano erupting, belching columns of smoke and flame. Bligh, smarting from his defeat at the hands of the Nomukans, continued his search for status at the expense of others.

There are anomalies of date in what happened next but these are explained by Royal Navy custom. At sea, the day began and ended at noon, so the afternoon, evening and night of a certain date preceded its morning. When remembering what happened on *Bounty* while becalmed within sight of erupting Tofua, some men used naval time and some civil.

Bligh's published accounts do not mention the following event, even though everyone else gave it great importance. Fryer and Morrison agree that some time on April 27 Bligh found further opportunity to torment his officers. It was his opinion that a pile of coconuts stacked between the guns had shrunk overnight, by which he meant some had been stolen. The coconuts are said to have been his personally, although they may have been purchased by him for ship's stores. Either way, Bligh would consider their loss an affront to his rank. Fryer was sent for because it had been his responsibility to stow them. Fryer agreed the store seemed less than before but thought the men could simply have flattened them by having to walk over them during the night. A simple explanation was not to Bligh's liking. He wanted names.

Bligh ordered every coconut on board to be brought on deck and subjected each man who owned some to the most mortifying cross-examination: *'How many coconuts did you buy?' How many did you eat?'* How could any man admit to theft under such circumstances, even if he were guilty?

Fletcher Christian was deeply wounded. *'I hope you don't think me to be so mean as to be guilty of stealing yours?'* Bligh retorted, *'Yes, you damned Hound, I do, you must have stolen them from me or you could give a better account of them.'* Multiplying his insult to Christian by lumping in the other officers, he damned them all, calling them thieves and scoundrels who joined the men to rob him.

When collecting evidence about the mutiny and the preceding events in London some years later, Edward Christian, spoke to many other men who had been aboard *Bounty,* none of whom were mutineers. The accounts they gave him confirm and extend the story and must be considered primary evidence. Of course, every source must be treated with at least a grain of suspicion, for memories often make heroes out of men who were only bystanders. Yet such defenders of Bligh as Madge Darby and Gavin Kennedy tend to dismiss Edward Christian outright, especially for concealing who told him what, as though he were fudging the evidence. What he was doing was protecting 'his sources', the men who spoke freely to him, because without this guarantee of anonymity they would not have spoken, still an important, judicially protected part of journalism today. It is very limited of understanding to criticise this.

Edward wanted memories of the event other than Bligh's published account. From these, Edward deduced a far more dramatic and wounding confrontation between his brother and Bligh. He says that, as officer of the morning watch from 4 am to 8 am, Fletcher Christian was in his hammock asleep when Bligh began his coconut tirade. Nonetheless, he was summoned. When he arrived, Bligh accosted him with *'Damn your blood you have stolen my coconuts.'* Christian answered: *'I was dry, I thought it of no consequence. I took one only and I am sure no one touched another.'* *'You lie, you scoundrel, you have stolen one half!'* was Bligh's rejoinder. Hurt and agitated Christian asked why he was treated thus. Bligh shook his hand in his face and said: *'No reply,'* but he continued to call him a thief and other abusive names.

Fletcher Christian was deeply wounded. 'I hope you don't think me to be so mean as to be guilty of stealing yours?' Bligh retorted, 'Yes, you damned Hound, I do. . .'

This version amplifies those of Fryer and Morrison. It was injury of the deepest and most unforgivable kind for his second-in-command to be accused of petty theft in such arbitrary and public fashion, even if he were a mere able seaman. Christian was allowed free access to Bligh's spirit supply simply by asking John Smith for the keys. If he were allowed alcohol, why not a coconut, which had been bought at the rate of twenty for an iron nail? It speaks badly of Bligh that he did not even hint at the coconut incident in his account of the events. He knew there was no way he could come out of it well so he ignored it, lied about it some might say.

The importance of the coconut incident is accepted by Fletcher's brother Charles as the final, irrevocable insult. In his autobiography he writes with much emotion: *'What scurrilous abuse! What provoking insult to one of the chief officers on Board for having taken a coconut from a heap to quench his Thirst when on Watch base, mean-minded wretch!!'*

Bligh's behaviour, first over the grapnel at Nomuka and then over the coconuts cannot be considered normal, even if it was an everyday event. Officers and gentlemen required public respect, whatever they might have done. Without it they were unlikely to be heeded by the seamen. By humiliating publicly his officers Bligh was encouraging the inefficiency he hated. Yet it seems that any inefficiency to which he could reduce his officers served to make him think himself the better sailor and more self-righteous man.

Not having acceptable boundaries himself, made it impossible for others to know or remember their boundaries. As happened after Adventure Bay, perhaps *Bounty's* crew was making their own? That's dangerous at sea.

During the coconut incident Bligh threatened that only half his officers and young gentlemen would return home, saying they would be made to jump overboard before they got through Endeavour Straits, or they would be left behind at Jamaica. It's childish bullying and terribly cruel to men at sea, again pushing for discipline rather than leading. When Bligh added that his officers were to have their grog rations stopped and their allowances of yams reduced, he was still not finished. He confiscated some of his officers' coconuts to replace those he thought stolen because, whoever had taken them, he considered it the incontrovertible fault of all his officers, not just Fletcher Christian. Then he told his officers that if they stole from him tomorrow, he would reduce their yam rations even further. In other versions this threat was made to the entire company.

If the momentum of events was not so dramatic this extraordinary moment would be laughable. There was Bligh nursing feelings of persecution about coconuts that had cost one nail for 20 of them. His officers murmured in astonished groups, having been threatened with forced suicide or abandonment. Some scurried about secreting food like nervy squirrels. England and sanity were further away than when *Bounty* sailed from Tahiti, only weeks before.

Bligh's records make it clear he imposed a divide and conquer regime that was contrary to every acceptable norm of Royal Navy discipline but that was a common thread in his management style throughout his career. Bligh had done himself no favours by flogging far less often than he might during *Bounty*'s long absence from Britain. When he did flog, he often ordered more than the 12 lashes at a time that a captain at sea was allowed to inflict. This added deliberate cruelty and ill-disciplined disrespect for the navy's regulations to his perceived weakness. His 'inconsistent' behaviour was often noted and was quite against the unwavering belief in absolute adherence to the law Bligh demanded of others.

Bligh's famous *'ungovernable'* temper always blew hot and cold, so that men were singled out who had not really offended and it is documented again and again how he would treat officers as common seamen, an insult to the confidence of both. To all that must be added his infamous bad language, not an accepted part of naval life as you might think. Articles of War XXIII and XXXIII expressly warn against *'reproachful or provoking speeches'* and behaviour *'unbecoming to an officer'*. Sir John Barrow, an administrator in the Admiralty during Bligh's time said of him: *'It is difficult to believe that an officer in His Majesty's service could condescend to make use of such language to the meanest crew, much less to gentlemen'*. It was ill-disciplined of a captain to do it, for not only did it diminish officers in the estimation of the lower ranks but the quality of the work of common sailors was hurt because, in the words of John Fryer, *Bounty*'s master, *'it could not be done with any pleasure'*.

After *Bounty* sailed from Tahiti, Bligh's pathological need to boost his sense of superiority by shattering the confidence of his officers and men was fully exercised. After April 12, when he punished John Sumner for neglect of duty, he worked everyone extra hard for no practical reason. *'Cleaning down below'* was ordered daily rather than the usual every two to four days a week.

As well as being targeted for personal abuse by Bligh, Christian was also expected to shoulder far more than his fair share of duties. Bligh's log shows that Christian was supposed also to oversee small-arms practice on a daily basis, something not mentioned once on the outward voyage. Cleaning and mending hammocks, normally ordered weekly, was now expected daily and as second-in-command it was Christian who had to supervise this.

It is right that Bligh should want his ship clean, tight and tidy, because he had neglected to keep up standards in Tahiti. The extra work he now ordered was well beyond that, seeming a spiteful punishment for everyone but himself and much of it was to be done by Christian.

It was not because her officers and crew were undisciplined that life on *Bounty* was profoundly unhappy. It was because Bligh himself was undisciplined and unable to demand respect. It was not Bligh's style of command but his lack of command. There were no boundaries given or taken.

The day of reckoning Fletcher Christian had foreseen was almost here. As long as Bligh did nothing worse, Fletcher Christian might just bear it. Any personal solution to what seemed like targeted persecution would mean the sacrifice of his honour, his family and his future.

CHAPTER 19

MUTINY

On April 27 1789 the atmosphere aboard *Bounty* was never worse and none was more oppressed or affected than 24-year old Fletcher Christian. His brother Charles had told him that each man has a breaking point and Fletcher was daily being pushed to reach his.

Later, Bligh once more abused his second-in-command. Fletcher ran forward with great tears welling from his eyes. Purcell stopped him, asking what had happened: 'Can you ask me and hear the treatment I receive?', Christian said. Purcell suggested he had received the same treatment but Christian pointed out a difference. As a warrant officer Purcell could not be flogged, so had a degree of protection when he defended himself to Bligh. Christian reminded him, ' . . . but if I should speak to him as you do he would probably break me, turn me before the mast and perhaps flog me, and if he did it would be the death of us both, for I am sure I should take him in my arms and jump overboard with him.' Purcell reminded Christian it was only for a short time longer but Bligh's threats about sending his gentlemen overboard in Endeavour Straits bothered Christian. Going through them would be added Hell.

Fletcher Christian was torn between a gentlemanly determination to do his duty and the manly requirement to defend himself. Although acting-lieutenant and second-in-command, he was still officially master's mate, no more than a superior midshipman, and could thus be flogged. He dared not to defend himself the way Purcell did. It was typical of Bligh to take advantage of Christian's vulnerability, driving him to the very limits of self-restraint. It was conscious and pernicious bullying. If Christian ever were to defy Bligh, he would lose his chance of promotion on their return, yet by not defying him he would seem weaker and thus a softer target for further abuse.

As young Fletcher wept, he protested: *'I would rather die ten thousand deaths than bear this treatment. I always do my duty as an officer and a man ought to do, yet I receive this scandalous usage.'* Bligh was to say later that Christian did not do his duty and that he had put the ship in great danger just a couple of weeks after leaving Tahiti.

`Flesh and blood cannot bear this treatment,' Christian cried. It was the only time men on board had seen him in tears *'He was no milksop,'* said one, something even his literary foes would concede.

Through inflexible responsibility to duty and in defence of his dignity Fletcher Christian was not prepared to descend to battle with Bligh. Instead, he decided to leave the ship. It was rash and it was desertion but it was brave and it allowed him to keep his pride intact while making an unmistakable comment upon Bligh. From my understanding of Georgian gentlemen and their codes of behaviour this would have been considered honourable, at least by officers who were gentlemen. Duelling with one's superior was out of the question at sea and England was too far away to issue a challenge that might be settled there.

Late in the afternoon, a sultry sullen time of a tropical day, Fletcher Christian gave away his Tahitian curios. He tore up his letters and papers and threw them overboard. He wanted no one to know the intimacies of his misery. If Christian had planned to mutiny well in advance of the event, as Bligh always preferred to believe, there was no need to do such a thing. He could have taken them to use as evidence. They must have been full of descriptions of his treatment and could have been distributed among his allies for use on the ship's return home. There is no evidence that he was bent on revenge as malicious as his own torture had been. So far, his plan was to escape with maximum dignity and as little added ill-feeling as possible.

The men to whom he turned for help were the carpenter Purcell, boatswain William Cole, acting-master's mate George Stewart and midshipman Thomas Hayward. Later events make Hayward seem a surprising choice. Strange situations make stranger unions, and confronted with an emotional and determined Christian, who was master of his watch and a mess-mate, Hayward would have been hard put to refuse. There was too much general sympathy for Christian for Hayward to dare side against him at this stage.

These men and others, who told Bligh and Courts of Justice they knew nothing of Fletcher Christian's initial plans to leave the ship, were lying to save their lives and future careers. Helping an officer desert his duty was almost as serious an offence against the Lords of the Admiralty and no one would freely admit it. In the super-charged atmosphere of insult and injury in the small, almost becalmed ship, Christian went about his business quietly because secrecy was impossible. The most noticeable of Christian's preparations was when he went in and out of the fore cockpit with George Stewart, a part of the ship neither would normally visit. He was collecting nails and other barter items from Purcell. He also collected wood and binding because now his plans were firm enough to let his confidantes know he was going to slip overboard and sail away on a raft.

This seems impractical but perhaps Fletcher Christian was simply 'doing something'? Arranging a method of escape could have been a therapeutic exercise. This achieved, he might well have been content. Such gestures are common in tense situations and the danger inherent in his supposed plans did not mean Christian was stupid or rash. He may never have been completely convinced about the plan.

As evening passed into night, Christian lashed two masts from the launch to a plank and hid some left-over pork and breadfruit. These rations were accidentally found by Tinkler, one of his messmates, who thought they had been secreted as part of a prank.

That night, Fletcher used the flimsy excuse to stay on board that there were too many men on deck for him to leave in secrecy. Sensing Christian's lack of conviction and knowing the grave danger of such a venture, George Stewart risked everything to help his friend. He humoured his friend Fletcher, supporting him when he most needed fellowship, keeping an eye on him, so that nothing really dangerous or suicidal would happen. There was enough practicality in his scheme to work and Stewart could never be certain Fletcher would not make the attempt. Christian was an excellent navigator and sailor and with his childhood life on a farm and his long stay on Tahiti, he was capable of self-sufficiency on any island.

Nearby Tofua was one of the Friendly Islands and Christian hoped he would be picked up by a local canoe and then, if the locals were suitably disposed, he could either stay with them or become one of the first South Pacific beachcombers. There was never any suggestion that either Tahiti or a woman was the reason for his desertion or his ultimate goal. No man aboard the ship ever entertained that idea except Bligh, who was always the last person to know what another man thought.

Bligh later poured scorn upon Christian's plans. What else could he do? He had no perception of his officer's misery and if Christian had left *Bounty* Bligh would have been permanently insulted because losing an officer would be quite as serious as losing a ship.

Bligh can't have heard what his second-in-command was planning. With extraordinary insensitivity to the coconut incident Bligh sent his servant John Smith to invite Fletcher Christian to join him for supper. Bligh also expected him to dine with him the following day. All Bligh's officers took turns at his dinner table, attending once every three days, dinner being served at midday. Fletcher was regularly invited to supper in the evening as well. It was out of the question that Fletcher should accept and he sent a message which politely saved Bligh's face, saying he was unwell. Bligh believed him, sensing nothing wrong. This was not the only snub to Bligh. All the officers agreed they would never again join Bligh at his table. Fryer and the late Dr Huggan had decided that more than a year before on the outward voyage.

Smith begged others to replace Fletcher at Bligh's table and when Hayward broke ranks to agree he was hissed by his brother officers. They were still smarting over the confiscation of their coconuts by Bligh but Hayward needed to flatter the captain and make up for his falling asleep on duty at Tahiti or he would never be promoted.

Night fell with tropical suddenness. There was only a little wind and any rain and clouds cleared by 10 pm. Bright moonlight then complemented the weird glow of the volcano. Bligh came up from his airless cabin to leave his orders for the night with Fryer, who comments that at the time he was on speaking terms with his captain, *'but I am sorry to say that that was but seldom'*. They discussed the breeze and the youthful moon. If all went well, they would reach the coast of New Holland with the waxing moon, greatly increasing their safety and speed of passage through the dreaded Endeavour Straits. At midnight, Peckover and his watch relieved Fryer's men. Those now responsible for the ship included Edward Young and George Stewart.

Fryer, Peter Heywood and the others went down to sleep but a man who should have been sleeping was not. Fletcher Christian, due on duty at 4 am was awake and restless. Had he really been watching for an opportunity to take his raft but been deterred by the number of men on deck, entertained by the volcanic eruption and cruising sharks? Or was he hesitating simply because he was perfectly aware of the dangers and hoped someone would talk him out of it?

Confused by lack of sleep, exhausted by the emotions of the last 24 hours, Christian still got up to command his watch. Duty had to be done.

When the men from Peckover's watch went to waken those of Christian's at 4 am, they found Fletcher had gone to bed less than an hour before. Confused by lack of sleep, exhausted by the emotions of the last 24 hours, Christian still got up to command his watch. Duty had to be done. Until this moment George Stewart had been able to watch his friend but now he wanted to sleep. First, he again soothed Fletcher, pointing out the unlikelihood of any chance of survival if he left the ship. In any case, he added, if Christian left, *Bounty* would be in a worse state, the men were *'ripe for anything'*.

Did George Stewart mean the men were ready for mutiny and that his friend should lead them? It is absolutely out of character and the only evidence that supports this is subsequent conjecture. Stewart had no motive to suggest mutiny. He was a strict disciplinarian, a stickler for duty, undoubtedly his reason for humouring Christian into staying at his post. Stewart did have a regular girlfriend on Tahiti to whom he was very much attached but a woman on a South Seas island was no reason for such a talented young man to abandon friends, family and career. Stewart would know that even to suggest such a thing was mutinous.

The simpler explanation is that he was cajoling, using any and every appeal to Christian's loyalty to his friends, telling him in return there was enormous sympathy for him on the ship. Christian was looked to for leadership and sympathy far more than Bligh and if he stayed the men would help put things right. There are always methods, and, apart from the sneaky Hayward, the officers had already quickly demonstrated united defiance by refusing to eat with Bligh, a remarkable comment on his behaviour and disrespect for his command that is rarely given the importance it should have.

Whether or not you are a Bligh or a Christian loyalist, it's a fact that no *Bounty* man ever blamed Christian for what he was about to do or felt personally injured by him. Plenty were willing to blame or to seek revenge upon Bligh but not one did the same to Christian, whatever their loss or distress, not even those subsequently tried and hanged.

Apparently satisfied he had once more mollified Christian, Stewart retired. Many men aboard Bounty later persuaded themselves or were persuaded that Stewart consciously suggested mutiny but his friend Peter Heywood always firmly declared that Stewart did not mention the subject even obliquely. Heywood and Stewart subsequently spent many months imprisoned together when they could have discussed the conversation and Heywood had plenty of opportunity to tell the truth after Stewart was dead. Heywood thought the imputation of Stewart's guilt distasteful, an insult to the memory of a fine sailor and an upright young man.

Yet, if Stewart did not think of mutiny, Christian certainly did. It was the alternative he had been seeking. It was the most invidious revenge he could wreak upon his tormentor. His breaking point had been passed and all thought of duty dissipated. On *Bounty's* almost empty, early morning deck the resolution suddenly became simple and practical. At some fatal moment he must have thought: *'Why should I go? Why shouldn't Bligh go?'*

Put like this it didn't even sound like mutiny and hadn't his own brother been forced to act against another captain? Familiarity with any crime can make the human mind see it as less prohibited, more acceptable.

The natural disinclination of Hallet and Hayward to work hastened Christian's resolve. Hayward was already asleep on the empty arms chest on deck. Hallet had not even appeared. With not even a warrant officer about to gainsay him, *Bounty* was for the taking. At first Christian carried out his normal duties, for he needed assistance and had to choose those most likely to be receptive. As well as the somnolent midshipmen Hayward and Hallet, the men on Christian's watch were the gunner's mate John Mills, the carpenter's mate Charles Norman, plus Isaac Martin, Thomas Burkitt, Thomas Ellison and Matthew Quintal, all ABs.

In fine weather, this morning watch is one of the most enjoyable at sea, combining a satisfying domesticity with the pleasure of daybreak. The South Pacific air is cool and silken, refreshing with its promises of breakfast, comforting as the rising sun quickly warms. Christian ordered the usual activities to commence. The lines were all neatly coiled and a gangplank unslung to gather water for washing the decks. The rhythm was interrupted only by the banalities of any crowded community. At this early hour Bligh's cook William Musspratt was chopping wood on the deck and Byrne fumbled his myopic way up to remonstrate against the thumping, which resounded through the quarters below. Then a shark was spotted and Hayward and Norman hung over the edge to watch the threat of its languid motion and malevolent upright fin.

Assessing the men of his watch, Christian settled upon Cornishman Quintal as the first to approach. Quintal was 5 feet 5 inches / 1.65 metres tall. Strong and muscular he is said to have been the only AB who had a permanent attachment in Tahiti. He was heavily tattooed on the backside and elsewhere. Already Christian must have realised the result of the mutiny would be exile in the South Seas and talking to Quintal he turned the conversation to those pleasant Tahitian times, eventually insinuating his plan of taking the ship and demonstrating his seriousness by showing he had slung a lead weight about his neck. If he failed, he would throw himself overboard to drown. Quintal refused to help but the black-bearded American, Isaac Martin, thought it a capital idea. He assumed responsibility for raising a party. Quintal quickly changed his mind and thereafter was an ardent supporter of Fletcher Christian.

As the light and warmth of the Pacific sun slowly spread, the heat of Christian's passion for revenge increased and kindled support throughout the creaking ship. Hidden amid the protests of sun-warped timbers, the response to whispers of revolt into the ears of drowsy men were as varied as those who made them. Some thought Bligh was to be put overboard to fend for himself on Tofua. Others believed he was to be confined and taken back to England with Christian in charge of the ship. Others hoped they were returning to Tahiti. Any inducement was used to recruit men and little of this was known to Christian, who would not have cared. He simply wanted men.

Once, Fletcher Christian could well have considered taking Bligh back to face his superiors in England but now he wanted to be rid of him, together with Hayward, Hallet and Samuel. The appeal of the South Pacific life certainly added impetus and fire to his resolve and to gain the supporters he needed to execute his simple plan. With enough men to guard the ladderways and keep everyone else below decks, he could achieve the exodus of Bligh and three other undesirables with speed and economy. Hayward and Hallet were universally unpopular for their arrogance. Samuel was Bligh's clerk and connected in everyone's mind with the cutting of rations.

Raising a gang went smoothly and caused no alarm but then they needed arms. Here they were helped by a small but significant breaking of the rules. Fryer, as master, was supposed to keep the keys of the arms chest but they were actually in the care of Coleman, the armourer. This was a sensible arrangement but against the rules. It was all too simple for Christian to shake Coleman awake and ask for the keys, saying he wanted to shoot a shark. Coleman suspected neither the man nor the hour. Christian was in charge of the watch, he was second-in-command and if Coleman were that suspicious, he could have stumbled into the morning light to see the fortuitous shark. Coleman handed over the keys and went back to sleep.

Christian went to the arms chest, where he found the sleeping Hallet, who was smartly hastened to his duty, suspecting nothing other than further chastisement for his lethargy. As if to reassume some semblance of dignity, Hallet started giving orders as soon as he was on deck. It was his turn to oversee the catering for his mess that week and he asked Burkitt to draw the three chickens that were hanging on the mainstay. Fresh poultry was one of the advantages the officers had and the coops were aft on the main deck.

Once they were crowded around the arms chest, Christian and his supporters quickly decided on a whispered course of action but a call for caution was agreed. The cutlasses, pistols, muskets and bayonets were only to be deterrents. There would be bloody threats but there was to be no bloodshed. Christian had exchanged his plan to leave the ship with little fuss for a plot to rid the ship of Bligh equally simply. It might take more force, that was all.

It would have been so easy for Christian to strike or murder Bligh that morning or at countless other times that accusations of his having little or no control of his temper cannot be supported. Such incidents, including mutiny, happened on naval and merchant ships all the time but even the bloodiest of these, where a captain was slowly flayed while bound to a mast, are rarely remembered. Although acting illegally and perhaps in the grip of a temporary mental breakdown, Fletcher Christian demonstrated great control. It was Bligh whom most of the ship thought was out of control, infecting others, including Christian, with ideas of abnormal action. It makes sense that a wide belief on *Bounty* that Bligh's mind was unbalanced is why plans to confine him or take some other unusual action were accepted without question by almost all those to whom they were put. The lack of protest or action by others after Fletcher Christian arrested him means they were equally unsurprised, even if unapproving.

As the insurrectionists mounted the fore ladderway, they were seen by Hayward, who asked why they were armed. He was peremptorily told that the captain had ordered them to exercise at dawn. Not believing this story, he started aft, apparently to warn the sleeping Bligh that something was amiss. William McCoy, tattooed and scarred from knife fights, had been loading his musket on deck. He banged it heavily three times on the deck to hurry his companions from below, warning them they were suspected.

Christian quickly emerged carrying a musket with a fixed bayonet and a cartouche box in his sweating left hand, a cutlass and pistol in his right. He ordered Burkitt to take the pistol. When the raw-boned heavily-tattooed man wavered, Christian yelled in fury: *`Damn your blood lay hold of it.'* From this moment on, Christian kept his dark-skinned face stern, *'darker than thunder'*, and by constantly threatening death and injury, kept everyone in fear of him, including his own party. Hurriedly the group caught up with Hayward. *'Damn your blood Hayward, Mamoo!!'* said Christian, using the Tahitian word to tell him to hold his tongue and threatening him with a drawn cutlass.

It hadn't taken long. At about 4.30am, Fletcher Christian descended the aft ladderway without disturbing Fryer, who slept directly opposite Bligh. Followed by Burkitt, Mills and Churchill, he burst through the habitually open door of Bligh's cabin, waking his victim with a flourish of naked steel and shouting, *`Bligh, you are my prisoner!'*

CHAPTER 20

A QUESTION ANSWERED

In the breathless few seconds it took to take in the scene, Bligh shouted, *'What's the matter? What's the matter?'* When he saw the cutlass blade at his throat he yelled: *'Murder!'* It was a natural conclusion. Christian was wild-eyed, beaded with sweat and dishevelled.

Bligh's shouts did not disturb the midshipmen who slept so close. Fryer was already awake and the prisoner of John Sumner and Quintal. His brace of pistols was taken but was of little use for he held no suitable ammunition.

Churchill shouted up the ladderway for rope to bind Bligh because Hayward's interference had made them forget it. No one moved. Churchill called up, *'You infernal buggers - hand down a seizing or I'll come up and play hell with you!'* Mills cut some line and sent it down. Unable even to don *`trowsers'* and with his hands bound tightly behind him Bligh was hustled on deck and demanded an explanation from Christian, *'What is the meaning of all this?'*

'Can you ask, Captain Bligh?' Christian answered. *'Can you ask when you know you have treated us officers and all these poor fellows like Turks?'*

Fletcher Christian made the cause of the mutiny perfectly clear to Bligh from the first moment he was asked to explain. Treating people like Turks meant working them like slaves and everyone on board would have understood this at once. Christian held Bligh just forward of the ship's wheel. In one hand he had the end of the rope binding his prisoner, in the other was a bayonet that he had exchanged for the cutlass and now pointed at Bligh's breast. To young Tom Ellison, just a few yards away at the wheel, Christian seemed like a madman. His long black hair hung loose and his shirt collar was wide open, exposing his tanned and tattooed chest. His eyes seemed to flame with vengeance.

The rope that bound Bligh's wrists had also caught the tail of his night shirt. Burkitt did not like to see his captain stand exposed, the stark ivory-whiteness of his skin contrasting deeply with his black pubic hair. Putting down a musket, Burkitt went to adjust Bligh's shirt, hauling it out of the tight lashings. Disregarding Christian's orders to take up his arms again, Burkitt called down to Sumner for some clothes for the captain but as John Smith came past he suggested he do this instead. Christian drew out a small pistol that Bligh had once carried and warned Burkitt to take care, that he was being watched. Afraid to do otherwise, Burkitt retrieved his musket and moved away to a less conspicuous place.

Christian ordered the small cutter to be put out and told Hayward and Hallet they were to disembark, which doubled their astonishment at the scenes they had witnessed. From this moment until they left, the two men kept up an incessant duet of tearful pleading, asking Christian what they had done to deserve such treatment.

Ellison, once puny but now well filled out, was as frightened and confused as the two sentenced midshipmen. Lashing the wheel and saying he needed to go to the heads, he asked Mills to take over his stint and then went in search of advice from Lebogue but he was in no mood to help and told Ellison to go to Hell. Not daring to ask anyone else, Ellison made up his own mind and joined Christian's party with a spirited offer to stand guard over Bligh. It must sorely have tried Bligh, for Ellison was a protege of his earlier employer, Duncan Campbell.

Fryer pleaded to be let out of his portside cabin so he could speak to Fletcher but was refused. When Christian relented and he was escorted up, Fryer said: *'Mr Christian, consider what you are about.'*

'Hold your tongue', was the sharp reply. *'I have been in Hell for weeks passed, you know Mr Fryer. Captain Bligh has brought this upon himself.'*

Fryer persisted, telling him his disagreement with Bligh need not result in the ship being taken. He suggested Bligh could be put in his cabin, adding he had no doubt they would soon all be friends again. Once more he was ordered to hold his tongue. Threatened if he said a thing more, Fryer risked injury by making one of the first pleas that Bligh be given something better than the leaky small cutter and made a motion to speak to Bligh. Christian aimed his bayonet at Fryer's heart and told him if he advanced further he would be run through.

No officer or crew member supported Fryer on the only real appeal to Christian that came from anyone on board. Fryer was returned to his cabin and John Millward was added as a third sentinel to guard the master and the aft ladderway.

The small cutter was so unseaworthy that the carpenter's mate Norman had to sit in it and constantly bail. Like the rotten sails, the condition of the boat could be blamed directly on Bligh, perhaps a reason he failed to mention this in his version of April 28 1789. Then Christian found others wished to leave *Bounty*. This was a complication and perhaps a disappointment to Christian but all morning he listened to the requests and suggestions of both the mutineers and those who wished to leave the ship, further indication that he was as unprepared for mutiny as anyone.

The large cutter was ordered out to replace the small one and Michael Byrne sat in it to keep it off the ship. Isolated by his poor eyesight and inability to hear what was happening on deck, he cried miserably, terrified he would be forgotten. It had taken time for the cutter to be launched and now a considerable number thought they would join Bligh's party, being loyal to themselves and to the king as well as to Bligh. There was no suggestion that Bligh would attempt crossing half the South Pacific to Timor and so the choice seemed simple. Should they renounce their duty and stay aboard or loyally sail to nearby Tofua with their captain? England was a long way off, perhaps never to be seen again, whichever course they took. The only men likely to have followed Bligh through true personal loyalty were his servants Smith and Samuel, although the latter had to be forced overboard. Bligh wrote that Samuel was forbidden, *on pain of death, to touch either map, ephemeris, book of astronomical observations, sextant, time-keeper or any of my surveys or drawings. (Mr. Samuel) attempted to save the time-keeper, and a box with my surveys, drawings, and remarks for fifteen years past, which were numerous; when he was hurried away, with "Damn your eyes you are well off to get what you have.".*

When told the large cutter was not big enough to hold those who wished to leave Fletcher Christian must have been sorely exasperated. What began as a small and simple act of personal revenge on Bligh and three others was getting out of hand. Only the launch remained, the most valuable part of the ship's furnishings and Fletcher Christian was adamant that it should not be given the sailing party. He most especially did not want Bligh to have it. After much pleading, discussion and hesitation, Christian eventually relented.

For the third time the troublesome and heavy task of hauling one cumbersome wooden boat back on board and then another lowering an even bigger one into the ocean was undertaken. To encourage the men, Christian ordered Bligh's servant to give a dram to each man under arms. This does not mean the mutineers quickly became drunk, incapable of decent or reasonable behaviour. A dram means a tipple, a mouthful. A dictionary defines it as only $1/8^{th}$ of a fluid ounce, just over 3.5mls, but it is hardly likely there was exact measuring on *Bounty* that morning. The men there said only one bottle was fetched and John Smith was unlikely to bring glasses. In any case, to men who were drinking rum every day, a quick refresher of spirits would be only a psychological lift. If Fletcher Christian had ordered a serving of grog it would have been a very different matter and then suggestions of drunkenness would be hard to dismiss. The imprecations hurled at Bligh and others were more likely to have been a result of exhilaration and the unfamiliar privilege of freedom of speech.

It took far too long but eventually the launch was out and it slowly filled with men and their possessions. Only Bligh took most of his and his selfish gathering of unnecessary bits and pieces was a major contributory factor to delay and to the boat's overloading, discomfort and to the subsequent danger of his fellow passengers.

When the time came for Fryer to be escorted to the ship's side, Bligh ordered him to stay aboard for the safety of the ship and in the hope that he might retake it. The master, anxious to obey, implied to Christian that without his skills Christian could never be master of the ship. Insulted, Christian once more brandished his bayonet at Fryer. *'Go into the boat or I will run you through.'* Fryer now thought of his young brother-in-law Tinkler and begged he also be allowed to leave. Churchill, who as master-at-arms should have been the first to defend Bligh, protested but Christian once more relented. When Fryer eventually reached the side and peered over into the full boat, he was one of the few amazed that so few armed men had been able to take the ship and turn so many out of her. He had forgotten that in Nomuka Bligh boasted he could take the ship with four men armed with stout sticks.

There was argument as to whether the carpenter Purcell should be allowed to go or stay. If Christian were perplexed or conciliatory during the morning, Churchill was not. The evidence shows he clearly understood the consequences of the mutiny and advised Christian of the practicalities required for their uncertain future. If the carpenter were allowed to take his tools, he could easily oversee the building of a vessel by the castaways. Compromising, Christian agreed to let Purcell go but kept most of his tools and both his assistants. He also kept armourer Coleman on board and he was a most useful man in the days to come. Churchill also ensured Peter Heywood and George Stewart stayed on board, seeing them as essential to the safe sailing of the ship, insurance against any future incapacity or loss of Fletcher Christian.

Throughout the long and unexpectedly complicated arrangements to get rid of Bligh, Christian had to withstand a constant tirade. Bligh was later proud he shouted himself hoarse trying to rally action and support on the ship even though there was no response. He alternated fruitless efforts with attempts to humour and dissuade Christian. Losing the breadfruit and *Bounty* was serious and anything Bligh could do to avert this must have seemed fair. Fairness to others wasn't one of his virtues and his protestations were ignored.

The day before, Bligh had pointedly refused to answer Christian when asked the reasons for his ill-treatment. Today, Christian clearly and repeatedly answered Bligh's same enquiry in a manner that left no doubt about his reason for the seizure. The only cause of the mutiny was Bligh and his behaviour towards Fletcher Christian. It was said so often in the evidence and reaffirmed so clearly by Rolf DuRietz in 1965 (in Studia *Bounty*ana), that I never fail to be amazed at writers who still search for the mutiny's cause.

The biggest reactions raised by Bligh's shouts for assistance were abuse and bad language, something he was used to giving rather than getting. There was to be no reversal of the situation and the defiance that men had suppressed so long now poured forth. Christian was advised to *'shoot the bugger'*, at least. Bligh was taunted with his handling of the food rations and there was satisfaction that he would have to try to survive on reduced rations and 12 ounces/340 grams of yams a day.

The show of solidarity must have pleased Christian. He was being unduly modest when he told men that something other than fear must have prevented anyone from acting against him. His determination and threats were enough and anyway there was enormous sympathy and understanding for him as well as widespread on-board unhappiness and resentment.

When Bligh wrote about the dramatic morning he omitted to mention that all three boats had been put out, for only the immediate putting out of the launch would support his belief in the story he wanted the world to accept, that he was the victim of a conspiracy, a well-laid scheme to which more than half the ship was party. Edward Christian thought readers of Bligh's Narrative would have the perspicacity to realise that it was not possible for 25 people, more than half those on board, to have conspired in such a plot beforehand without being overheard. Bligh was convinced otherwise and published a memorable paragraph: *'Notwithstanding the roughness with which I was treated, the remembrance of past kindness produced some signs of remorse in Christian . . . I asked him if this treatment was a proper return for the many instances he had received of my friendship. He appeared disturbed at my questions and answered with much emotion, "That Captain Bligh that is the thing: I am in hell I am in hell.".*

As is always the case with Bligh's defensive published recollection of events that might have dishonoured him, there are other memories of that morning. Most of those that follow were collected from men who were not party to the ship's seizure. Purcell said of the confrontations between the two men, one so pale, one so swarthy: *'Captain Bligh attempted to speak to Christian, who said, "Hold your tongue and I'll not hurt you. It is too late to consider now. I have been in hell for weeks past with you.'*

From others on deck at the time, Edward Christian collected accounts of a similar conversation and the details confirm Purcell's memories. Bligh said: *`Consider Mr Christian, I have a wife and four children in England and you have danced my children on your knee.'*

`You should have thought of them sooner than yourself, Captain Bligh. It is too late to consider now, I have been in hell for weeks past with you.'

Thomas Burkitt, who was to be hanged for his part in the insurrection, gave an even more poignant account. While the second of the three boats was being hoisted out, he said he heard Bligh say the following (and like the rest of these reports it is more likely to be in the words of the storyteller than of Christian and Bligh):

'Consider what you are about, Mr Christian. For God's sake drop it and there shall be no more come of it,' said Bligh.

'Tis too late, Captain Bligh!'

'No, Mr Christian it is not too late yet. I'll forfeit my honour if I ever speak of it. I'll give you my bond that there shall never be any more come of it.' For Bligh to speak of honour was dangerous. In defence of his duty to the king he could do whatever he thought fit. No sooner was he free than he could have pursued Christian with every means at his disposal and would not have had the slightest qualms about forfeiting his *'honour'*. Christian was right not to trust his word. Men who did so later in Bligh's career regretted it deeply. When Bligh was deposed as the Governor of New South Wales, he signed a solemn agreement to quit the colony but, as soon as he was aboard the ship to sail him to England, he commanded it to stay in the colony's waters for almost a year.

If pride had driven Christian to mutiny, pride was going to see him carry it through.

A second appeal for Fletcher to change his mind, came from Cole and Purcell, party to Christian's initial plan. Christian reminded them that they well knew how he had been used. When Cole answered, he settled once and for all the cause of the mutiny.

'I know it very well, Mr Christian,' he said. *'We all know it, but drop it for God's sake.'* Cole identified that almost the entire ship's company knew and understood Fletcher Christian's personal situation, even if they did not agree with his solution. If pride had driven Christian to mutiny, pride was going to see him carry it through. To Bligh, Christian seemed to be *'meditating'* destruction on himself and everyone else. Bligh played for time by asking for arms, a request greeted by laughter. Domineering Churchill forced the climax, when he told Christian that the heavily laden boat waited only for Bligh.

'Come, Captain Bligh,' said Fletcher Christian. *'Your officers and men are now in the boat and you must go with them. If you attempt to make the least resistance you will instantly be put to death.'*

Without further ceremony, a group of armed mutineers untied his hands and forced him over the side. Once Bligh was in the boat the jeering and ridicule increased. There was some bargaining for possessions. Christian gave Bligh his own compass, further food and clothing were thrown down plus four cutlasses. There was a plan to tow them closer to Tofua but Bligh was terrified that the noisy men on board would shoot into the boat. After pitiless hours the Captain Bligh and his loyalists were finally cast adrift on the open ocean. Ellison climbed up to unfurl *Bounty's* main top-gallant sail and Bligh says he saw George Stewart come on deck and dance in the Otaheitian manner. In Bligh's words *Bounty* was now in the hands of 25 men *'the most able men of the ship's company'*. He fancied he heard cries of *'Huzzah! for Otaheite'* and may well have. For whatever side a man was on, in the strange, strained exhilaration of those first independent hours, the indolence and flesh of Tahiti would have been especially attractive, perhaps the only concrete idea of what might happen next.

Bligh's immediate and consoling thought was that Tahiti's allures were the cause of the mutiny and that there had been a conspiracy by the majority to throw out a loyal minority. It never occurred to him that he might have been the cause of Fletcher Christian's mutiny. The motive had to be less personal, far more dramatic. Even though it was clear not everyone still aboard was an active mutineer, Bligh had to include everyone or it was too shameful for words. Only a conspiracy of the majority served as the perfect salve to his private pride and salvation of his public reputation.

Bligh's overriding characteristic was a type of vanity that prevented him understanding the effect of his behaviour or language on any other man or woman. It was perpetual, self-righteous, and consuming. This is why he could not delegate authority. He simply could not bear to see others excel and thus had to wound constantly, to bolster his view of himself as incontestably superior.

William Bligh's initial admiration of Fletcher Christian's shipwide popularity, obvious physical strength, navigational skills, superior education and higher social position turned to envy when Fletcher grew more independent of him when on shore on Tahiti. Bligh wanted, needed perhaps, Fletcher's dependence on him but Fletcher had grown out of need for him.

Professor J. C. Beaglehole, who was Emeritus Professor of British Commonwealth History at the Victoria University of Wellington, gave an enlightening lecture on August 3 1967, *'Captain Cook and Captain Bligh'*, in which he contrasted one with the other. Cook punished more often, was more cruel physically and had a smouldering irascibility that made both him and others jump up and down when he was in a rage but he was a *'character'*. His men, whom he flogged until their ribs were exposed, loved him, calling him *'the old boy'*. Well over 6 feet/1.8 metres tall, this made him especially admired on Tahiti, where height was akin to godliness. Cook had charisma, presence and a natural ability to command respect, which recommended him to the men before the mast, where manliness was highly regarded.

Bligh was short. With the unusual combination of blue eyes, delicate white skin and black hair he looked like a doll. He never understood the finer points of human behaviour and naively believed aping of Cook would bring similar adulation. Charisma cannot be learned like navigation charts, so when Bligh went into a Cook-like rage he merely looked silly. His language made it worse. Bligh must have developed his legendary linguistic ability consciously, as a weapon against the attacks on his vanity he so feared. Every spirited coward attacks before he is threatened and Bligh so developed his arsenal of insult that the most hardened users of foul language were stunned by his inventiveness. Bligh did not just use esoteric and arcane obscenity. He constructed intricate combinations of physical defects, such complexities of unnatural relationships. It was startling to hear filth pouring from such decorative lips. The contrast made him more, rather than less, ridiculous.

Much of Bligh's conspiracy theory relied on his absolute belief that his constant aspersions on the abilities and reputations of his officers and men were justified. Professor Beaglehole points out: *`Bligh habitually talked to his officers, and wrote about them afterwards, as if it had been the special purpose of the Divine Power, for some unrevealed reason, to inflict upon him for every voyage a unique collection of fools and knaves as his subordinates. It is unlikely that this is the case.'*

Beaglehole particularly pointed out that the men about whom Bligh was most vitriolic all went on to pursue blameless and distinguished naval careers. Bligh however, continued to be accused of the same faults wherever he went and whatever position he held. He added that Bligh never mellowed and that after his second breadfruit voyage in 1793, `. . . *(he) made no more discoveries except, one is tempted to say, of his own limitations, and of those he was always incredulous'*.

The William Bligh, who found himself sitting cramped and half-clad among his hastily packed possessions and with men for whom he had daily expressed contempt, had a great deal of work to do to patch up his pride. Almost immediately he cross-questioned the boatload to reinforce his suspicions. He was convinced of a conspiracy and that both George Stewart and Peter Heywood were hard-core plotters, relying for evidence solely on their companionship with Fletcher Christian.

Fletcher Christian took this Chinese porcelain bowl with him in Bounty, perhaps a touchstone of the lifestyle he enjoyed ashore. It was given by Mauatua to Captain Mayhew Folger of Topaz, the US sealer which discovered Fletcher's hideout and is now owned by the Nantucket Historical Association and displayed at the Nantucket Whaling Museum.

REPERCUSSIONS I

April 28 1789 to June 14 1789
By open-boat with Bligh to Coupang

March 23 1791 – June 19 1792
Cruelty and death in *Pandora's* box

August 3 1791 – August 2 1793
Bligh's 2nd breadfruit voyage & Bond's revelations

September 12 -18 1792
The mutineers' trials – without Bligh

1794
Fletcher Christian's defence by his brother Edward

CHAPTER 21

'...INSTEAD OF INK, IT WAS MY HEART'S BLOOD...'

The last sight Fletcher Christian had of Captain Bligh was as he and his 18 companions sailed for erupting Tofua. Here they were made anything but welcome and quartermaster Norton was killed in a rock-throwing battle. *Bounty's* cast-off men jettisoned extra clothes and equipment as temptations to their pursuers, thus deflecting them. News of Bligh's treatment of the Nomukan chiefs is thought to have preceded him and they were lucky to lose only one man. Accepting the dangers of Pacific islands, Bligh announced he saw no safe succour closer than Timor, 1,200 leagues/almost 7000 kms away.

This is Foul-Weather Bligh at his most magnificent. According to his assiduously kept notes he had on board 150 pounds of ship's biscuits, 28 gallons of water, 20 pounds of pork (presumably salted), 5 quarts of rum, 3 bottles of wine, some coconuts and some breadfruit. In modern terms, this is 68 kgs of ship's biscuit, 127 litres of water, 9 kgs of salted pork, just over 5 litres of almost 100% proof rum (about 50% alcohol) as well as the wine, coconuts and breadfruit.

Conditions were so cramped that no one could lie down to rest, not even one at a time but they could have rested back on one another, perhaps even encircled by another's arms, bodily contact welcomed not just for the comfort it gave but also for consolation in shared danger that can be so sustaining. Many men who had barely touched another man sought and found such consolation in the trenches of WWI.

Bligh never mentions what the men did to increase their comfort in the challenging closeness, how they coped with performing personal functions or any description of individual trauma except for bowel pain and constipation. Bligh writes that: *. . . all agreed to live on one ounce (25g) of bread and a quarter pint (fewer than 150mls) of water a day. Therefore, after examining our stock of provisions, and recommending this as a sacred promise for ever to their memory, we bore away across the sea, where the navigation is but little known and in a small boat twenty-three feet long from stem to stern, deep loaded with eighteen men; without a chart, and nothing but my own recollection and general knowledge of the situation of places, assisted by a book of longitudes and latitudes to guide us. I was happy, however, to see everyone better satisfied with our situation in this particular than myself.*

The confidence of these starving men in their humiliated commander was well placed, for he did have knowledge of the area and was an ace navigator. One sailing professional said to me, if you overlook the distance, all he had to do was go across quite a lot and then up a bit. Nevertheless, it was a truly heroic decision that few would be brave or confident enough to do.

The misery of this boat voyage can scarcely be imagined. Bedevilled by rain for 21 of 43 days, they laboured towards the eastern Australian coast. Bligh meticulously divided each day's rations in makeshift scales of coconut shells with a bullet as a weight, which may be seen today in the National Maritime Museum, Greenwich, London. Although they constantly trailed a line over the stern, they caught none of the fish they knew abounded beneath them.

With few words, Bligh summons up pictures of the dreadful conditions: *. . . so covered with rain and sea that we can scarce see or make use of our eyes . . . Sleep, though we long for it, is horrible . . . we suffer extreme cold and everyone dreads the approach of night . . . the least error in the helm would in a moment be our destruction. The misery . . . has exceeded the preceeding . . . The sea flew over us with great force and kept us bailing with horror and anxiety . . . another such night would produce the end of several.*

On May 25 1789 after almost a month on the open sea, Bligh had to cut rations even further, which he described as feeling like taking life itself from his haggard companions. Four miserable days later, after negotiating the treacherous Great Barrier Reef, they landed on the Australian coast. As with any interlude in a difficult sea voyage, a change of mood overcame the men when they were on land. There were disagreements about the proportions served of an oyster stew and someone stole some of the remaining pork. There were even factions aboard the little boat. The majority supported Bligh but the rest looked to Fryer for leadership. There were constant complaints about who was collecting the most food, who was eating too much, the excusable petty niggles of men who, famished and dispirited, faced anonymous death for reasons not of their own making.

Further up the coast on May 31, Purcell's temper erupted into mutiny when he said to Bligh he was as good a man as he and that, anyway, they would not be here if it had not been for Bligh. Bligh wrote: *'I saw there was no carrying command with any certainty or Order but by power, for some had totally forgotten every degree of obedience'.* Bligh had sympathy, yet it is natural that the men might doubt that his every command was right. No man likes to feel his life wholly dependent on just one other when the situation is so dramatic. If Bligh had such severe disciplinary problems with weak, undernourished men lacking firearms, it is not surprising that Fletcher Christian, on the other side of the Pacific, had to rule his stronger group by pistol. It is the way of seafaring men when on land, and the only wonder is those who wonder at it. Henceforth, Bligh always carried a cutlass.

Bligh's manly courage at a time when he could have been murdered did little to encourage tighter order or respect. Yet, as well as having to contain passions that would turn the most phlegmatic into an apoplectic, he had the self-discipline to write regular reports, observations and sketches, noting details of current and coastline and fixing his daily position.

Once the boat sailed away from the north-eastern tip of Australia and he and his companions were on the open sea again, Bligh noted it seemed as if they *'had only embarked with me to proceed to Timor, and were in a Vessel equally calculated for their safety and convenience'.* Days later the condition of his men had weakened alarmingly, legs were swollen and there was much of the terrible sleepiness that often precedes the body's predisposition to surrender. Bligh, too, was seriously ill.

On June 12 Timor was at last sighted. Bligh wrote that it was scarcely believable that this should have taken only 41 days. By his reckoning they had made about 90 miles/145 kms a day. If land was in sight, co-operation and comradeship were not. Bligh refused when Fryer and Purcell demanded that they should land immediately, perhaps unwise to do on an island both unknown and inhabited by tribesmen. All the men in the boat were filthy, starving, exhausted and probably in pain, so to be denied respite when so close to land must have been agonising. In a nice turn of phrase and using further reserves of courage, Bligh severely reprimanded Fryer, telling him *'he would be dangerously troublesome if it were not for his ignorance and lack of resolution'.* Something changed Bligh's mind. A few hours later Peckover and Cole were landed in a small settlement, where they learned that Coupang was just along the coast. A local agreed to tow the boat there once he had sight of Bligh's parcel of Rix dollars, the money the Admiralty had given him to buy additional plants for the West Indies and Kew Gardens from this Dutch dominated part of the world. That puts paid to the charge that Fletcher Christian stole *Bounty's* cash reserves.

The pitiful cargo was towed into Coupang, now Kupang, on the south-west coast of Timor, as day broke on June 14 and Bligh's attention to detail and duty now assumed the trappings of a mature mania. Even though he and his men were desperately ill, he raised a pennant of distress and prepared to wait at sea for formal permission to land. Fortunately, the signal was quickly spotted and there was little delay in landing, upon which Bligh was astonished to be greeted in English by a sailor from his own country. The sailor's commander, a Captain Spikerman, organised an English breakfast with pots of tea and asked Bligh to invite his men up, too. Extraordinarily, Bligh ordered Fryer to remain on board to guard the boat and his belongings and so Fryer demanded that Bligh's servant John Smith should also remain.

The two deserted and starving men were noticed and brought a kettle of tea and some small cakes by a soldier. Fryer, moved to tears by the man's thoughtfulness, began to thank him in Dutch but was surprised when the man answered in English. His father was English and he liked people from England, he said. It was some hours before Dutch soldiers told Captain Spikerman that there were two men still in the boat. He was amazed, thinking the breakfast party in his house represented the boat's full complement. Tea and bread were immediately sent out but only the tea was taken as the two men felt that to eat too much would endanger their health further. There now being little else to do but wait, Fryer shaved himself. Learning that Smith did not think himself capable of doing the same to his own very large black growth, Fryer set to and within a half hour had given the man a smooth and clear face. He adds, *'all this time I might have gone to the devil for my good friend Captain Bligh'.*

In describing the events that found him in a part of the world he never expected to be, Bligh noted his disappointment in Ellison, saying he felt he had been run down by his own dogs. He makes enigmatic reference to a problem with Hallet and Hayward but never expanded. It is clear that nothing deflected him from his impression that the mutiny had been planned in advance and by a large group.

Governor Adrian van Este was sick but put a big house at Bligh's disposal. Bligh had to invite his haggard crew to share it, for there was no other comfortable space available. Bligh continued his manic industry, quickly laying down the foundations of his defence as he began making statements to his hosts, writing a full report of the mutiny, descriptions of the mutineers and letters of excuse. He had plenty of time. Inquiries showed they would have to sail to Batavia to ensure passage to England and Batavia was 1,800 miles/2900 kms away. Such a sea voyage was inconceivable with his men in the condition they were. For two months the men convalesced and Bligh crammed his logbook with the minutiae of life in the settlement; history, growing rice, Chinese burial customs, the slave trade and market prices. On July 20 the botanist David Nelson, who was collecting breadfruit shoots locally, died of fever. Bligh seems to have been touched by his death.

Bligh only got around to writing to his wife Betsy the day before they sailed for Batavia. His expectation of sympathy in this letter is obvious but understandable. His priorities of money and position are neatly woven into a missive that is a masterful manipulation of events and the reader's emotions. Bligh was more than fighting for his reputation, he was fighting for his future. In Coupang he bought a 34 foot/10.4 metres schooner named *Resource* that was armed with four swivel guns and stands of small arms because there were pirates in these waters. When he sailed on August 20 1789, the ship towed *Bounty's* launch and carried a collection of plants, including more breadfruit, for which the Rix dollars had originally been allocated.

After a month at sea in *Resource* they reached Surabaya on the north-east coast of Java and Purcell and Fryer were misbehaving again. This time Bligh marched them below at bayonet point. It didn't help when Bligh sent for help from Dutch officials but learned that calumny had been spread in the township, suggesting Bligh would be hanged or blown from a cannon's mouth when he returned to England.

Next day a court of inquiry listened to both sides. Fryer accused Bligh of overcharging the Admiralty for items purchased in Coupang and produced a paper that seemed to implicate Governor van Este in this slander. Bligh countered by producing his receipts and vouchers that had been signed by Fryer and the boatswain as well as two respectable residents of Coupang. Being so discovered, Fryer begged forgiveness, ending a note to Bligh, *'if matters can be made up, I beg you will forward it'*. This was pointless, because Fryer had also blackened Bligh back in Coupang, suggesting no bill would be honoured by the Admiralty if it had only Bligh's signature. The two men were reduced to the silliness of communicating solely by letter. Fryer wanted assurance that Bligh would forgive him but Blight replied saying he was too busy to see him. Purcell and Fryer were put onto separate Dutch ships that would also be sailing on to Batavia. Bligh did visit Fryer eventually, only to hear more disavowals of the infamous reports he had spread, then ordered the convoy to sail. On September 22 the ships called at Samarang on the north coast of Java and there Fryer apologised properly and was released. Purcell remained in irons until they reached Batavia on October 1.

Here, Bligh became feverish. His condition deteriorated, so he obtained a medical certificate to prove it, which explained his haste in departing and leaving the rest of the men in Fryer's charge. All he could raise for the *Resource* and the launch was $Rix295.

It was October 16 before he sailed with his servants Samuel and Smith on the Dutch East Indiaman, *Vlydte*. By now Bounty's cook Thomas Hall had died and before the rest of the group departed master's mate Elphinstone and quartermaster Linkletter also perished, both possibly of malaria. The butcher Robert Lamb died during his passage home but the fate of acting-surgeon Ledward is not known. Of the 19 men Fletcher Christian turned off *Bounty*, 12 returned to England.

On February 17 1790 the Cumberland Pacquet reported that

Bounty, although *'retarded beyond the proposed period'* was bound for the West Indies via Cape Horn, news based on journalistic surmise and on ignorance of the return route *Bounty* was supposed to have taken. Nobody panicked when a sailing ship was overdue in those days, so there was no anxiety when 1789 dawned and passed without word from the ship.

On March 14 1790 Bligh landed on the Isle of Wight and days later arrived in London. The interest in his story was enormous but publicity was not necessarily what Bligh wanted. It was more important that Sir Joseph was still prepared to support him. Banks was persuaded to put the best face on it, no happy man as long as the breadfruit plants were in the South Pacific. The press had a heyday and soon most people knew about the mutiny and the boat voyage. Bligh had only adulation and sympathy, a reasonable initial reaction.

Quite what Fletcher Christian's immediate family made of Bligh's story is largely unrecorded . . .

Quite what Fletcher Christian's immediate family made of Bligh's story is largely unrecorded but one of the supposed results that has been widely advertised is quite untrue. His first cousin John Christian XVII did not change his name to John Christian Curwen because of the mutiny. Ten days before Bligh landed, the Cumberland Pacquet for March 3 1790 records: *'Mr Christian of Workington will shortly take the name of Curwen. His Majesty's Royal Licence and permission respecting same will probably be announced in the next Gazette'* which was the official court newspaper. On March 17th it was announced that the change had been celebrated *'by a large party of principal inhabitants at the Indian King on Monday last'*. John had been preparing for the change for some time and had formalised the Christians' coat of arms in 1788. John Christian was too well known as a politician to hide behind any name and anyway always retained Christian as part of his surname. He was simply getting things straight between his two ancient families, the Christians and the Curwens.

Three weeks after the name change was announced, the Cumberland Pacquet on March 24 told the news of the mutiny. Although John Christian Curwen is mentioned in the same issue there is no linking of the two and of all the Cumberland Pacquets I read that mentioned *Bounty* or Fletcher Christian, none made a connection with John, even though as a rich and famous MP, John had quite as much news value.

The first-hand account we have of a family member's reaction comes from Charles Christian, Fletcher's mutinous surgeon brother. By 1790 he was in Hull, where he says he was well known for *'the successful Extirpation of two Women's Breasts, one just broken into an open cancerous state, the other an immense excrutiating painful schirrous Enlargement . . .'* He continued though to have trouble with the jealousies of others and writes of the latest intrigue which involved *'the circulation of . . . malicious poetic productions'* to the effect that Charles had gotten a widow pregnant, then arranged her marriage to another man so he would not have to care for the child. He writes:

When my mind was in a state of extreme soreness from the invidious and malignant attack by a villain whom I had never spoken to, came the heart-rending account of the Bounty's mutiny.

I was struck with horror and weighed down with a Sorrow to so extreme a pitch that I became stupified. It was hard to bear, but I thank God, strength was given to me equal to the burden.

I knew that this unfortunate occurrence, following so close on the heels of my late eventful and disastrous voyage [in Middlesex], would occasion the lies which had been spread abroad in consequence to assume the aspect of Truth. I have in bed perspired with agony of mind till I thought my nostrils were impressed with a smell of Death such was the peculiar sensation I experienced.

There was similarly melodramatic, but more serious, reaction by another brother. According to genealogical notes written in the 19th-century transcript of the diary of the boys' first cousin Jane Christian Blamire, Humphrey, *'then at a station on the Barbary coast of North Africa, being in very bad health, died shortly after reading the account of the mutiny'*. Edward, by now a Fellow of St John's and confirmed as a Professor of the Laws of England at Downing, appears to have made no immediate comment.

When Charles Christian recovered, he wrote to Dr Betham, William Bligh's father-in-law and *'. . . firmly prophecied that it would be found that there had been some Cause not then known that had driven Fletcher to this desperate step. I was enabled to form the just presage from what I had so recently observed possible to occur on Board of Ship where Strife and Discord prevailed. I told Betham that my feelings were so harrowed up with this unlooked for and unhappy intelligence that I would have him consider that instead of Ink, it was my Heart's Blood I wrote with.'*

This was in vain. Betham had died of *'dropsy of the breast and terminated his existence by suffocation'* almost a year earlier, on June 3 1789. This was one champion Bligh would sorely miss.

With the undue haste that scandal excites in the thespian world, the Royalty Theatre in London dramatised the events with *The Pirates! or The Calamities of Captain Bligh*. You could see *Bounty* sailing down the Thames, an Otaheitian Dance, the seizure of Bligh, the distress of the open boat at sea, their miraculous arrival and friendly reception. In the interests of drama, the latter two events are said to have happened at the Cape of Good Hope, which would indeed have been a journey to write about, and so the final scene was one of the Hottentot dances and ceremonies on their departure. Most intriguing of all was the final line of the bill poster, *'Rehearsed under the immediate Instruction of a Person who was on-board the Bounty,*

Store-ship'. A short time later a Parisian company went further and set the whole spectacle to music.

In June, cashing in on the best story of the century, too, Bligh's first book, *A Narrative of the Mutiny on board His Majesty's Ship Bounty,* was published by the Admiralty. By July 21, the news had sped across the Atlantic and the *Columbian Centinel,* printed and published on State Street, Boston, Massachusetts, devoted more than a column to the story.

In October 1790, the London-based artist and engraver Robert Dodd had painted and then engraved what was to become one of the most famous images of the period. William Bligh with hair streaming down his back stands in the launch, a protective hand raised as he looks at a man whom we can only presume is Christian, standing high on the stern of *Bounty* with his cutlass laid back on his right shoulder. An unidentified mutineer has just thrown Bligh one of the four cutlasses he was allowed and this is in mid-air.

The engraving is dedicated to the West India Planters and Merchants who had started the whole venture and the purchaser is referred for further information to Bligh's Narrative, so it has some claim to being an official, or at least approved, image. The face of Bligh is not unlike some of his portraits and considering the interest in the story, Dodd seems likely to have consulted Bligh for his blessing and advice. Allowing for the vagaries of the engraver's instrument, are we also seeing Bligh's recollection of Fletcher Christian? If so, it is the only image we have. I look at my hand-coloured original of the engraving that is still in its 18th-century frame and glass and wonder if I am looking at my great-great-great-great-grandfather.

Bligh had to face a formal court martial for the loss of *Bounty,* which was held at Spithead on October 22 1790. He was then obliged to state whether or not he wished to charge any of the officers or men among the 11 other survivors now in England. Details of his low opinion of *Bounty's* master Fryer and carpenter Purcell had been published in his Narrative and thus it might be thought he would have charged Fryer but he did not, seriously diminishing his published criticisms of the man, which were common knowledge and yet again demonstrating how quickly Bligh could retreat when challenged. Purcell, less highly ranked, was charged on six counts but the prosecution was so mild he earned only a reprimand. Was Bligh being conciliatory? Had he mellowed after his harrowing experiences? It was more likely he was protecting his career by avoiding long trials in which the accused might make counter accusations. The advice of others, including Banks must have swayed him.

Captain Edward Edwards of *Pandora,* who was sent to the Pacific to find *Bounty,* subsequently arresting some of its crew on

Tahiti, made abstracts from Heywood's and Stewart's journals and also talked to Heywood at some length. Heywood's story does not waver and the men who accused Heywood of complicity at their trials were later to express regret for presenting suspicion as fact. There is another very important document from Captain Edwards that supports Heywood's innocence, a draft of a letter he wrote to 'C. Christian.'

Apparently, the addressee and a woman called Mrs Christian wrote to Edwards as soon as he returned to England, inquiring about Peter Heywood, whom they knew to be a friend of Fletcher Christian. The only C. Christian I can identify at the time who had a living wife or mother was Charles Christian, Fletcher's brother and the woman would thus be his mother, living on the Isle of Man. Or perhaps she had fled to her son in Hull at the news of the mutiny and they had stayed together for some time. This is unimportant compared to what Edwards says in one of the few letters of which he made a copy.

On July 17, 1792 he begins by letting them know that *'the unfortunate young man Peter Heywood whom you mention'* was aboard a guardship in Portsmouth. He goes on: *'I apprehend he did not take an active part against Captain Bligh. How far he may be thought reprehensible for not taking an active or decided part in his favour in the early part of the business will depend on the construction the court may put on the evidence and the allowance that may be made in consideration of his youth, should that also be made to appear. I have had some conversation on the subject with Cmmnd Pasley with whose family the Young Man has some connection . . . he has been informed the Young Man was only seventeen years old at the time of the mutiny I have only to observe that he appeared to be much older and I understand he passed for and was considered to be so on board the Bounty. Whatever . . . his conduct . . . he certainly came on board Pandora of his own accord almost immediately . . . it is greatly to be lamented that youth through their own indiscretion or bad example should be involved in such difficulties and bring ignominy on themselves and distress to their friends'.*

Coming from a man whose reputation for cruelty and insensitivity surpasses that of Bligh, this is praise indeed. It is scraping the barrel to say Heywood's crime was not to assist his captain, as this is true of two dozen others. On April 28 1789 Bligh had little support among his officers and crew and that is what made it simple for Fletcher Christian to mutiny.

Relatives of the men who had not returned with Bligh wrote asking for news of their sons, husbands and fathers. In Bligh's replies that survive, wounded vanity is venomously compressed into a few words. Peter Heywood's mother, recently widowed, received the following, dated London, April 2 1790.

Madam, I received your letter this day and feel for you very much, being perfectly sensible of the extreme distress you must suffer from the conduct of your son Peter. His baseness is beyond all description, but I hope you will endeavour to prevent the loss of him, heavy as the misfortune is, from afflicting you too severely. I imagine he is with the rest of the mutineers, returned to Otaheite

To Colonel Holwell, an uncle of Peter Heywood, he wrote: . . . *your nephew Peter Heywood is amongst the mutineers. His ingratitude to me is of the blackest dye, for I was a father to him in every respect, and he never once had an angry word from me through the whole course of the voyage, as his conduct always gave me much pleasure and satisfaction. I very much regret that so much baseness formed the character of a young man I had real regard for, and it will give me much pleasure to hear his friends can bear the loss of him without much concern.*

In both letters he moans for sympathy rather than giving information, ever the wounded, never the wounder. Worse, he was vilifying a young man unable to defend himself. When face to face, self-defence was a possibility but, as with Fryer and Purcell, Bligh said nothing or little.

On October 24 1790 Bligh wrote to reassure Banks he had been honourably acquitted at the court martial. He said he had held back some of the evidence against Purcell, *'for it threatened his life'*, surprising in view of Bligh's vindictiveness in print.

Lord Chatham assured Bligh he would be promoted to post captain as soon as Chatham had seen the king and Bligh proudly told Banks that he was to be presented to His Majesty on Wednesday. An enigmatic section in the letter shows that Banks and Bligh were not on the best terms. Bligh writes, *'I am concerned at losing your kind assistance just at this time . . .'* but nevertheless makes suggestions about how Banks might help his promotion. Bligh was presented to King George III, who was enjoying a spell of popularity and clarity between losing the American colonies permanently and misplacing his reason temporarily and was then promoted to commander and then to post captain, the usual conditions of service being waived.

Bligh and Betsy and the girls lived for five more months on the generous £600/£45,000 today given him by the Jamaican Government in commiseration, plus half pay. On April 16 1791 Banks had him appointed commander of a new breadfruit expedition, thus ensuring and preventing both offenders and offended from receiving fair and balanced hearings in courts, ward rooms or drawing rooms.

Mutiny was common at the time and it was just as common for mutineers to counter charge their captain and for him to be punished as harshly as he had expected the mutineers to be. The Royal Navy was the Senior Service and, although not always the case, an appearance of fairness and justice was vital for the security of the nation. If Bligh had been present at subsequent trials, history and his future are likely to have been markedly different. The judgement of Sir Joseph Banks would also be called into question.

I think Banks was aware of the danger of counter-charging, which would be as troublesome for him as for Bligh. *Bounty* should have been bigger, so it had other officers and marines and should probably have had fewer young midshipmen aboard. If blame were attached to the organisers rather than to Bligh, the opprobrium would land on Banks, just as he had predicted before *Bounty* sailed. It was better if Bligh were out of the way and not unlikely he was persuaded in return for continued patronage by Banks, including further promotion.

It did not work out like that. Bligh found his reputation was worse after his third trip to the South Seas, the second breadfruit expedition, when public opinion had swung heavily against him. It never reverted and it was largely his own fault. Bligh thought he had told the truth about the mutiny but it was not the whole truth, nor was it nothing but the truth and he was exposed. While he was away, Edward Christian had been talking to those who were also aboard *Bounty*.

When his defence of Fletcher Christian was published what happened before his mutiny would finally be told, events the judges of the alleged mutineers did not hear and could not take into account.

CHAPTER 22

PANDORA'S BOX

On March 24 1790, ten days after Bligh returned with his story of South Seas mutiny, King George III ordered Fletcher Christian to be pursued. The Admiralty appointed Captain Edward Edwards, a commander who in no way equalled the marine talents of Bligh. In classic deprecatory mode, Bligh doubted Edwards' ability even to find his way to and from Tahiti.

On November 7 1790 Edwards' expedition sailed in *Pandora*, a frigate of 24 guns and 160 men. After months of searching in the South Pacific it apprehended a few of the fugitives on Tahiti but overall the voyage cost almost as many men as had originally sailed on *Bounty*. It was a far more horrid illustration of barbarity than anything aboard *Bounty* had been.

Edwards is not a reliable source for details of the voyage. Both Morrison and Heywood furnish us with accounts of their treatment at the hands of this man, treatment that John Barrow, Second Secretary to the Admiralty, described as having *'a rigour which could not be justified on any ground of necessity or prudence'*. Edwards based his justification on something other than necessity or prudence, fear of another mutiny. He had already been the victim of one aboard HMS *Narcissus* in 1782. Six of those mutineers were hanged, one sentenced to 500 lashes and another to 200.

Pandora anchored in Matavai Bay on March 23 1791, after rounding the Horn and sailing close enough to Pitcairn's Island to have warranted investigation but Edwards ignored it or perhaps couldn't find it.

The success of his mission seemed assured in the first hours after Edwards arrived in Tahiti. Coleman put out in a canoe and, although he was almost drowned when the canoe capsized, he climbed aboard. To his astonishment he was arrested. None of his ex-shipmates had the slightest idea that Edwards was ordered to treat all *Bounty's* men he found as mutineers.

Next George Stewart and Peter Heywood arrived in a double canoe. They asked for Thomas Hayward, *Bounty*'s lazy midshipman, now a third lieutenant, whom they had discovered was on board and expected him to agree their innocence but they were disappointed. Heywood wrote later, *'he (like all worldlings when raised a little in life) received us very coolly and pretended ignorance of our affairs!'* Stewart and Heywood were clapped in irons. Then the unsuspecting Skinner arrived and joined them in bondage.

Ten more *Bounty* men remained on Tahiti. They were all in Papara, acting as mercenaries for Teina and helping prepare for a major massacre. Just as Christian's force had been embroiled in Tubuaian politics, the men he had left behind on Tahiti became involved in the inter-clan squabbles there. With arms and ammunition, they were a prize beyond price for any man with ambition and helped permanently change the social structure of Tahiti.

At first the Europeans panicked and sailed out of the lagoon aboard home-made *Resolution*. Only Byrne refused to go and fuddled his way back to Matavai Bay. The vessel was unprepared for a lengthy voyage and soon sailed ignominiously back, where the Tahitians were now torn between saving the extraordinarily useful ship and saving their friends. They also kept an eye on *Pandora* in case it could offer bigger and better rewards. Some of *Bounty*'s men hid in the mountains but Morrison, Norman and young Tom Ellison remained on *Resolution*. They were quickly arrested by Tahitians, who did not want their warship damaged.

Morrison's trio escaped but were found by one of *Pandora's* boats and were as surprised as the rest to be put into irons. Norman and Coleman expected Bligh to have spoken up for them. He had, but not strongly enough and Edwards was ordered to make no exceptions.

Edwards knew enough about Tahiti to note there was no important man among those welcoming him. Teina, father of Tu, was nowhere to be seen. He was in hiding, in fear of his life. His brother Ariipaea was in charge of his district and acting as regent for Tu, still only seven years old.

Eventually, the opportunist Teina came to offer his help in rounding up the rest of the Europeans but he was not needed. He saw that his hegemonistic aspirations were dashed and begged Edwards to take him away. The captain refused and Teina headed to hide in the hills. Tahiti returned to its previous state of petty clandoms but young Tu and his advisers had learned from the experience. With the advantage of guns left behind by *Bounty*, careful cultivation of the English captains and missionaries who subsequently came to the island and the building of a loyal body of mercenaries, Tu eventually did conquer the entire island and founded the Pomare dynasty, which later sold out to the French 'for protection against the English'.

The men from *Bounty* were treated insufferably. Perhaps because Edwards was terrified that his own crew would be persuaded to mutiny, they were chained and forbidden to speak in English or Tahitian under pain of instant death. The hammocks they were given were verminous and they were unable to change their clothes because they were bound so tightly. The foul air and enervating temperature of their floating prison must have been torture enough to men who had so recently been free in the perfumed air of Tahiti and enforced silence and lice made it intolerable.

Still not certain the prisoners were fully secure, Edwards ordered the construction of his infamous *Pandora's* box. On the after part of the ship's quarterdeck, it was 18 feet by 11 feet/5 metres by 3.4 metres. Only just taller than a man, the box was entered by a scuttle less than 2 feet/ 61cms square and had only two holes 9 inches/23cms square for ventilation. Iron grates in these effectively halved the amount of air that could pass. This unshaded box held 14 men who were constantly protected by a musical comedy trio. Two guards stood on top and a midshipman marched around the four sides.

When the weather was sunny the heat was so intense that sweat ran from the suffering men in streams to the scuppers. When it rained, they were soaked. The prisoners could have the food friends brought them but the foul conditions must have killed much of their appetites. There were two open basins for their wastes.

Two guards stood on top and a midshipman marched around the four sides.

George Stewart was so shaken by the tormented cries of his wife he begged she should never be allowed to come aboard again. When McIntosh's leg slipped out of irons as he slept, everyone's irons were tightened. When their wrists began to swell, they were told the handcuffs were *'not meant to fit like gloves'*.

The torture continued in Tahiti for two months before *Pandora* sailed on May 8 1791 but the prisoners were still confined in the box. Edwards hoped to find Fletcher Christian and *Bounty* but had only faint suspicions and rumours to go on. In early August he gave up and ran for home. When near the Great Barrier Reef, he foolishly made inadequate safety precautions one night and *Pandora* ran aground, tumbling her manacled prisoners upon one another. Somehow, they managed to shatter their irons but Edwards ordered them manacled again. Though the water rose higher and higher, more guards were placed over the prisoners and Edwards ordered they be shot or hanged if their irons were broken again. By six-thirty in the morning it was apparent that *Pandora* was lost. Edwards gave the order to abandon ship and jumped over the side with his officers, at last saying the prisoners could be released.

An armourer's mate scrambled into the box to unlock their shackles. Muspratt, Skinner and Byrne struggled out but then *Pandora's* master-at-arms, more heartless than Edwards and unaware of his orders, secured the hatchway once more. The moment he did this, the ship lurched and threw him overboard. The prisoners were locked in again with the man who was to release them. The ship was sinking but the armourer's mate worked in a frenzy to unshackle the men. At the last minute a sailor named William Moulter shouted that he would release the prisoners or go down with them and wrenched the bolt off the hatchway. Amazingly, everyone escaped but Hillbrant, who drowned while still fully shackled.

When Edwards arranged a roll-call on a sandbank three miles from the wreck he found 35 of his men had been lost including the prisoners George Stewart, Henry Hillbrant, Richard Skinner and John Sumner but his vile treatment continued. The escapees from *Pandora's* box were naked but he refused them permission to shelter from the tropical sun under a sail, so to avoid sunburn and sunstroke they buried themselves in sand. As scapegoats for the entire disaster, *Bounty*'s men were cruelly and ruthlessly treated as Edwards then led a hungry and thirsty flotilla of four open boats containing 99 men to Coupang. For the odious Hayward it was his second such arrival there.

Only in March 1792 when the prisoners were aboard HMS *Gorgon* in Cape Town, were they treated with any humanity. They arrived at Spithead on June 19 and were rowed through the foul waters to HMS *Hector* in Portsmouth Harbour.

CHAPTER 23

'...A SMALL PORTION OF THE TORMENT...'

Of the 25 men left on *Bounty* when Bligh was turned off, only ten were to be tried. Some should not have been but Bligh had forgotten to speak up, or did not do so loudly enough. For three months the suspected mutineers were confined aboard HMS *Hector* in Portsmouth harbour and there they were allowed writing paper, ink and visitors. Morrison composed his Memorandums, wrote his own defence and helped others construct theirs. With help from his companions he was also working on his Journal, a detailed record of his experiences since December 1787. Heywood worked on his dictionary of the Tahitian language.

Two days after Edwards' court martial for the loss of Pandora, which clearly showed there was no sympathy for the men of Bounty, not even the ones Bligh conceded were innocent, the ten survivors were brought to trial aboard HMS *Duke* on September 12 1792. If they were found guilty, death was the sentence but it was the practice that recommendations for mercy could and would be heeded. On, each was charged with *'mutinously running away with the said armed vessel (Bounty) and deserting from His Majesty's Service'*. There was no mention of piracy and, as is often overlooked, there was no interest in any contributing causes of the mutiny. The trial was only about what happened on April 28 1789.

It was over a year since Bligh had sailed away in *Providence* on August 3 1791, so he could not be cross-examined, which made it difficult for the prisoners or their counsel to plan suitable defences. The prosecution said Bligh's Narrative was admissible as evidence because it was *`official'* and, not being able to contradict or to add what had been omitted, this meant the prisoners could easily be disbelieved on major issues. Public opinion was decidedly against them. Peter Heywood, who had turned 20 in June was the only officer to be tried, so expected to be made an example.

What was said during the trial was taken very seriously at the time, as it should be now. Much detail from the proceedings has been incorporated into my narrative of the mutiny.

Peter Heywood was doubly damned by acts of commission and omission, especially when the Captain Bligh did not publish some lines from his original manuscript in his public versions of the mutiny. The most relevant omission was: *'As for the officers . . . they endeavoured to come to my assistance, but were not allowed to put their heads above the hatchway.'* If Bligh had published that sentence, his story that everyone left on board was associated with the mutiny could not have been sustained. Suppression of the sentence was as spiteful as the letter he had written to Heywood's widowed mother.

Heywood was further damned by the evidence of Thomas Hayward who, overlooking his own tearful begging to stay aboard, said he supposed Peter Heywood must have been a mutineer, simply because he stayed. Hallet said he saw Heywood laughing when Bligh spoke to him on the brief occasion that Heywood had come on deck to see what was happening. Both these men later expressed remorse for their statements.

Neither was Heywood helped by the defence statement written by his counsel and read on his behalf. It was mawkish, relying largely on appeals of youthful indiscretion rather than hard evidence that clearly showed he was forced to remain below. There was just enough suspicion of collusion, based mainly on the extraordinary idea that this comparative boy should have led a counter-revolt, for him to be found guilty but recommended for mercy. An enormous effort was made on his behalf, led by his sister Nessy, something criticised as the upper-classes conniving to protect and save one of their own but what is surprising about that?

The king granted Peter Heywood a free pardon on October 24. Later legal wrangles resulted in the declaration that the King's Warrant was not so much a pardon for a crime committed as a quashing of the conviction. The pardoning of Heywood has been put down to family influence but it is in no way unusual for a young man to wish to live, or for his family to support him in this. Although the death sentence was mandatory for many civil and martial crimes, there were thousands of precedents for it not being carried out. It has been estimated that fewer than half of those sentenced to death in the civil courts were actually executed. All that was needed for a reprieve were connections but not necessarily in high places. A mother, a brother, or anyone else of honest and industrious standing in the community could vouch that they had some need or care for the condemned person. This was all the 'connection' the courts looked for. The only people who could be assured of execution were those for whom no one spoke up. Money and titles helped but thousands of men and women who had no access to either escaped the death penalty simply because they had friends. Wouldn't you have done the same to protect a family member if you had the resources, the friends, or both?

Morrison was found guilty but also given a free pardon. It is likely that copies of his Memorandum, subtitled Vedi et Scio (I saw and I know), which was written while he was awaiting trial, was circulated discreetly amongst senior members of the Royal Navy during the trials and was one of the 'various reasons' contributing to his pardon. There was plenty of precedent for counter charges at mutiny trials and it was precisely this kind of nastiness that Banks would want to avoid. Read in conjunction with the trials' evidence, Morrison's outline of what could be said in reply (even if not substantiated), showed it would be troublesome to convict men about whom there was any possible doubt. They, then having nothing to lose, would counter claim. In some cases, the captains against whom counter charges had been made were found guilty and dismissed from the service. Banks could not afford to let this happen to Bligh because it would ruin them both. Morrison's Narrative was not published until 1935 but even so details of what Bligh had not said were repeated in and out of the Royal Navy.

Norman, Coleman and McIntosh were acquitted, as was the fiddler Byrne, but none was given compensation for their appalling treatment by Edwards aboard *Pandora*.

Royal Navy hangings were not a sudden drop to break your neck.

There was too much evidence against young Ellison, so the court did not feel they could take his age into consideration. The verdict of guilty was not accompanied by a recommendation for mercy. The most convincing statement against him was made by Hayward, who said he saw Ellison arm himself and run towards Bligh saying, *'Damn him I'll be sentry over him.'*

Burkitt, who had equivocated and covered Bligh's exposed genitals, found that this was not enough to outweigh his having been with Christian when Bligh was seized. He was convicted and so was Millward, who had stood guard over Fryer while the master was detained in his cabin. Muspratt was also convicted but he had been defended by Stephen Barney, who managed to get him discharged because of a technical irregularity.

No friend was prepared to speak up for Ellison, Burkitt or Millward. On October 29 1792, the three men were hanged publicly aboard HMS *Brunswick*. Seamen from every other Royal Navy ship in the port were present. Millward is said to have made the following stirring and penitential speech to the assembled tars. It sounds unlikely but several people vouch for its veracity and perhaps it was penned by Morrison.

Brother seamen, you see before you three lusty young fellows about to suffer a shameful death for the dreadful crime of mutiny and desertion. Take warning by our example never to desert your officers, and should they behave ill to you, remember it is not their cause, it is the cause of your country you are bound to support.

The shameful death the onlookers were about to see was an example indeed. Royal Navy hangings were not a sudden drop to break your neck. Instead, a cohort hoisted you aloft from deck level, so that the slow choking and the facial discolouration, the bowel and bladder voiding and the futile struggling against suffocation were public and prolonged.

Once free, Heywood quickly told Fletcher Christian's family the truth about *Bounty,* as he had promised Fletcher he would in Matavai Bay. Having himself been a victim of Bligh's post-mutiny malice, he did everything he could to prevent his friend being damned more than necessary. He sent a moving letter to Edward Christian that was also published in the Cumberland Pacquet, the first intimation the public received that they might not have been told the whole truth. Writing from Great Russell Street in London on November 5 1792, he said:

SIR, I am sorry to say I have been informed you were inclined to judge too harshly of your truly unfortunate brother; and to think of him in such a manner as I am conscious, from the knowledge I had of his most worthy disposition and character (both public and private), he merits not in the slightest degree; therefore I think it my duty to undeceive you, and to rekindle the flame of brotherly love (or pity) now towards him, which, I fear, the false reports of slander and vile suspicion may nearly have extinguished.

Excuse my freedom, Sir: If it would not be disagreeable to you, I will do myself the pleasure of waiting upon you; and endeavour to prove that your brother was not that vile wretch, void of all gratitude, which the world had the unkindness to think him; but, on the contrary, a most worthy character, ruined only by the misfortune (if it can be so called) of being a young man of strict honour, and adorned with every virtue; and beloved by all (except one, whose ill report is his greatest praise) who had the pleasure of his acquaintance.

I am Sir, with esteem
Your most obedient humble servant
P. Heywood

Edward Christian must have been astonished. Although his brother Charles thought there were extenuating circumstances, Edward, like everyone else, must have believed Bligh's version. Now he was told that *'the dreadful mutiny on the Bounty originated from motives, and was attended with circumstances different from those which had been presented to the world'*. Edward's legal chambers were in Gray's Inn, where he had been since July 5 1782 but he went to see Mr Romilly of Lincoln's Inn and then took advice from a more senior legal man who had been present at the trial, probably Sir Archibald McDonald, Attorney-General.

Bligh's servant John Smith went to see Edward Christian of his own accord, clearly wanting to tell a different story. *Bounty's* sailing master John Fryer visited Joseph Christian of No. 10 Strand and he said Fryer also told a new and different story. There may have been an element of revenge in the actions of Heywood, Smith and Fryer but might they not have been equally interested in putting the story straight? Bligh's reputation was a casualty, not the target. They acted with speed but Edward Christian acted with caution. The new intelligence he had was extremely dangerous if mishandled. He apprised the public at large of his intention by releasing Heywood's letter to the Cumberland Pacquet. The Pacquet published it and other papers reprinted it.

Imagine the feelings of Sir Joseph Banks in late November when he read this preface to Heywood's letter. Bligh was far away at sea, so he cut the reprinted piece out of a newspaper, the name and date of which he omitted to note, and kept it to show him. This is part of the preface: '*Though there may be certain actions which even the torture and extremity of provocation cannot justify yet a sudden act of phrenzy, so circumstanced, is far removed in reason and mercy from the foul deliberate contempt of every religious and virtuous sentiment and obligation excited by selfish and base gratifications. For the honour of this county, we are happy to assure our readers that one of its natives FLETCHER CHRISTIAN is not that detestable and horrid monster of wickedness and depravity, which with extreme and perhaps unexampled injustice and barbarity to him and his relations he has long been represented but a character for whom every feeling heart must now sincerely grieve and lament . . .*'

Heywood's letter then followed but, apart from saying he was an officer on *Bounty*, his identity was not revealed. The article went on to say that Edward Christian had since spoken to three officers and two seamen from *Bounty* who were in the London area. McIntosh alone is named, having said about Fletcher: '*Oh! he was a gentleman, and a brave man, and every officer and sailor on board the ship would have gone through fire and water to have served him.*' The newspaper article concluded by saying the mystery was soon to be solved and that shame and infamy would be distributed in the just proportions it had been earned.

It was to be almost 18 months before Edward Christian's revelations were published. He had no need to hurry, for others were equally interested in telling versions other than Bligh's. It is naive to call everything written by other men simply an attack on Bligh. Why shouldn't Fryer and Morrison and Heywood wish to defend themselves more publicly than the trial allowed? If Bligh was entitled to publish his opinions, why should not others be heard?

In the meantime, the court martial (and perhaps Morrison's Memorandum) had radically changed opinion about Bligh throughout the Royal Navy, where it counted. He found he was neither welcomed nor given audiences at the Admiralty. His lieutenants aboard *Providence* found his recommendations for promotion were often useless.

Whatever was being said in public was different from that said in private. Charles Christian spoke to a captain of marines from *Providence*, who said '*from Bligh's odious behaviour during the voyage, he would as soon shoot him as a dog, if it were not for the law*'. Lieutenant Bond wrote much more in that vein but he did not do so in public, so it had no immediate effect.

Sir Joseph Banks received a welter of correspondence and documents concerning the *Bounty* affair from Edward Christian. Banks asked Bligh to answer the allegations and questions they brought up. For 18 months after returning in *Providence,* Bligh had little else to occupy his time. He was on half pay, with no offer of a command. It is not clear why this should have been so. He had promised Betsy that if he was returned safe from the *Providence* voyage, he would forswear the sea but that was ages ago and promises were the blight of his life. His long furlough may also have been an unofficial punishment or expression of displeasure after the publication of alternative versions of Bligh's Narrative.

The Mitchell Library in Sydney holds Bligh's draft replies to Banks concerning three separate sets of documents. I cannot think why these notes have not been given more importance, except they are not indexed and so take days to read and few authors on the subject want to spend so much time.

First, Bligh sketched answers to three letters addressed in December 1791 to Banks by Edward Christian in which Christian revealed the information he had collected from ex-*Bounty* men in London to which the newspaper article alluded and that he wished to publish. It was an act of fairness to present them to Banks before publishing, not the actions of a man who wanted to defend his brother Fletcher at any price.

Wishing to keep the interest of Banks, Bligh carefully moves for the first time towards allowing the possible veracity of other men's stories, indeed in places he positively contradicts his public statements. Perhaps he realised that to retain his position he had to resort to the truth, except Bligh asserts he was Fletcher Christian's sole patron and had employed him in the Merchant Service for three years, a convenient lapse as the correct length was nine months.

Bligh scratchily makes the major confession that the mutiny might possibly have been caused solely by Fletcher Christian's desperation but then dismisses this because he was not aware of it: *'What Mr Christian calls his Brother's being drove to desperation must have been a very sudden impulse, for it is certain that Capt. Bligh's usual invitation to dinner and supper with him was sent to him the day before the mutiny happened.'*

Fletcher Christian's mutiny against him was a sudden impulse but Bligh saying so is utterly inconsistent with his determined view that the mutiny was planned in advance. He dismisses the idea by believing a dinner invitation is proof enough that Fletcher Christian was not driven to despair. Elsewhere in the notes, Bligh finally admits he had words with his officers and men about coconuts but does not elaborate, another case where silence was his only defence. His inconsistency on shore equals that at sea.

Bligh also reveals that he argued with Christian after they left

Tahiti, another important new admission and further proof his Narrative was by no means exhaustive or balanced. He shifts his ground on other points too but keeps his paranoia firmly in place. He constantly says to Banks that many of the men now criticising him should have done so at his court martial for the loss of *Bounty* in 1790. The fact they did not proved to Bligh they had no just complaints and had subsequently been suborned. On the other hand, Bligh had not prosecuted Fryer or others when he could but he ignores this. He also moves the theory of concerted action against him first from Tahiti to *Bounty* and then to England.

Once the prisoners from *Pandora* arrived back in Britain, Bligh claims they and their friends formed connections to prevent the mutineers from being hanged. And why not? It would have been very silly for Fryer and Purcell, for instance, to have attacked Bligh at the first court martial without the supporting evidence of others who were involved. It would be strange if there were not a degree of collusion among the defendants and possibly the prosecutors. It's what accused men do, especially when threatened with death if found guilty.

There are two more points about the trial. First, Bligh did not accept those men had any right to defend themselves, citing his Narrative as *'sacred truths'*, even though he knew this was not so. He had defended himself and dissembled and he should have expected others to do so, except he might then have been revealed as a liar or withholder of the truth.

Bligh also overlooked that if the prisoners were plotting to save their lives, they could have done so by uniting to blame everything on Fletcher Christian but nowhere was Christian blamed. They could have made him a temperamental, inconsistent and disloyal second-in-command and Bligh would have had to support them, making him appear on their side. If as desperate and groundless as Bligh thought, they might also have played the homosexual card.

The penalty for homosexual behaviour on a Royal Navy ship was death and even the suspicion of such acts would have been fatal to Bligh's career. If there had been a homosexual relationship between Christian and Bligh it's just as likely one or the other would have tried to net others into his sexual lair. There is no evidence for this but it would have been a powerful tool against Bligh that would have reaped them a rich and long revenge.

There is another theory, that Christian was addicted to opium or laudanum, a liquid draught containing the drug. Although touted by Bligh apologists, the suggestion is as much an insult to Bligh's command and extraordinary attention to detail, as it is to Christian. The only source of these drugs was the medical supplies in the ship's stores and these were the eventual responsibility of Bligh. If anyone used these supplies to feed addiction, it suggests Bligh allowed this to happen or had been negligent, which would be a serious abrogation of his duty to ensure the safety of ship and crew. In *Bounty's* cramped confines and amidst an increasingly tense and hostile crew, how could Fletcher Christian possibly have been addicted and escaped notice or subsequent mention? The unlikelihood is compounded when tight control of the ship's supplies is what would make Bligh's important profit from the voyage.

Bligh said Peter Heywood's evidence from the court martial incriminated Fletcher Christian in the fullest manner. That Heywood went on to write the letter to Edward Christian simply confirmed Bligh's belief that evidence had been tampered with. Bligh was sure that Edward Christian and Peter Heywood believed Fletcher was innocent but this is totally untrue.

Edward Christian knew his brother was a mutineer, said so in print and did not defend this. He, his brother Charles and many others wished only to know the reasons the crime had been committed. If Bligh's reputation was damaged, that was unavoidable and had not Bligh already done this to almost everyone aboard *Bounty*? Professor Beaglehole is not the only one to say that the low opinions Bligh had of everyone who sailed with him were subsequently shared by very few others, yet that the low opinion of Bligh was extensive in and out of the Royal Navy for the rest of his life, when he was commonly that '*Bounty* bastard'.

In Bligh's letters to Banks about the court martial, he refers once more to the conspiracy he thought was planned on Tahiti, certain that Fletcher did have a particular female on Tahiti. He said Coleman remembered her and that Lebogue, who had gone out again with Bligh in *Providence,* had actually seen her at Otaheite on that visit. *'She went with Christian always untill his last Departure, which was sudden and unknown.'*. If Lebogue and Bligh agree that Fletcher's favourite was abandoned, this casts doubt on why Mauatua went with Christian to Tubuai and then to Pitcairn.

It is noteworthy that Bligh never asked this 'favourite' or any Tahitian for information about Fletcher Christian or any other ex-*Bounty* man when he was again in Tahiti.

In the Mitchell Library notes to Banks, Bligh adds new information about Fletcher. First, he writes that aboard *Britannia, `Christian stood godfather to Peter Roger's child my boatswain who was a very low man and was connected in plundering sugar belonging to Mr Hibbert £17 as Mr Lamb reported.'* Presumably Bligh is attempting to show that Christian was also 'low' because he stood godfather to the son of a dishonest man. Christian's act was a form of patronage and a telling example of how well he was regarded. Bligh was possibly jealous, never having been asked to stand as godfather to anyone's child. Bligh adds that *`Christian also had a hoard of provisions aboard Britannia'*, perhaps supporting his belief that Fletcher was likely to steal coconuts. A private store of comforting provisions is almost the first thing any prospective long-distance sailor does. Why Bligh thought this worth mentioning is a desperate curiosity.

Having seen Edward Christian's evidence and answered Banks' questions, it was typical of Bligh's constant inability to judge other men that he did not foresee what was about to happen. In the middle of 1794 Edward Christian published Stephen Barney's *Minutes of the Proceedings of the Court-Martial* and to this he attached *The Appendix*. The rumours, gossip and contradictions in private and naval circulation were suddenly public. Society was transfixed because, except for mutineer Muspratt, all the nice things said to Edward Christian about his brother Fletcher were from Bligh loyalists.

He was a gentleman; a brave man; and every officer and seaman on board the ship would have gone through fire and water to serve him.

I would still wade up to the arm-pits in blood to serve him

As much as I have lost and suffered by him, if he could be restored to his country, I should be the first to go without wages in search of him

Everybody under his command did their duty at a look from Mr Christian

Mr Christian was always good-natured, I never heard him say 'Damn you' to any man on board the ship.

CHAPTER 24

FLETCHER CHRISTIAN'S DEFENCE BY BROTHER EDWARD

Edward Christian had asked permission to reprint the Admiralty's official minutes of the court martial but was refused on the grounds that public records could not be released for private publication. So, Edward presented himself to Stephen Barney, the counsel for Muspratt. He had taken copious notes throughout the proceedings and although they were incomplete, with discrepancies and mistakes, they were better than nothing for Edward Christian's project. He would be able to publish men's words spoken under oath.

To these Edward Christian attached a pamphlet known *The Appendix*. The introductory page is dated May 15 1794 and Edward thanks Barney, acknowledging that his notes had been taken with no thought of future publication. Edward declares that he alone is responsible for the accuracy of what he says in *The Appendix* and that the information was obtained exactly how and from whom he says.

The Appendix is unequivocal about how abhorrent was the crime of mutiny and Edward pointed out the trial's concern was to establish only who did what on the morning of the mutiny. There was no interest in preceding events that might have been contributory. This unpublished background to his brother Fletcher's mutiny is what Edward was now presenting.

As a legal man, Edward Christian had collected a group of men of impeccable reputation and almost every time he spoke to someone from *Bounty*, one or more of these men were present. Just as sworn witnesses in court, these eminent and successful men were prepared to be named and to bear testimony that what they heard first-hand was faithfully represented in *The Appendix*. The list further averts any charge of falsehood because each could be contacted for verification, as their addresses or official capacities were printed.

John Farhill, Esq., 38 Mortimer Street
Samuel Romilly, Esq., Lincoln's Inn
Mr Gilpin, No. 432 Strand
The Rev. Dr Fisher, Canon of Windsor
The Rev. Mr Cookson, Canon of Windsor
Captain Wordsworth, *Abergavenny* [an East Indiaman]
Rev. Mr Antrobus, Chaplain to the Bishop of London
John France, Esq., Temple

James Losh, Esq., Temple
Rev. Dr Frewen, Colchester
John Atkinson, Esq., Somerset Herald at the College of Arms

Most of the men had connections with St John's, Cambridge and with William Wilberforce, the great opponent of slavery, whom Edward befriended there. There is considerable circumstantial evidence in C. S. Wilkinson's *The Wake of the Bounty* to support the claim that Edward Christian was able to gather these men because they were opposed to the slavery-related business interests of Bligh's earlier patron and relative by marriage, Duncan Campbell. It may have been one of the things they had in common but what was the relevance here? If an anti-slavery campaign was their motive, they would have said so. Many of the men had connections with the Wordsworths, too. Edward had won their famous case against the Earl of Lonsdale in 1791 and William Wordsworth was also a St John's man. Both the Canons of Windsor were very close to the king, far too sensitive a position in which to join in perjury. Dr Fisher was known, according to one source, as the 'King's Fisher'. Antrobus was from Cockermouth and Romilly, later Sir Samuel, became a famous law reformer. You will remember that the Dictionary of National Biography was dismissive of Edward's later days. Even if true, this should not colour views on who or what he was when investigating the causes of the mutiny on *Bounty*.

Edward Christian's window in the Great Hall of Gray's Inn, London, commemorating his appointment as Thesaurarius or Treasurer, the Inn's highest honour.

Admitted to Gray's Inn in 1782, Edward was called to the Bar in 1786, after which he was permitted to act as a barrister. In 1788, aged just 30, he was named Downing Professor of the Laws of England, although the college was not to be founded until 1800, after which he was a Fellow of Downing from 1800 to 1823. He was appointed Law Professor at the East India Company College from 1806-18 and then was Chief Justice on the Isle of Ely.

In June 1791, the year of his successful Wordsworth / Lowther trial, Edward renewed his lease of chambers at 3 Coney Court at 3 Gray's Inn Square but by December 8th assigned the balance to another member of the Inn. From now on he must have spent much time in researching and writing in defence of his brother Fletcher, including his extensive correspondence with Sir Joseph Banks. If there were anything suspicious, devious or dishonest about Edward's research and writing that was published in 1794 as part of Stephen Barney's *Minutes of the trial of the Bounty mutineers,* it is unlikely that Gray's Inn would have appointed him to their Bench, its managing committee in 1809 or to their most senior rank, that of *Thesaurarius* or Treasurer in 1810. A stained-glass image of his unicorn-crested arms commemorates this in the west window of Gray's Inn Hall. If he were to have been twisting facts it is equally unlikely he would have retained his position with Downing College.

What else would Edward Christian have done but gather those

he felt would support his cause? What balances the supposed imbalance found by sniping modern writers is that he published the names and addresses of them all, inviting Bligh and others to confront them directly if they disagreed. To my knowledge nobody did.

Edward's equally qualified older brother John might have helped but he was dead. As a second wife he married the widow of a sugar merchant, 23-years older than him and, presumably, a great deal richer for they lived in Pall Mall, steps away from St James's Palace. He was only 39 when he left her a widow a second time in 1791. According to a document kindly sent to me by Caroline Alexander (*The Bounty: 2003,)* the cause was 'a gradual decay'.

Edward's investigating committee must have been formidable to meet and perhaps some of *Bounty*'s sailors did alter their stories slightly 'to please the gentlemen'. It was not just able seamen who were interviewed and Edward Christian also published the whereabouts of each man to whom he spoke. The only thing he hid was who said what and there were very good reasons for that, as shown when McIntosh was threatened in the street. The *Bounty* men interviewed who had returned with Bligh were:

John Fryer, the master
Thomas Hayward, midshipman
William Peckover, gunner [who lived with some nicety in Gun Alley, Wapping]
William Purcell, carpenter
John Smith, personal cook to Bligh
Lawrence Lebogue, sailmaker
Joseph Coleman, armourer, tried and acquitted
Thomas McIntosh, carpenter's mate, tried and acquitted
Michael Byrne, musician, tried and acquitted
Peter Heywood, midshipman, pardoned
William Muspratt, able seaman, might have been convicted but for a legal error
James Morrison, acquitted and who wrote to Edward Christian.

Edward Christian's Appendix re-tells the whole story, which because it is first-hand has been used in this book as authoritative. A particular paragraph in which Edward Christian discusses his brother's relationship with women has always created problems. The sentence continuously misinterpreted is one in answer to Bligh's assertion that Christian had a favourite female and that his mutiny was based entirely on his desire to return to her.

Fletcher Christian was on shore all the time *Bounty* was in Tahiti, yet not one man on the same duty agreed he had a female favourite or any attachment or particular connection among the women. Even if Fletcher did, Bligh then tells us that this was not the woman who went away with him.

So, either Fletcher had nothing to do with women on Tahiti or, as

Edward said, Fletcher did but had no regular or favourite female companion and that thus for either of these reasons could not have mutinied for the sake of a woman.

Edward's last paragraphs are impressive: *The writer of this Appendix would think himself an accomplice in the crime which has been committed, if he designedly should give the slightest shade to any word or fact different from its true and just representation; and lest he should be supposed to be actuated by a vindictive spirit, he has studiously forborn to make more comments than were absolutely necessary upon any statement which he has been obliged to bring forward. He felt it a duty to himself, to the connections of all the unfortunate men and to society to collect and lay before the Public these extraordinary circumstances.*

The sufferings of Captain Bligh and his companions in the boat, however severe they may have been, are perhaps but a small portion of the torments occasioned by this dreadful event: and whilst these prove the melancholy and extensive consequences of the crime of Mutiny, the crime itself in this instance may afford an awful lesson to the Navy, and to Mankind, that there is a pressure, beyond which the best formed and principled mind must either break or recoil. And though public justice and the public safety can allow no vindication of any species of mutiny, yet reason and humanity will distinguish the sudden unpremeditated act of desperation and phrenzy from the foul deliberate contempt of every religious duty and honourable sentiment; and will deplore the uncertainty of human prospects, when they reflect that a young man is condemned to perpetual infamy, who, if he had served on board any other ship, or had perhaps been absent from the Bounty a single day, or one ill-fated hour, might still have been an honour to his country and a glory and comfort to his friends!

Bligh's best response would have been suitably edited and published extracts from the letters I discovered from him to Banks in response to Edward Christian's questions. Instead, he published a series of miscellaneous correspondence and orders that have no cohesion and upon which he does not comment. Some of them even tell against him, such as Heywood's letter to Edward Christian, which Bligh does not counteract by comparing it with Heywood's evidence in court. The few comments Bligh made were meant to debunk Edward Christian's method of collecting statements. Bligh suggested he withheld facts about who said what to obfuscate the truth and to prevent perjurers being brought to justice. This was a dangerous thing to claim about a Gray's Inn barrister and reckless to say about two Canons of Windsor or a Somerset Herald. Bligh was publicly calling them liars and plotters but Edward was only protecting his sources, then and now an acceptable part of collecting fact for publication.

Next, in 1795 Edward Christian published *A Short Reply to Capt. William Bligh's Answer.* Even Christian's foes admit it is brilliantly argued for it removes absolutely any suspicion that Edward

Christian kept his quotations anonymous for any but the most noble reasons; without a guarantee of anonymity the men would not have spoken at all. McIntosh had been threatened for what he had said to Christian and whoever else was present.

The Short Reply is one of the rarest and most expensive pamphlets in the world. Only 150 were printed and today (1982) only three are known to exist. When *A Book of the BOUNTY* was published in 1938, it collected all contemporary documents, including Bligh's account of *Bounty* and his open-boat voyage, the minutes of the mutineers' trials, The Appendix and Bligh's Answer. Edward Christian's *A Short Reply* is not included and this must be why so many authors have overlooked its important substantiation of everything in The Appendix. Could it be they just don't want to know?

Edward Christian and his eminent colleagues worked together to publish facts that otherwise would have been suppressed. Journalists get prizes for that today.

Kennedy and others suggest that *A Short Reply* was meant to sting Bligh into taking legal action. Bligh could then have been cross-examined, hoping to disprove the allegations in *The Appendix* and their corroboration in *A Short Reply*. They could not be disproven, so once again William Bligh said nothing.

In April, just months after publication, Bligh was given command of HMS *Calcutta* and supervised her conversion from the East Indiaman *Warley* to a 54-gun fourth rate and in October sailed in her with 200 troops to put down a mutiny aboard HMS *Defiance*.

Once again, he was far less trouble to the authorities when he was at sea than ashore, yet he had learned nothing from his South Seas ordeals.

CHAPTER 25

BOND EXPOSES BLIGH

When I embarked on this project of bringing Fletcher Christian to life, I expected to have to blacken Fletcher Christian's reputation and to defend Bligh. It was amazing how short a time I was able to maintain that expectation. The available evidence seemed to prove overwhelmingly the traditional harsh view of Bligh, except that he was never as physically cruel as films made him out to be; the keel-hauling seen in the Gable / Laughton movie was never part of Royal Navy punishment. I was rather worried about this. I had found so much new material about Fletcher Christian, I wanted to say something new about Bligh, good things, ideally. Recent works gave no clues to new sources. All the research seemed to have been done. Then, when I was at the Australian National Library in Canberra, an assiduous librarian pointed out extraordinary material. Many of the original documents are in the Greenwich Maritime Museum and copies were available on microfilm around the world but, although discovered and published in 1953 and 1960, they had been largely ignored by Bligh's apologists. Gavin Kennedy's *Bligh* published in 1978 didn't know about these papers, so writers who followed ignored them, too. The first scholar who realised and strongly argued their significance was the Swedish historian Rolf E. DuRietz, in 1963 and 1965.

The material in question is the draft of a letter plus journal notes written by Lieutenant Francis (Frank) Godolphin Bond, first lieutenant on HMS *Providence*, the ship commanded by Bligh on the second breadfruit voyage. Bond's position aboard *Providence* was the same as that to which Fletcher Christian was promoted at sea in *Bounty*. The differences were that he had not sailed with Bligh before and there was no passionate friendship although he was a relative through marriage, the son of Bligh's half-sister Catherine Pearse, the daughter of his mothers' first marriage to Richard Pearse. His middle name of Godolphin is widely used by relatives of the dukes of Leeds, but as Francis is the only one of his brothers and sisters with this, the connection must have been fanciful or wishful. The officers included an excellent water-colour artist Lieutenant Tobin, Lieutenant Guthrie, whose health was broken on the expedition and who died in Innsbruck soon after returning, and Lieutenant Portlock. Also aboard was midshipman Matthew Flinders, off for his first look at a part of the world he would make very much his own. Comfortingly there were discipline-keeping marines and the brig *Assistant* as a tender.

The second breadfruit expedition sailed on August 2 1791, not only well before the mutineers' trials but before anyone knew if *Pandora's* mission to find *Bounty* had been successful. The departure was hardly the relief to Bligh that it was to Banks. He was suffering terribly from intense and recurring headaches and fever and should have stayed at home in bed. It was the health of his patronage and expectation of promotion that must have persuaded him.

Providence sailed for Tahiti on August 2 1791, after Bligh's *Narrative* was swiftly published in 1790 but well before the 1792 trials of the mutineers, when everyone heard differing stories.

The document found in Canberra is a draft of a letter Bond wrote to his brother in 1792, during the late days of this second breadfruit expedition, when *Providence* was returning through the Atlantic towards the West Indies. In its original form it is extremely dense, so I have divided it into paragraphs. Bond writes:

'To say a southern voyage is quite delectable is also to say you have every domestic comfort; but on this score I must be silent, for at present I mean to say but little of our Major Domo (i.e. Mr. Bligh) . . . Yes Tom, our relation has the credit of being a tyrant in his last expedition, where his misfortunes and good fortune have elevated him to a situation he is incapable of supporting with decent modesty.

'The very high opinion he has of himself makes him hold everyone of our profession with contempt, perhaps envy: nay the Navy is but [a] sphere for fops and lubbers to swarm in, without one gem to vie in brilliancy with himself. I don't mean to depreciate his extensive knowledge as a seaman and nautical astronomer, but condemn that want of modesty in self estimation. To be less prolix I will inform you he has treated me (nay all on board) with the insolence and arrogance of a Jacobs: and notwithstanding his passion is partly to be attributed to a nervous fever, with which he has been attacked most of the voyage, the chief part of his conduct must have arisen from the fury of an ungovernable temper.

'Soon after leaving England I wished to receive instruction from this imperious master, until I found he publickly exposed any deficiency on my part in the Nautical Art etc. A series of this conduct determined me to trust to myself, which I hope will in some measure repay me for the trouble of a disagreeable voyage in itself pleasant, but made otherwise by being worried at every opportunity.

'His maxims are of the nature that at once pronounce him an enemy to the lovers of Natural Philosophy; for to make use of his own words, 'No person can do the duty of a 1st lieut who does no more than write the day's work of his publick journal'. This is so inimical to the sentiments I always hope to retain, that I find the utmost difficulty in keeping on tolerable terms with him. The general orders which have been given me are to that purport I am constantly to keep on my legs from 8 o'th'morning to 12, or noon, altho' I keep the usual watch. The Officer of the morning watch

attends to the cleaning of the Decks; yet I am also to be present, not only to get it done, but be even menially active on those and all other occasions.

'He expects me to be acquainted with every transaction on board, notwithstanding he himself will give the necessary orders to the Warrant Officers, before I can put it into execution. Every dogma of power and consequence has been taken from the Lieutenants, to establish, as he thinks, his own reputation what imbecility for a post Capn! The inferior Warrants have had orders from the beginning of the expedition, not to issue the least article to a Lieut. without his orders so that a cleat, fathom of log line, or indeed a hand swab, must have the commander's sanction. One of the last and most beneficent commands was that the Carpenter's Crew should not drive a nail for me without I should first ask his permission but my heart is filled with the proper materials always to disdain this humiliation.

'Among many circumstances of envy and jealousy he used to deride my keeping a private journal and would often ironically say he supposed I meant to publish. My messmates have remarked he never spoke of my possessing one virtue tho' by the bye has never dared to say I have none. Every officer who has nautical information, a knowledge of natural history, a taste for drawing, or anything to constitute him proper for circumnavigating, becomes odious; for great as he is in his own good opinion, he must have entertained fears some of the ship's company meant to submit a spurious Narrative to the judgement and perusal of the publick.

'Among the many misunderstandings that have taken place, that of my Observing has given most offence, for since I have not made the least application to him for information on that head, he has at all times found illiberal means of abusing my pursuit; saying at the same time, what I absolutely knew was from him. Tir'd heartily with my present situation, and even the subject I am treating of, I will conclude it by inserting the most recent and illegal order. Every Officer is expected to deliver in their private Logs ere we anchor at St Helena. As our expedition has not been on discoveries, should suppose this an artibrary command, altho the words, King's Request, Good of the Country; Orders of the Admiralty &c &c &c are frequently in his mouth—but unparrelled [sic] pride is the principal ingredients in his composition.

'The future will determine whether promotion will be the reward of this voyage: I still flatter myself it will, notwithstanding what I have said. Consistent with self respect I still remain tolerably passive; and if nothing takes place very contrary to my feelings, all may end well: but this will totally depend on circumstances; one of which is the secrecy requested of you concerning the tenor of this letter. My time is so effectually taken up by Duty that to keep peace I neglect all kind of study; yet the company of a set of well informed messmates makes my moments pass very agreeably, so that I am by no means in purgatory . . . The 2nd August [1791] we left England and had pleasant w[eather] to Teneriffa, where Captain B. was taken very ill, and from particular traits in his conduct believe he was insane at times.'

Those are strong, contemporary and first-hand observations and show that men thought Bligh mentally unstable on both *Providence* and *Bounty*.

Lieutenant Bond's son, the Rev. F. H. Bond, had a paper that summarised what he had been told by his father about Bligh and supported this with quotations from notes, which have not been made public. Most of what follows are now the words of F. H. Bond.

'Though a prime seaman, however his [i.e. Bligh's] passionate temper and violent language were so uncontrolled that he was hardly ever employed without exasperating his officers and ship's company. The story of the Mutiny on the Bounty, the immediate causes of which were an outburst of temper on his part and grossly insulting words to one of his officers, and the marvellous voyage of the Bounty's launch for 4,000 miles are so well known that they need only be referred to here. This extraordinary feat of seamanship was now on everybody's lips and Bligh was universally commiserated. He had of course told his story in his own way and was, like many violent tempered men, perhaps really unconscious of the amount of provocation he had given. He had represented the mutiny as the result of his crew's experience of the delicious climate and the life of the island of Otaheite

'The most perfect harmony was luckily maintained among the Lieuts throughout [the Providence voyage], a matter of immense importance for a reason which must unfortunately be presently noted . . .'

During the voyage to the Cape of Good Hope, Bligh transferred Bond to *Assistant*, a much smaller ship and something of an insult. Once at the Cape, Bond was returned to *Providence*. His comments in notes made at the time were full and hasty and his son declines to quote them as Lieutenant Bond had written under the influence of strong feeling. He merely wrote of the notes:

'It will be enough to show that there was great cause for discomfort. Hardly had the voyage commenced when Cp. Bligh's arbitrary disposition and exasperating language began again to render his ship a most unfortunate one for his officers and especially for his First Lieutenant [Bond] who from his position was brought into closer contact with him. Orders of an unusual nature were given with haste and in a manner so uncalled for and so devoid of feeling and tact as to occasion very great irritation.

'The short exchange with the Assistant was felt at the time quite a relief and his resumption of duties as First of the Providence was attended with discomfort which he speaks of as frightful. A dictatorial insistence on trifles, ever-lasting fault finding, slights shown in matters of common courtesy, strong and passionate condemnation of little errors of judgement all these stung the hearts of his subordinates and worked them up into a state of wrath which would probably have much surprised Bligh himself had he known it.

'Instances are given of his want of courtesy. At a ceremonial visit to the Governor of the Cape, Bligh takes the opportunity of snubbing Lieuts

Portlock and Bond by presenting them after a junior Lieut, and the Commander of the Assistant last, quite against the rule of etiquette. At Teneriffe he refused to present two of his officers to the Governor, who thereupon corrected the intentional blunder and presented themselves. The Governor, it is added, received them well.

Refusals of leave to land, apparently without cause, which annoyed at the Cape, were felt still more strongly at Otaheite, when frequent leave was naturally expected during their 3 month stay. One other point was very trying to Bligh's nephew, the great inconsistency of his conduct. He says that in prosperity Cap. Bligh was all arrogance and insult, despotic insistence without explanation, advice or show of kindness; often an hauteur and distance which utterly ignored the nephew as well as the rank of his First Lieut.

In time of real danger what a change to cordiality and kindness! The Devil's Hole for example! 'Oh Frank! What a situation; into what a danger have I brought you! God grant that we may get safe out of it.'

I replied, 'No sir, we shall do very well; I don't see that there's any real danger to the ship.'

The event which called for this conversation is not given but there is the hint that the danger was caused by the helm being put the wrong way through mistake.

. . . The serenity of the weather gave us the most flattering hopes of a safe passage; but several affairs have lately occurred to prevent the cordiality which should have existed between my commander and myself, and his remarks tended to deprive me of self-confidence. I was e.g. reproached and threatened because my men on the Fore Tops yard were beaten by Tobin's and Guthrie's, the carpenter was abused for acting on my orders and ordered to take the skippers out; the boatswain was similarly treated. The usual etiquette in our respective positions was quite set aside.'

Bond did not mutiny on Providence *because he was not in purgatory. Unlike Fletcher Christian he had peers aboard . . .*

There is that charge of inconsistency again, that lack of universal boundaries on which sailors and their safety rely. Who can doubt that Lieutenant Bond tells us the truth about Bligh and gives us a concise picture that is also relevant to his behaviour on *Bounty*?

Bond did not mutiny on *Providence* because he was not in purgatory. Unlike Fletcher Christian he had peers aboard and he had no close ties of friendship with Bligh. On later ships there were continuing complaints about Bligh's manner of command, his constant inconsistency and insensitivity but he never learned, never changed. When commanding HMS *Warrior* in 1804 he was court-martialled for using abusive language to an officer and behaving in *'a tyrannical and oppressive and unofficerlike behaviour contrary to the rules and discipline of the Navy'*. The charges were found largely proven and he was reprimanded.

Remoteness might be the key to why there was not as much turmoil on *Providence* as there was on *Bounty*. The logs show Bligh was not in direct contact with his officers on a daily basis because he was so seriously ill and confined to bed for most of the time. Other men would probably have refused the command, perhaps even disembarked before it was too late. Not William Bligh. When he had orders, he fulfilled them whatever the personal cost. It was hell on his inferiors but exactly what Joseph Banks wanted for two reasons. Bligh could be relied on to do the job and it kept him well away from the trial of the *Bounty* mutineers and any chance of cross-examination or of counter claims being accepted by the court. Instead, Bligh was once more cursing and swearing his way to and from the South Pacific.

On August 7th 1793 the second breadfruit expedition arrived back from the West Indies and docked in Deptford. The venture was a resounding botanical success, apart from the exasperating refusal of the West Indian slaves to eat the stuff. When *Providence* and *Assistant* were paid off at Woolwich, the Kentish Register of September 6 1793 reported that Bligh was cheered, which made a good impression on everyone. Sailors know it pays to be nice to superior officers in public and, as far as the complements of *Providence* and *Assistant* knew, Bligh was still a hero.

The appointment of clearly unwell Bligh to this second breadfruit expedition was selfish and defensive of Banks, yet it helps answer another perplexing question about Bligh's later career.

Why was he later appointed Governor of New South Wales at a salary of £2000? That's about £88,000 today and twice what the previous Governor had been paid.

CHAPTER 26

'THE FURY OF AN UNGOVERNABLE TEMPER'

In 1806 William Bligh was in Sydney as Governor of New South Wales, living in considerable style and wielding tremendous power. Some say the British Establishment just wanted him as far away as possible, some that the appointment was an overdue reward for what happened on *Bounty*. Others believe it to be a punishment for *Bounty* and other incidents.

Curiously, it is more likely Bligh was appointed *because* of *Bounty* and his other misdeeds, not in spite of them. For this important insight I am grateful to Paul Brunton, then Curator of Manuscripts at the State Library of New South Wales and a noted Bligh scholar. Brunton explained that the governorship of the colony, which was in deep strife at the time, was in the personal gift of Sir Joseph Banks. Whoever Banks recommended for the post, the king would accept. Banks, of all people, knew Bligh could be relied upon to tackle challenges fearlessly, for precisely the qualities and reasons that made him unpopular; he had an absolute belief in the written law and was insensitive to the opinions of others.

Whatever the rights or wrongs of his motives, Bligh was nothing if not magnificently courageous and would wade in where others would dare even to tread. As far as Banks was concerned, the man whom much of the Royal Navy establishment called 'that *Bounty* bastard' could be as horrible as he liked, if he did the job. He'd also be well out of sight and sound.

Banks tempted Bligh with a greatly increased salary and the guarantee that his place in the Navy List would not be compromised. Like every other naval officer, Bligh's promotion to higher ranks, to Vice-Admiral in his case, would come through simple seniority, by outliving contemporaries. You didn't need to go to sea again, so such rank was not specifically tied to reward for duty done or even to any proven degree of ability, thus it's understandable that this protection of his precedence was important.

It was cruel of Joseph Banks to send the sick Bligh on the second breadfruit expedition but this minimised harm to Banks that might have been the result in Bligh being cross-examined in the *Bounty* trials. It was deeper, calculating self-interest of Banks to send him to New South Wales because if this colony failed, Banks himself would fail. It is likely that Banks believed Bligh alone had the ruthless focus to confront and tame the factions tearing the colony apart.

Anti-British sentiment combined with the greed of privileged settlers and the venality of the powerful New South Wales Corps had created a hell for freemen and for transported prisoners alike.

Bligh never trusted the opinions of other men. Every action he took was based entirely on the written word, as long as it agreed with what he thought. The unwavering support of Banks and his official orders provided the armoury he needed to begin the taming of New South Wales. He did what he was sent to do. He stood up to the troublemakers and began the long process of turning the penal colony into a calm and profitable settlement, even though he was jeered when he walked in Sydney, sometimes because he ordered the full ceremonial of pipes, drums and guards of honour.

On January 26 1808 Bligh was arrested for tyranny by the NSW Corps and its supporters. It was the second mutiny against him and marked the beginning of the so-called Rum Rebellion. Even if it is true that he was found under a bed when the NSW Corps arrested him, it would not have been through cowardice but because he was determined to protect the written records of his administration, which he had clasped to his breast. These represented vital proof that he had done his duty, quite as important as the journals he doggedly kept after the *Bounty* mutiny. Once imprisoned, Bligh made no attempt to avoid conflict or change his behaviour. He was so difficult about the conditions under which he would agree to leave the colony that he remained a prisoner in Sydney for over a year. It took until February 1809 before he signed an agreement with the NSW Corps to quit New South Wales.

In *Distracted Settlement* published in 1998, Dr Anne-Maree Whitaker introduces the previously unpublished journal of Lt James Finucane, who was in the colony at the time. He said Bligh *'eventually entered into a most solemn engagement to leave the Colony on the 20th of this month [February 1809] . . . and proceed to England with all possible expedition, not to touch at nor return to any part of this territory without receiving His Majesty's orders . . . and not to interfere in any manner or under any pretext . . . with the Government of the Colony'.*

On March 17 1809 Bligh sailed for England in the sloop HMS *Porpoise,* his flagship as governor. As soon as *Porpoise* was through Sydney Heads, he went back on every aspect of the agreement, arguing that it was extracted by force and thus not binding. He sent letters proclaiming the officers of the NSW Corps and others were all traitors and rebels and continued *'cruizing in sight of land'* for almost a year, most of it in and around Tasmania. Just as with the mutiny on *Bounty,* he believed none of this was his doing and that he should be reinstated as governor. The man who so vehemently condemned others for failings of duty and honour had diminished his triumphs in New South Wales by breaking his word.

Meanwhile, HMS *Hindostan* had arrived from Britain with Lach-

Ian Macquarie as the new governor and the 73rd Foot to replace the NSW Corps, which had been renamed the 102nd Regiment of Foot and recalled to Britain. On May 12 1810 *Hindostan* sailed from Sydney with both Bligh and the men who had mutinied against him aboard, discomforting for both.

All that Lieutenant Bond says about Bligh's behaviour in his draft letter, and what his son subsequently wrote, dramatically confirms every point made by Morrison, Fryer, Edward Christian, Peter Heywood and other men who were aboard *Bounty*.

When Bligh returned from Australia, Charles, Edward, Fletcher's mother, John and Isabella Christian Curwen had heard about Pitcairn Island and *Bounty's* destiny.

Yet, nobody could say what had happened to Fletcher Christian. No-one could even agree if he were dead or alive.

This sign shows where Fort George was built but there are no physical remains as the walls have been bulldozed and the moat has been filled in.

Overview of the Fort George site, now a market garden

REPERCUSSIONS II

April 29 1789 – September 22 1789
Fletcher chooses Tubuai and builds Fort George

September 23 1789
Fletcher's final departure from Tahiti

CHAPTER 27

A UNIFORM GOAL

Fletcher Christian might not first have thought of mutiny as a solution to his anguish but he had certainly mused many times about what he would do if he were in charge of *Bounty*. To do so is typical of the second-in-command anywhere on land or sea. Many of his early actions as the new commander of *Bounty* clearly show a measure of detailed forethought about the situation, without meaning he had thought about how it might arise.

Once *Bounty* was his, Fletcher Christian made gestures large and small that demonstrated his ability to assume command, to comfort and to deserve the approbation of the men aboard. He also needed to order his own mind as full comprehension of the enormity of his act dawned. As you will read later, this understanding might have come because a brief psychosis ended and reality returned, a reality that would never have changed if he *'if he had served on board any other ship, or had perhaps been absent from the Bounty a single day, or one ill-fated hour, might still have been an honour to his country and a glory and comfort to his friends!'*

It is said Fletcher sat below, his head on his arms, and gave orders in monosyllables. Yet there was no slackening of discipline. Avoiding comparison with Bligh's autocracy, he encouraged meetings of the men to decide broad issues. He even introduced voting to ensure that everyone's voice was now being heeded. This new style of management meant that George Stewart was chosen as second-in-command.

Stewart was not a mutineer and he was unpopular on board because of his severity but he was acknowledged as particularly efficient. When even hard-line mutineers were given the choice, they demonstrated they understood the dangers of a sailing ship and chose the best man for the job, not the most popular. Some would have preferred non-mutineer Peter Heywood to be second-in-command but as the crew now had to stand only two watches, Christian wisely thought him far too young to assume such responsibility.

The 25 men aboard were certainly *Bounty*'s most able men. They were a community of craftsmen and tradesmen like no other on earth. Most were tattooed, most were scarred. If their faces were not deeply pitted by smallpox, like Adams, Burkitt, McIntosh and Norman, they had wounds of old fights, of abscesses and accidents. They used Tahitian words and phrases, shouting *mamoo* rather than telling someone to shut up and calling women *vahine*.

Henry Hillbrant was Hanoverian and spoke little English and

that with a heavy accent. Heywood had a broad Manx accent, Martin was American, John Williams was from Guernsey and spoke French as easily as accented English. Edward Young had played an active but quiet part in the mutiny, standing armed with a sword behind Christian and his prisoner. Always said to be one of the first to follow Christian, he was dark complexioned, probably because he had some West Indian blood, had lost most of his front teeth and those that remained were rotten. His slurred speech was flavoured with the argot of the West Indies.

Heywood was described by Bligh as fair-skinned and well-proportioned although still growing. Dark-skinned, perspiring Fletcher Christian seemed bow-legged and Stewart was bottle-shouldered and probably had an Orkney burr.

Their tattoos were as varied as their personalities. Men who had quickly been seduced by Tahiti were the more heavily decorated. Adams was tattooed on his body, legs, arms and even his feet. Boatswain's mate James Morrison, who was never despised for having to perform the floggings Bligh ordered, was tattooed with the insignia of the Order of the Garter, as were several others, including Fletcher Christian. Adams had the star on his breast, a garter on his leg and the priestly *tahu'a ta'tau* had punctured *Honi soi qui mal y pense* into his leg.

They were a rum lot but perhaps no worse than most sailors of the day, except that *Bounty* was theirs by mutiny.

There are those who rather too loudly claim that Christian and his activists were not mutineers but pirates, even saying there was never a mutiny. Well, of course there was a mutiny. If one is to believe dictionaries, what then followed was not piracy because the definition is the taking of *another's* ship. To be fair, the use of 'pirates' and 'piracy' began early, whatever the true meaning. Bligh wrote of piracy in his first account of the mutiny and when he wrote on October 7 1790 to explain he no longer had the Kendall chronometer, he said *'it had been left in the said ship when pirated from my command'*. Allowing that Christian had taken the king's ship, then I suppose this makes him and those who followed him pirates of a sort and it seems to have been exciting and romantic to call them such, as was seen in the publications and theatrical productions based on the mutiny when news got to London. Pirates and piracy are still seductive to some, even when there are none about.

Christian's idea of democracy on board has also been touted as a direct result of his time in the Caribbean, inevitably learning about the pirates of Jamaica's Port Royal. The profound fault with this stance is that piracy was an occupation of the 17th century and that by the late 18th century sugar was Jamaica's major industry. There were no pirates when Fletcher was there and few memories of them because their Port Royal stronghold at the mouth of Kingston

Harbour had been destroyed by earthquake in June 7th 1692. The nastiest threat in Kingston was malaria and other fevers that regularly raged amongst sailors living in their accustomed cramped, airless and foul conditions. Add the town's miasmas and the sewage-strewn harbour and it's a wonder William Bligh and Fletcher Christian came back at all.

The Caribbean's past pirates did operate their ships in a very orderly way, with voting common on both major and minor issues, a sensible way to act considering they had no naval regulations to fall back on for disciplinary matters. Although meant by today's critics as an insult, I can't think that calling *Bounty's* new crew pirates is pejorative. Might not Fletcher Christian's idea of pirate-like power-sharing demonstrate an admirably keen understanding of how best to get the most from men when official rules and regulations no longer exist? The source of ideas is less important than their efficacy. Pirates had nothing to do with the mutiny on *Bounty* but because by the late 18th century pirates had become something romantic, it might have been thrilling to think of them as such, even if you knew little of reality.

Fletcher's ideas of working by agreement, voting and equality are much more likely to have come from ideas sprung from the Age of Enlightenment he heard in the salons of London Society visited with his cousins Jane and Bridget. He certainly knew the views of William Wilberforce, a university friend of his brother Edward, and Jane and Bridget's brother John Christian XVII was a great social reformer. By the time Fletcher sailed in December 1787, the American states had revolted and there was much revolutionary talk of France's future and of a possible new kind of world order.

Unusually, Fletcher's mutiny was bloodless in execution and that has always made it easier to romanticise. To demean Fletcher Christian in the 21st century by saying there was no mutiny and that he and his ship-mates were merely wicked pirates is wearisome, not to say ill-educated.

Some academics like to divide *Bounty's* men after the event into two opposing parties, mutineers and loyalists but with Bligh gone everyone had to unite in a cohesive need for survival at sea. This is not to say there was no mumbling and dissension but neither was there any sharp division. Warring parties on-board would be suicidal, the sort of thing that leads to shipwreck or mutiny.

Sources for the post-mutiny episodes have generally been accounts written by Adams on Pitcairn and by Peter Heywood and Morrison on their return to England. Other contemporary records include Captain Edwards' abstracts from journals kept on a daily basis by both Peter Heywood and George Stewart, which were lost with the foundering of *Pandora*.

The quotations from Heywood's journal are detailed and give new slants on many events, whereas Stewart's are mainly a recital of dates. They both generally support the content of Morrison's journal and it is not surprising that Morrison, Heywood and Stewart all get dates wrong, perhaps because one or the other might have used the naval system. Such inconsistency does not affect the content or veracity of any account. If readers find my version varies from what they know, it is because I have used the abstracts from Heywood's journal as my touchstone.

On April 30th, only two days after the mutiny, Fletcher Christian ordered the royals, the smallest sails used at the very top of masts, cut up so they could be made into uniforms for all hands. British sails were most likely to be made from linen, which softens with age and use, so should have been comfortable to wear. With a nicety that reflected previous thought, he gave his navy-blue woollen officer's jacket so it could be cut into strips to use as edging, a design entirely new and creative. At that time only naval officers wore uniforms and these were dark blue with wide, white facings turned back over the chest. Seamen wore simple working clothes, known as slops. Morrison tells us Christian observed *'that nothing had more effect on the mind of the Indians than a uniformity of Dress, which by the by had its effect among Europeans as it always betokens discipline especially on board British men-of-war.'*

. . . an admirably keen understanding of how best to get the most from men when official rules and regulations no longer exist?

The second part of the statement is Morrison's private observation but it is acute and relevant. Christian's uniforms would bind his crew of mutineers and loyalists together both as equals on board and in the eyes of those encountered. It was an extraordinary idea on a Royal Navy ship, where difference had always been thought essential and was a further brave example of Fletcher's attachment to the ideals of the Age of Enlightenment. George Stewart adds that on May 2 they cut the mizzen and main staysails to make more jackets. This newly free assortment of tanned and tattooed young gentlemen, tradesmen and seamen, sitting cross-legged in semi-circles to sew jackets under the probable supervision of tailor William Muspratt as *Bounty* sailed east in the pleasant days of the early dry season, provides one of the few amusing pictures of the entire saga. Added to Fletcher's use of voting, *Bounty* was unlike any British ship that had ever sailed, naval or otherwise.

On May 1 all but a few of the breadfruit plants were thrown overboard. On the 6th and 7th the men divided *'the pleasing apparel of the People and Officers'* who had preceded the breadfruit overboard. The curios, woven mats and tapa cloth were also divided and Morrison says that those of Christian's party always got the biggest and best. The loot, there is no other name, was stored in the emptied greenhouse. Fletcher Christian moved into Bligh's small book-lined cabin but he was more than just a new commander faced with the safe and healthy transportation of men and goods from one part of the globe to another.

He also had to master and use to his advantage high emotion and unfettered speech, shipmates no 18th-century captain would freely choose.

As commander of *Bounty*, Christian had to make sure no man out-manoeuvred him. The decisions facing him would have been cruelly complex for an experienced sea captain. He was still only 24 and there was absolutely no precedent for what he had done or for the situation in which he now found himself.

Bounty was the first European ship to be totally free upon the Pacific. No other band of European men had ever had the choice of where they went and when. They were masters of all they could survey and most of that had never been seen by an Englishman. It was at once intoxicating and beyond simple comprehension. Yet, if any of the men had supported the mutiny so they could return to Tahiti, their seizure of the ship meant it was forever forbidden them. It was no enviable task for Fletcher Christian to tell men who had broken the law that they had done so in vain. He must have kept them working and the two small boats must have been repaired almost at once because Fletcher would have known they would be needed to get to and from shore wherever they next anchored.

There were navigational guide maps and accounts of earlier voyages to the South Pacific in Bligh's hot cabin and these now became the tools of his escape from retribution. Alone with these reminders of his benefactor and teacher, Fletcher made his first unilateral decision. *Bounty* was to sail to Tubuai, 350 miles/565 kms south of Tahiti.

The announcement quickly polarised some of the more spirited men. Having proof that they were expected to disappear into the Pacific, they planned to overthrow Christian and his followers. The plotters vaguely talked of getting rid *'of those we did not like by putting them on shore and that in all probability our design might be favoured by an extra allowance of grog'*. If it were so easy, it is astonishing no attempt was made to do it.

The tale comes from Morrison. He defended himself at his trial by saying he stayed aboard *Bounty* so he could retake her in the name of Justice, but Justice needs braver fellows. One author dismissively wrote that Christian *'was incapable of keeping control of his men, just as he could not control his passions'*, citing the plot as evidence but to call it a plot is to dignify it more than it deserves. Saying something like that totally misunderstands what it was like to be free men on a ship taken by mutiny and also ignores what actually happened. Anyway, Christian was clearly in control and continued to be so, as there is no evidence he was superceded or dismissed.

Much to the plotters *'unspeakable surprise'*, the whispers were discovered by Christian, who began carrying Bligh's pistol in his pocket at all times and also ordered his followers to arm themselves. Churchill slept on the arms chest and, Biblically, whenever two or more of the plotters were gathered together, they were joined by one of Christian's men. The situation was never serious and its discovery was a relief. Christian now knew who might be a threat.

On May 24 *Bounty* made Tubuai and the next day prepared to move into the lagoon and anchor. Few works other than Professor Maude's *In Search of a Home* have written in full about Tubuai and most who do twist the first-hand account by Morrison into a sustained and malicious attack on Christian, which demonstrates little or no thought about reality on board *Bounty* or what faced 18th-century Europeans making first contact with South Pacific islands.

Bounty *was the first European ship to be totally free upon the Pacific.*

Of course there was dissension on *Bounty* and Christian's revolutionary power-sharing was dangerous to his position. Yet he did establish the first English colony of free men in the South Seas, just a year after the first fleet of prisoner ships arrived in Sydney, so far west. When Tubuai was abandoned, he was alive and maintained mastery and ownership of *Bounty*, clear proof of the power and endurance of his command.

If Cook had landed when he discovered the island 12 years earlier, Christian would probably not have. Knowing nothing of the customs and conditions on shore, he hove to outside the reef and ordered the small cutter to be launched so that George Stewart might locate an opening. There is only one of consequence and once inside the reef a sailing ship is still far from safe, for the lagoon is treacherously uneven, with a sunken patchwork of ragged shallows and sudden shoals amid depths of over 40 fathoms, almost 75 metres. If clouds scatter the water with shadows, or if the sun is not high and bright, disaster is always a possibility, as had been seen in Toaroah harbour.

The inhabitants of Tubuai had seen sailing ships but were clearly not pleased to see another and paddled across the lagoon to attack the cutter. To Fletcher Christian it must have looked as though Stewart was doomed. The Tubuaians' canoes were like nothing anything he'd experienced. Each was 30 to 40 feet/9 to 12 metres long, with a high prow carved into an animal's head and with a tall scroll at the stern. Painted red and decorated with pitch and with glittering fish scales and shells, each carried up to 20 warriors with long spears of dark wood.

After mobbing the cutter, some Tubuaian warriors boarded it, thus giving the Europeans a strange safety because at close range the islanders' 18 foot/5.5 metre spears were useless. The cutter's men had foolishly left the ship with only a brace of pistols, one of which did not work. In the melee, one of the boat's crew was wounded and the working pistol misfired but wounded a Tubuaian. The cowed boarding party tumbled back into their canoes, capturing only a jacket.

The attack was not enough to deter Fletcher and next day *Bounty* was carefully worked further into the lagoon, to anchor in 16 fathoms/30 metres, a quarter of a mile/400 metres off what Morrison calls a sandy bay but which is more properly only a slightly deeper curve in the unbroken ribbon of white sand that circumscribes most of the island. Tubuaians assembled from every district, crowding the beach and flocking about the ship in their canoes, wailing and hooting with great conch shells. Six months in Tahiti had made Fletcher Christian and several of the ship's complement passably fluent in Tahitian and Tubuaian was close enough for the two parties to understand one another. No inveiglement persuaded the islanders to come on board but by their dress Fletcher knew they were warriors and that battle was on their minds.

The fighting costumes were largely red and white and clearly took some time to don. First, pieces of red-dyed bark cloth or of woven coconut fibre were wound around the body and held in place at the waist with a plaited sash with coconut fibre tassels. The shoulders were bare and unencumbered and the folds in the cloth were used to carry stone projectiles, just as theatre tickets were once tucked into cummerbunds. Across the chest was suspended a pectoral, sometimes of pearl shell, always highly decorated. On his head each man wore a helmet of woven and matted coconut fibre shaped like a beehive. Some were covered with white cloth and crowned with black man-of-war bird feathers. Those could ward off a cutlass swipe but those with a pearl shell and a semi-circle of wild duck wings were more vulnerable.

This was not the welcome Fletcher had hoped. As the sun rose next morning, he could see from *Bounty* that the number of canoes pulled up on the white sands had increased considerably. At last an old man, probably a chief, came aboard. *'He appear'd to view everything with astonishment and appear'd frightened at the Hogs, Goats, Dogs etc, starting back as any of them turned to him.'* Fletcher discovered this was because Tubuai had no native mammals other than rats, so to his visitor the goats and hogs aboard were just as likely to be gods as the men aboard the ship that sailed with no paddles.

Fletcher Christian made gifts to the man, who was not so overwhelmed that he did not carefully and too obviously count those aboard. He promised to return and in the meantime the ship's arms were got to hand, for such an occasion was not likely to be one of welcome, something easily inferred from the islanders' weapons. As well as the long spears, they also displayed a far more dangerous weapon. About 10 feet/3 metres long, it was finely worked into a club at one end and flattened and pointed at the other. A warrior clasped the weapon with his arms spread and swung it. Alternately clubbing and stabbing, he could maim and kill with ease.

At noon on May 27 the canoes were launched. Among them was an unusual double canoe, carrying women decorated with flowers and pearl shells. They were young and handsome, with black hair that hung to their waists in waving ringlets. As they approached, the girls stood and beat time while one of them, seemingly a chief's daughter, sang a siren's song.

Christian had ordered all hands to change into the new uniforms and mounted a guard spaced evenly around *Bounty*. The 18 women and five of the men who had paddled them came on board so readily that his suspicions were further aroused. Sure enough, about 50 canoes containing perhaps 1,000 men, perhaps most Tubuains, had glided up to the other side of the ship and began blowing conch horns. The defensive precautions Christian had taken were heeded and the war party chose not to board, even though the odds were massively in their favour.

The young women were treated civilly and given presents. Stilted conversations of compliment and deceit were exchanged because Fletcher Christian knew the pretence had to be played out. The men who accompanied the women were pests, stealing anything they could. The glass of the compass was broken. One man stole the compass card but was noticed by Fletcher Christian and in the struggle for its return it was torn. The Tubuaian disliked being thwarted and a hand-to-hand scuffle followed. Christian's well-documented strength got the best of the thief, who was sent smartly off to his canoe with several stripes from a rope end and then the women and remaining canoeists quickly followed. Once in their canoes, they brought out and brandished their hidden weapons.

In the moments of relief that followed the departure of the boarding party, Fletcher Christian noticed one of the buoys that marked an anchor being cut away. He fired a musket at the offender and ordered the firing of a cannon that had been primed and loaded with grape shot. Grapeshot was a canvas bag filled with grape-sized metal balls, that begin to spread as soon as fired and that also included a fuse and explosive designed to detonate when it hit its ultimate target. Frightened and sustaining many injuries, the canoeists hastened ashore but Christian pressed his advantage further. *Bounty*'s two boats were quickly manned with armed men and pulled to shore in pursuit. The boats were pelted with stones until muskets were fired into the crowd. In a few moments the beach was deserted.

Although they had seen only two men fall, 11 men and women had been killed by *Bounty*'s men, remarkably few considering grapeshot had been fired into open canoes at very close quarters. Was Fletcher Christian being bad tempered or undisciplined in firing at the Tubuaians? With so few men aboard, *Bounty* was extremely vulnerable and survival was more important than relationships with unfriendly people.

There were no published guidelines for Fletcher Christian on making initial contacts with Pacific islanders. Some were friendly and trustworthy. Others were bellicose, deceitful and trusted only power. When Tubuaians threatened the ship and damaged its equipment, *Bounty* had to assert itself. Even Captain Samuel Wallis, first European discoverer of Tahiti, had first to defend himself with firearms.

Determined to find some basis for discussion, Christian ordered the nine canoes left on the beach to be collected and made fast behind *Bounty*, hoping that in negotiating their return the islanders might be persuaded to talk in broader terms. Cords were found in each of them, which the *Bounty* men presumed were intended to bind them. If Fletcher Christian had not been offensive as well as defensive, many of those depending on him might have died.

That night the wind swung to the north-west, filling the canoes with water and so they were released and floated to the shore. From this time on, the consensus was that the place was to be called Bloody Bay.

For two more days Fletcher attempted to make contact. Bounty's two small boats sailed around the eastern end of the island showing a Union Jack and a white flag. He landed in several places and pushed through the thick fringe of undergrowth and trees to find houses and to leave gifts of hatchets but he saw no one. Unbeknown to him, many of the 3,000 islanders had come to the uncultivated swamp behind Bloody Bay. They watched *Bounty* there for several days and *'for want of their usual bedding they caught Colds, Agues, and Sore Eyes, Running at the Noses'* which was interpreted as some sort of punishment from the men on the ship. Fletcher Christian's firm action was having a salutary but negative effect, a classic story of first contact in the Pacific that was to be repeated hundreds of times by Europeans.

Fletcher liked what he saw on Tubuai apart from the natives and his determination was so fixed that he *`dream't of nothing but Settling on Toobouai'*. He thought, incorrectly, the island was sparsely populated, persuading himself that it could be induced to friendship, either by persuasion or force. As to the safety of its position, he considered the difficulty of anchorage would deter ships, which could go easily on to Tahiti. He was certain he could rest peaceably and permanently here. Heywood says that there were discussions about the lack of women as well as suspicions of plans to take the ship but by whom was not transcribed by Edwards. Having landed a sickly young goat and two hogs, *Bounty* weighed anchor and headed for the warmer welcome of Tahiti and the friends made there.

Teina welcomed Titreano and *Bounty* back with such pleasure that it dulled his ordinary sense of curiosity and critical ability. Christian lied that they had unexpectedly met Captain Cook, whom Bligh had persuaded them was still alive and discovering islands. He was said to be furnishing a new settlement on the island of Aitutaki and had taken Bligh and his breadfruit to help. *Bounty*, now with *ra'atira Titreano* in command, had ostensibly been sent back to top up supplies and then return.

Teina believed all he heard and was exceedingly active in attending to *Bounty*'s requirements. According to Heywood, *Bounty* was quickly loaded with 312 pigs, 38 goats, 9 chickens and the bull and cow left by Cook, which had either never mated or had done so to no effect. Dogs and cats were adopted and many of the empty pots in the great cabin were planted with flowers and tubers. If he could not have the real thing, Christian was certainly taking much of Tahiti with him, although nowhere on Tahiti would there have been the farmyard crush and stink there was now on *Bounty*.

After an initial scare, when *Bounty* seemed likely to ground on Dolphin Bank off Point Venus, necessitating the cutting of a cable and the loss of an anchor, the ship headed south for Tubuai. The dates for all these events are given by Stewart a day later than in Heywood's account, saying they arrived on June 7 and left on the 20th.

Two weeks in Matavai Bay had not been the orgy of eating and sex that might be imagined. To protect his plans, Christian kept an armed guard at all times and his people were forbidden to tell the truth on pain of death. Presumably his own faction was occasionally allowed on shore but even if not, women were enjoyed. They came on board, but not to stay. Some seemed to have had second thoughts about Europeans and the idea of sailing away to live permanently with them was far less appealing than spending the occasional night afloat. There was no point having a white lover and collecting large quantities of iron nails in return if their families and friends could not see their enhanced wealth and status. So, *Bounty* sailed with only nine women. Adams had Teehuteatuaonoa, known as Jenny, McIntosh a woman he called Mary, and Fletcher Christian now had Mauatua, whom he renamed Isabella, presumably after the rich cousin he would like to have married. Also on board were eight men and ten 'boys', probably energetic teenagers keen on adventure.

Heywood thought most Tahitians on board came voluntarily, so possibly Fletcher and Mauatua were continuing a relationship and this is the basis for my historical novel MRS CHRISTIAN – 'BOUNTY' MUTINEER. She and Jenny were easily the two most influential Tahitian women who later went to Pitcairn and I find it interesting that they should so early have been the consorts of the two men who successively took the leadership on Pitcairn, to me illustrating how much they wanted change from their lives on Tahiti.

Fletcher Christian's voyage back to Tubuai held several surprises and much discomfort. Stowaways were discovered including young chief Hitihiti from Bora Bora, who had once been especially friendly with Bligh and previously sailed as far as New Zealand with James Cook. Morrison says he was amazed that neither Hitihiti nor his retinue ever showed the slightest regret at leaving their friends.

There should be no surprise at this or that Mauatua, Jenny and others were keen to follow Fletcher Christian. They shared a heritage of centuries of adventurous sailing into the unknown that is deep in the blood of the South Seas. Without it, the vast triangle formed by Hawai'i, Easter Island and New Zealand, that later became known as Polynesia and is almost as big as the land mass of Africa, would never have been settled, first by groups of star-navigating adventurers who once sailed into the unknown East from south-east China and the island of Taiwan.

The sea was more troublesome than *Bounty's* unexpected companions. Pigs and goats were so tossed about they trampled upon each other but only five were lost. The bull, which could not keep its footing and obstinately refused to lie down, fell several times and died. They heaved the carcass overboard because it was easier to dispose of it than to attempt to salvage the better parts of its overfed flesh.

By June 26th *Bounty* was once more anchored at Tubuai and this time she was welcomed but any sense of security was soon dispelled by landing the cow and 200 of the pigs. The locals were far more terrified of these than they had been of the firearms, the more so because the animals were let to run loose on their unfenced land, uprooting, guzzling and even dining on the low hanging thatched roofs of their homes. It took two days to ferry the other animals to the small keys in the lagoon to the east of Bloody Bay, where they were easier to husband and caused fewer problems. From Captains Edwards' abstracts of the journals of Heywood and Stewart, we learn that considerable discontent broke out soon after *Bounty*'s second arrival. The mutineers probably thought their freedom, long overdue, was now upon them. On July 5 trouble began and on the 6th two were put into irons by a majority vote. Fletcher was still a believer in collective action. The problems were drunkenness, fighting and threatening of each other's lives, so that Heywood said those abaft were obliged to arm themselves with pistols. Normally those 'abaft' would be the officers but on *Bounty* and on Tubuai these would probably be the hard-core mutineers. By July 7 matters were so bad that a sort of truce was declared. Fletcher Christian and Churchill drew up articles that specified mutual forgiveness of past grievances, which every man was obliged to sign. Only Matthew Thompson refused. A list of rules is something else that pirates did and how sensible it was.

Fletcher Christian must have been working from dawn to dusk, believing that a permanent settlement on shore would alleviate the tensions on board. It would indeed have been wonderful if a man so young, so beset by the threats of both mutineers and Bligh loyalists, perhaps also with the pangs of conscience, was able to establish a settlement that satisfied everyone. At first glance, the island of Tubuai did seem to offer Fletcher Christian the possibility of a settled future.

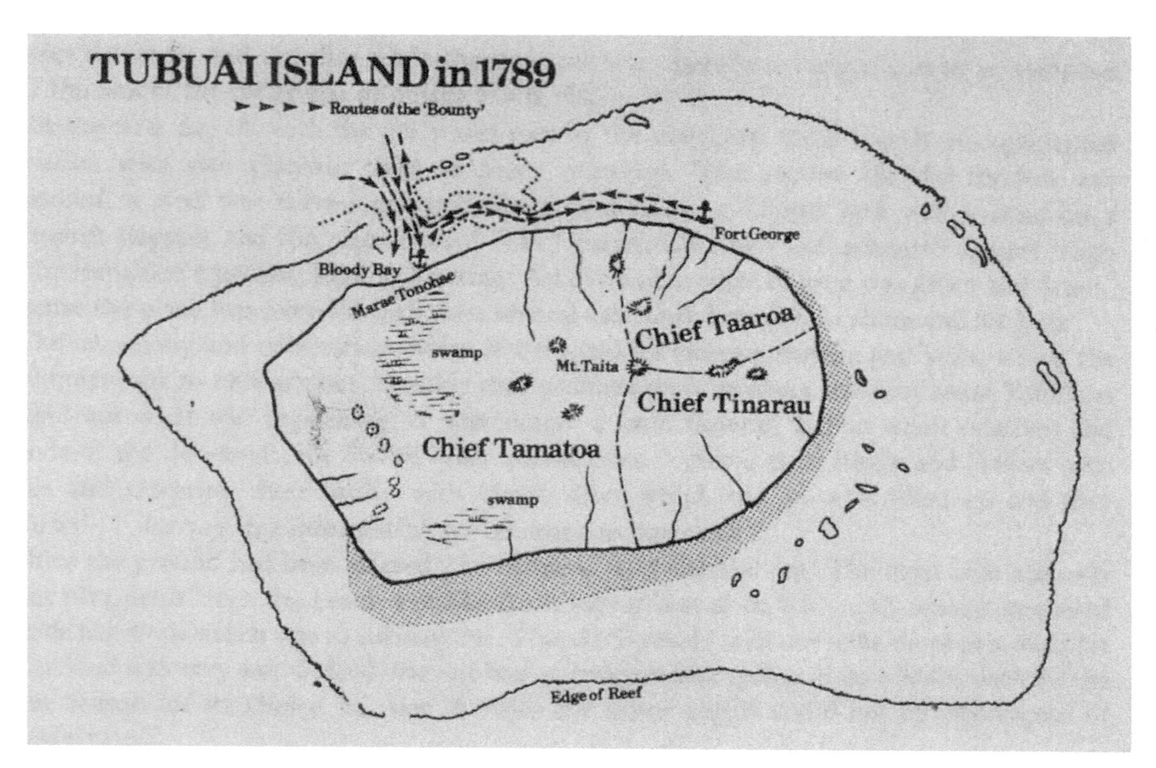

It's commonly said Fletcher Christian should have tried to settle in Tamatoa's bigger chiefdom but my 1980 expedition discovered his practical reason for choosing Taaroa's district.

CHAPTER 28

'HE ALWAYS TOOK A PART'

Tubuai looks as if it might be two islands, one flat with a great central outcrop of rock, the other high and rocky. This is an illusion caused by a combination of volcanic and coral origins and the single island is enclosed by an encircling reef that considerably varies in its distance from the shore, sometimes as far away as three miles/almost five kms. At its furthest points Tubuai is five miles by three miles/eight by five kms. The dazzling, narrow white beach that surrounds most of the island was then the best access from one place to the other, for a very deep strip of timbered lowland encircled the island and the thick undergrowth made foot passage difficult.

The western end is the flatter and more fertile and in most lowland places the soil is a rich black mould, changing towards the foothills into a red earth that supports ferns, bamboos and reeds. The interior of the west leaps up in three naked rocks of hard stone. The east is generally more mountainous, emerging from the verdant lowlands to look like the pastel peaks of Cumberland.

The island produced perfumed flowers and nuts, breadfruit, coconut, yams, taro, plantains and everything else common to Tahiti. Within the reef there were fish, shellfish and large sacred turtles, which women were forbidden to eat. Throughout the island were innumerable rivulets which, having been banked up for the cultivation of the swamp taro, gave shelter to wild ducks and a profusion of eels, prawns, shrimps and freshwater fish. Morrison's estimate of 3,000 inhabitants seems to have been accurate.

The Tubuaians were more robust and savage looking than the Tahitians, a look heightened by their use of turmeric and oil to dye their clothes which gave them *'a Yellow disagreeable look'*. The women were considered more handsome than any the *Bounty* men had seen and this might have been because their faces had not been deformed to make the flat noses and widespread eyes considered ideal on Tahiti. They were considered good at dancing but there was no lewdness.

If the men were more warlike than the Tahitians, their society was, by our standards, more humane. There was no infanticide and parents alone looked after their children. There was no acknowledged form of marriage so, while a man and woman agreed, they lived together. If they parted the men took the boys and the women took the girls and neither was a hindrance to further alliance. The islanders were also less sophisticated, dirtier and smellier than Tahitians. Tubuai had no tattooing, no supercision of the foreskin and no societies of *arioi.* They did pluck their body hair but they bathed infrequently, rivers being too shallow for that purpose. Wherever they went they carried a length of purple bark cloth, glazed to make it waterproof. So efficient was it that the Tubuaians scorned gifts of European cloth. They also preferred their own tools of stone.

Their oval houses looked like haystacks thatched right to the ground on one of the 40 to 80 feet/12 to 24 metres long sides. Thatch did the same at both ends and to within six feet/1.89 metres of the ground along the front, which was provided with several shuttered openings, carved and painted red. A tier of stones divided the men from the women of the house. At the men's end was another area fenced with standing stones. Here the male heads of the house were buried and their images were kept, carved and decorated with the hair, teeth and nails of the dead.

The natural and physical glories of the island by day were scarcely matched by its nocturnal features. None has ever described the misery better than Morrison: *'and when they go to sleep they beat the musquettoes out and make a fire at each Door to keep them out as they are very troublesome and together with Fleas and lice keep them employed till sleep gets the better of them and the Rats run over them all night in droves, but as we left several cats it is possible they may reduce their numbers.'*

Although so different in comfort, hygiene and custom from Tahiti, Tubuai was similar in the way it was ruled with no over-all chief but divided into three clandoms. That which included Bloody Bay, which is almost on the dividing line between east and west, was the biggest, comprising most of the western end. The chief was Tamatoa and he made great demonstrations of friendship. Fletcher was taken to Tamatoa's marae of Tonohae and seated on a large parcel of cloth. After being presented with a young plantain tree as a symbol of peace, he was given a root of *yava* and exchanged names with the chief. The chief's relations gave Fletcher plantain, *yava* and lengths of cloth and then 50 of the landed *ra'atira* followed, each with servants carrying more cloth and each with two baskets of provisions, including baked and raw fish, breadfruit and all the produce of the prolific island. Once the wives of the chiefs made similar presentations, the huge pile of gifts and Tamatoa were taken to *Bounty.*

Tamatoa remained on board all night, most of which he spent in prayer at Fletcher Christian's bedside. In the morning, Fletcher gave hatchets, red feathers, Tahitian cloth and matting. The feathers were the most popular, a point quickly noted. The pleasantries and prayers over, Fletcher and Tamatoa returned to shore to fix on the site for a settlement. Finding none suitable in Tamatoa's western district, Fletcher Christian then made the move which doomed the project. He went to the north-eastern clandom, smaller and poorer, and ruled by Taaroa, who was not allied with Tamatoa and quickly seized the opportunity to score over his adversary by first welcoming Fletcher Christian and telling him to choose the land he liked. Then he too performed a name-changing ceremony.

Tamatoa was furious. Unable to get Fletcher Christian to change his mind, he and Tinarau, chief of the south-eastern district, joined forces and prohibited the people in their districts, the majority of the population, from going to the ship or having anything to do with the visitors.

Fletcher was so intent on planning the settlement he did not at first notice the effect this ban was having on supplies. When he did, he tried approaching Tinarau but he and his family always fled. He decided the best plan was to be firmly settled before solving the problem. He ordered frugal rationing of the ship's provisions and greater care of the stock, something that scarcely endeared him to those who had mutinied against precisely such restrictions.

The site Fletcher had chosen for his settlement was some miles east of the reef opening and Bloody Bay and to further his plans the ship was moved closer to the site. This proved laborious and dangerous, far more difficult than moving *Bounty* from Matavai to Pare. There are so many coral shoals on the course that it is impossible to move in a direct line, making progress under sail out of the question and for hour the ship could not move because of strong sea breezes. *Bounty* had to be kedged, first throwing out anchors and then winching in the line, hoping the anchor stayed attached to the lagoon bottom. Some of the lagoon is so shallow on the way, *Bounty* had to be lightened, first by laboriously hand-pumping out the water used as ballast and then by jettisoning the fresh water. The spare booms and spars were put overboard and moored to a buoy but they went adrift and were lost, to the horror of some but unconcern of Fletcher, who thought it no great cost as he never intended to go to sea again.

By July 10, according to Morrison's reckoning, *Bounty* was anchored only 100 yards/90 metres from the shore. Fletcher then chose the exact site for his fort, abreast of the ship and was given permission to do with the land as he thought proper, probably having purchased the site with red feathers.

It was no time for sighs of relief. When Christian returned aboard, Morrison says he found Sumner and Quintal had gone ashore without permission and it was not until next morning they returned. When they were called to explain their absence, they said, *'the ship is moored and we are now our own Masters'*. Christian clapped the pistol he carried to the head of one, saying *'I'll let you know who is Master'*, and they were both put into leg irons. When they were brought up next morning the resoluteness of Christian's behaviour had convinced them he was *'not to be played with'*. They begged pardon and promised to behave better.

It is possible that this disagreement is confused with one on May 6, when two men were put in irons by a majority of votes. Heywood says it was May 9, when the ship was kedged up to the fort site, close enough for there to be confusion. One incident or two? I think two, as I do not feel Morrison would confuse a joint decision with one made solely by Christian.

The reporting of this Quintal/Sumner incident is one which illustrates how other authors colour events to favour or denigrate a character. In *Bligh* by Gavin Kennedy, which makes it clear that Morrison is the source of the anecdote, the original telling of the story and the author's version are quite at variance. Morrison says, *'He [Christian] called them aft and enquired how they come to go on shore without his leave.'* In Kennedy's book this is rendered as *'Christian demanded to know on what authority they had gone ashore.'* He goes on that Fletcher went into a rage but there is no suggestion of demand or rage in Morrison's journal. Fletcher Christian may well have been very angry but no-one who was there said this. It is just as likely that he enquired quietly, in the way of a stronger, and to use Morrison's word, a more *'resolute'* man.

Fletcher acted to help prevent such frustration happening again by giving leave for two hands to sleep on shore each night, as had been Bligh's custom at Tahiti. As many as pleased could go ashore on Sunday on condition that the livestock on the keys was checked.

The next man Fletcher had to deal with was Tom Ellison, who now seems to have been his servant for he is described as *'Thos. Ellison, who waited on him and was frequently there'*. Knowing the

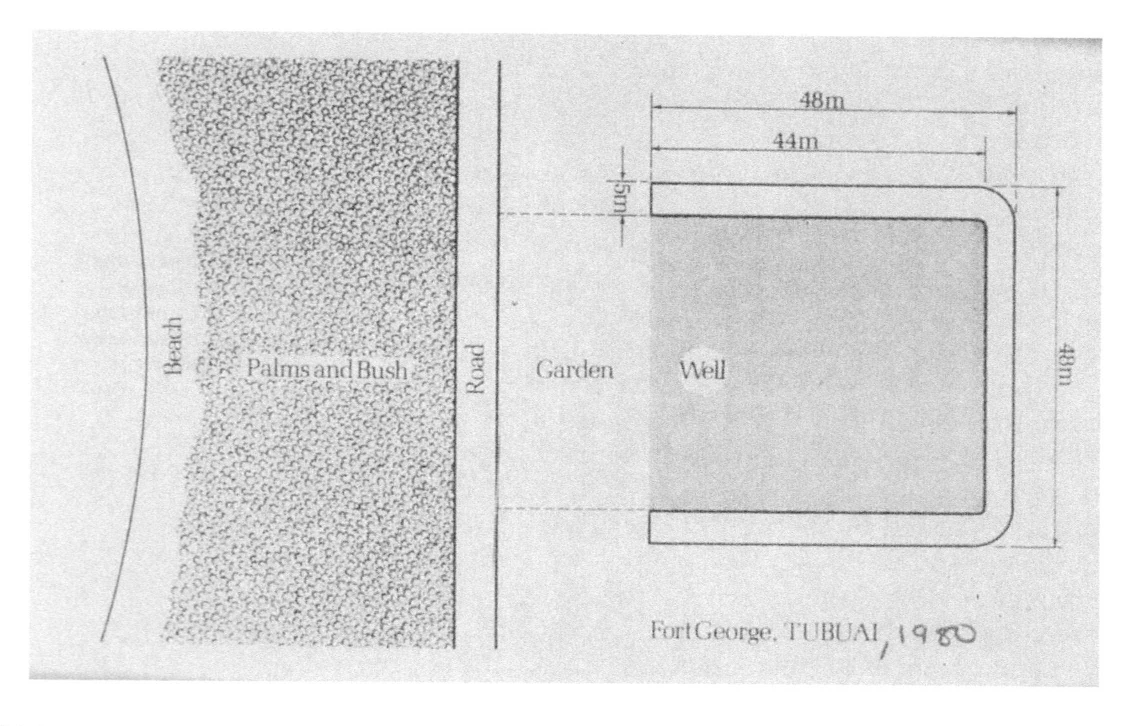

popularity of red feathers on Tubuai, Fletcher distributed some to all hands, which gave each of them status and bargaining power when on shore, a thoughtful and placatory gesture. When some of Christian's feathers were suddenly missed from his cabin, Ellison was accused of being the thief. He was stripped and tied ready for flogging but was persistent in declaring his innocence. As no other man could say he had seen the boy with what might have been stolen feathers, he was released.

Work on the land base began. The forge was set up and armourer Coleman made iron rammers for the muskets and then altered axes for felling trees. Botanist William Brown, assisted by a Tahitian, was to clear a piece of ground to plant yams. Coleman and McCoy were to work at the forge making spades, hoes and mattocks. Hillbrant stayed on board as appointed cook, Byrne and Ellison with some of the Tahitian men cared for the boats but the rest were sent ashore to work. On shore, arms were to be carried but left under the care of a sentinel while ground was being cleared. To protect the ship while the men worked, one of the cutters was anchored near the beach, the other was to stay close to *Bounty*.

On the first day of work the party was met by Taaroa and friends, who presented Christian with two plantain trees as peace offerings. The ground for the fortress was measured, a clod was turned as symbolic of possession, a Union Jack was hoisted on a makeshift flagpole and the place named Fort George for, although Fletcher had mutinied against Bligh, he remained a patriot, loyal to his king. Because the place was overrun with rats, several cats were brought on shore and let loose and then an extra allowance of grog was given and drunk.

The ceremony and celebrations were interrupted by hideous shrieks and yells, which the celebrants took to be war cries. Quickly they grabbed their muskets and sent some Tahitians to find out what was happening. It was a local funeral, a time when relatives and friends of the deceased rent the air with horrid cries, *'cutting their heads and breasts with shells and smearing their bodies with blood; after which the grave is filled up and they departed . . . having this information we returned to our work'.*

Once the ground had been cleared, the fortress was laid out. The front was only about 50 yards/45 metres from the beach and Morrison says it was to have been 100 yards/about 90 metres square, measured outside the ditch which was to surround it. This ditch would turn out to be more of a moat, for the ground was very wet but whether salt or fresh water was the likely content is unknown.

Critics claim Fletcher Christian chose this site through ineptitude, misunderstanding the political balance and internecine tension on the island but none that I know of ever bothered to go there. When I took the trouble and expense to visit Fort George in 1980, I discovered he had chosen this site most wisely, because it featured an independent spring in the middle of it. With no exposed contributing stream, the settlement's supply of fresh water could not be interrupted or contaminated by others.

The ditch that became a moat was to be 18 feet/5.5 metres wide, enough to ensure it was not easily crossed, wet or dry. The walls were to be a monumental 20 feet/6 metres high, measured from the bottom of the ditch, 18 feet/5.5 metres wide at the base and 12 feet/3.5 metres wide at the top.

Heywood's journal cast fresh light on these specifications, for he says the fort was only to be half what Morrison says and this is closer to the remains we measured in 1980. The exact height the walls reached has never been established but in 1902 they were 6 feet 7 inches/2 metres high and the ditch about the same depth, about two-thirds of the original target. The ditch seems never to have been more than 8 feet/2.4 metres wide.

Defence consisted of a drawbridge on the north side facing the beach, with a cannon on each corner and two swivel guns on each side, with two of the latter always in reserve for reinforcement. Each direction was thus protected by at least two cannons and four swivel guns.

There was a professed dearth of knowledge about fortification but work proceeded. Stakes and battens were cut, sods were dug and carried. Barrows were made and timber was cut for the gates and drawbridge and for the one long and two short huts that the Tahitian Jenny said were built. This was not a Wild West fort of split timber as some later illustrations have shown, although there might have been a palisade planned atop the earth walls.

Fort George was the most splendid opportunity imaginable for a romantically inclined adventurer brought up with the ancient fortifications of Moorland Close and Fletcher was no idle spectator. *'He always took a part in the most laborious part of the Work'* and, to encourage his men to keep up with him amid the heat, rats and insects, he allowed an extra half-pint of porter, a strong, dark bitter beer, twice a day.

From this time on, about the beginning of August, the reports vary a great deal in their dates but more or less coincide concerning actual events. I think it likely that Captain Edwards was as slack in copying correct dates as he was in applying the third person singular only to Heywood. The timing of what follows, which includes new incidents reported only by Edwards/Heywood, is entirely mine.

At the end of July, some of Fletcher's party had been waylaid on their way to feed the cow and shot an islander in the back, a murder that Kennedy attributes to Christian even though there is no such suggestion in Edwards' abstracts. Early in August a party went ashore to take women by force and meeting with expected opposition shot a Tubuaian and wounded another with a bayonet. Fletcher's strict rationing of time ashore must have been relaxed for Morrison says, *'We remained quiet some days, but as the people were fond of sleeping ashore, some of them were decoyed by the Weomen into Tinnarow's district, where they were strippd.'*

This was vexatious and when Adams was imprisoned in Chief Tinarau's house, Fletcher was obliged to assert authority on behalf of his men. He marched south with others into Tinarau's district but the chief fled. The woman who had enticed Adams brought him to Christian wearing only his shirt. The rest of Adams' clothes had been taken by Tinarau's men, so Fletcher sent messengers asking for their return and suggesting that both sides should make friends. It was a game of patience and Fletcher waited hours, sending repeated placatory messages. All were refused.

Fletcher ordered Tinarau's house to be burned, a serious action made more insulting to the chief by the removal of his revered ancestor images. Just as Fletcher had hoped the canoes that he took on the first visit might stimulate conciliatory conversation, he thought these immensely valuable objects would perform the same task. The party returned to *Bounty* accompanied by the woman who had lured Adams because she was now afraid of being punished for releasing him. Once aboard she found friends, so it is apparent that some of the men had learned it was safer to lure women into their hammocks.

By September, work on the fort had slowed but the gate posts were fixed and three-quarters of the walls completed. Tinarau wanted his household gods returned and on September 2 crossed the island with an entourage carrying baskets of food as a peace offering. Fletcher agreed to return them provided women were allowed to go freely with the *Bounty* men. He also wanted his men's stolen clothing returned and a promise there would be no more inveiglement or ill-treatment in Tinarau's district. Tinarau agreed and proposed they should drink *yava* together to seal the bargain.

When Christian refused the beverage, Tinarau was furious and swept away in a passion not because his offer was spurned but because he knew the trick behind his visit had been discovered. Earlier, one of the Tahitian boys had seen the food carriers secrete weapons close to the fort. When the Europeans were under the narcotic effect of the *yava*, they were to have be exterminated. Although knowing this, Fletcher went through the greeting ceremonies but had ordered armed men to take positions atop the works. Discovering this, Tinarau left in a hurry and without formality.

The Tahitian boy who had brought the earlier information had quickly been sent out to the ship with orders for Coleman. As soon as the retreating Tubuaians appeared on the beach he was to fire grapeshot at them. He did and their exit was considerably hastened. No damage to flesh was inflicted but some of the shot passed through a house and cut away a rafter on which a man was hanging a gourd of water. He could not see the ship and did not see where the shot went. He thought it something supernatural and he never returned to his house.

Taaroa's 'royal' family of the district in which the fort was situated was always loyal and friendly to Fletcher. They made a special visit a few days after Tinarau's failed ploy. Taaroa brought his old father and the usual retinue of attendants bearing provisions. His daughters sang and danced for Fletcher, then the compliment was repaid by the Tahitian women. Not feeling he had entertained enough, Fletcher invited his noble guests to return the next day for a Tahitian *heiva,* a far grander and more formal entertainment. Special attention was paid to the costume of the entertainers. Two women were *'neatly dressed'* and two men donned the astonishing mourning dress of the Tahitians, the *parae,* which *Bounty* had loaded as gifts for King George. Yet the mood had changed and while the Tahitians sang and danced the entertainments of their homeland, the Europeans were under arms and mutiny bubbled to the surface again.

So advanced was the fortress and so single-minded its architect, there was finally talk of removing the masts from *Bounty,* thinking that these and the ship's timbers might be used for proper houses. Morrison secretly entertained the idea of escaping back to Tahiti in the ship's large cutter. Stewart said Heywood and he had formulated the same plan. To prevent their being pursued, Morrison suggested he would sabotage the ship's lifting blocks after the masts had been removed, thus preventing their replacement. It was first decided to make the cutter shipshape but Fletcher had ordered that the boats were not to be repaired until all were settled safely ashore.

The trio of plotters, feeling others might the same plan, thought they might be better off simply risking the voyage with the boat as it was. Morrison wrote: *`tho the passage was short and it might perhaps be made with safety in 5 or 6 days, yet had we the Chance to Meet with bad weather our Crazey boat would certainly have made us a coffin . . .'* The situation on Tubuai was soon to change and they were unable to make the attempt.

From the recollections of both Adams and Jenny there seems to have been at least one more battle with locals, perhaps two, not mentioned by Morrison, Heywood or Stewart. Adams says the Tubuaians believed a rumour that suggested the fort's ditch was to be a mass grave for Tubuai's entire population. Jenny says there was also conspiracy between Tubuaians and one of the Tahitians to do the reverse, that is to murder all the Europeans and divide their goods.

Mauatua is said to have told Fletcher of the conspiracy, neglecting to mention it was one of her countrymen who was the traitor.

It is possible that both Adams and Jenny are speaking of the same battle for there are reasons enough in their tales for both sides to wish to attack the other in the interest of their own survival. The lack of women had never been solved. They would not join *Bounty's* men at the fort, although they were happy to sleep with them at their own houses, which was not thought good enough. One plan suggested the Europeans would make slaves of the Tahitian women, men and boys, who would be distributed by casting lots. Another wanted Fletcher to lead a party to capture women by force.

Fletcher would have none of these strategies, even though he was told those disaffected would do no more work until each man had a wife. Fletcher's objective was to persuade rather than to force the Tubuaians, which is hardly the decision of a man *'who could not control his passions or men'*. One reason for not turning the Ma'ohi men and women into slaves was personal: he would lose Mauatua. The unrest was so general that work did stop and three days were spent debating. Exercising their tongues was as thirst-making as labouring with their bodies and the men demanded more grog but Fletcher refused what he considered yet another absurd demand, so they broke the lock of the spirit room on the ship and took it by force.

Perhaps playing for time while they cooled their tempers, Christian ordered a double allowance of grog to be served every day but to no purpose. He called yet another meeting of all hands to gather opinions. At first the consensus was to move to Tahiti where they might get women without force but he was overruled. Next day the proposal was revived and on a show of hands the idea was accepted.

Fletcher had now lost his fort on Tubuai. Knowing he could never go anywhere he might be captured and taken back home to shame his family further, he asked for *Bounty* in a speech that resounds with black despair *'I'm resigned to sailing where the wind takes me. Somewhere, anywhere . . . but it will be to oblivion. Then one day, if it pleases God, you will forget me and what I have brought you to.'*

The best description of Fletcher Christian was written by Bligh in the open boat. Demoted back to master's mate, he was: *aged 24 years, 5 feet 9 inches*/1.75 metres *high, blackish, or very dark brown complexion. Dark brown hair, strong made; a star tatowed on his left breast tatowed on his backside; his knees stand a little out, and he may be called rather bow legged. He is subject to violent perspirations, and particularly in his hands, so that he soils any thing he handles.* The darkish complexion described was probably exaggerated by tanning from months under the Tahitian sun but now Fletcher Christian would never again pale in English weather.

It took less than a minute for every man aboard to agree to Fletcher's request for the ship, probably based on certainty that regaining Tahiti was worth any price. He was supervising the daytime addition of water and lime juice to the high-proof navy-issue rum when Matthew Quintal stepped forward, saying that he would be proud sail with Fletcher.

Cornish Quintal, barely as tall as Fletcher's shoulder, was 21 and *'very much tatowed on the backside and several other places'.* The first man Bligh flogged for insolence and mutinous behaviour, he was also first to join Fletcher against Bligh. Now, he too opted for a future with no boundaries except to wander the South Seas, forever hoping to avoid retribution.

When William McCoy said he would sail wherever his mate Quintal went, Fletcher Christian saw a different future shaping. McCoy was 25 and only slightly taller than Quintal. Fair-skinned with light brown hair and strongly built, he had a scar where he had survived a stab in the belly and another under his chin and *'had tatows in different parts'.*

Then Isaac Martin volunteered, by now Jenny's new lover. Martin was taller than Fletcher by two inches/5cms and another victim of Bligh's contradictory orders. He took 24 lashes for striking a Tahitian thief while he wrestled back things stolen from *Bounty*. One would expect him to have been rewarded for saving the ship's property, so Bligh's punishment can only be explained as inconsistency and muddled thinking. Martin vacillated during the mutiny, first refusing to support Fletcher and then agreeing. Sallow skinned, brown haired and raw boned, he too had a Garter star on his left breast.

Fletcher next took the proffered hand of William Brown, the assistant botanist. A little shorter than Fletcher and from Leicester, he had been a career naval officer, working up to be an acting lieutenant and a second-in-command but had then elected to become a botanist. He once refused to join the daily dancing on *Bounty* but wasn't flogged and merely had his daily grog ration stopped. He was tattooed and had a remarkable scar that contracted his eyelid and ran right down to his throat, the result of scrofula.

John Mills joined Fletcher next. Slightly taller than Fletcher, the gunner's mate was a Scot from Aberdeen and at 40 was one of the oldest aboard. He went with Fletcher to arrest Bligh, who described him as strong made and raw boned. He was a bully, the self-deluding sort who excused sadistic taunting or physical abuse as 'only having a bit of a laugh'.

Any thoughts by Fletcher of noble self-immolation unseen on the open seas of the South Pacific must now have evaporated if these were to be his companions. As with the mutiny, he hadn't expected so much support from other men.

John Adams came forward next. He was 22 and stood only as tall as Quintal and described by Bligh as strong made, with a brown complexion, brown hair and a face pitted by smallpox. Widely tattooed, he had a scarred right foot, a feature attributed in Richard Bean's 2014 play *Pitcairn* to Fletcher Christian, amidst other extraordinary notions that cannot be defended as dramatic license; for instance, Tubuai is called Tubai, and Tahiti's *manahune* become *manahane*. Adams stood beside Fletcher from the earliest moments of the mutiny. In Tahiti he took up island life easily and more fulsomely than any of the company. He was first to be tattooed and first to have a permanent Tahitian mistress and a *taio*.

Fletcher then accepted the offer of Williams, the French-speaking native of Guernsey aged 25 and another active mutineer. Black haired, slender and tattooed, he was as short as Quintal, Adams and McCoy. Williams received six lashes in False Bay, South Africa, when Bligh found fault with the way he heaved the lead, by which the ship knew how much water was beneath her hull. Fletcher suddenly had a company of seven. What did Mauatua think through this? When did she decide she would not leave Fletcher? Did she envision something wonderful with Fletcher or dread an unknown future with men like those? Being men, neither Heywood nor Morrison who were present mentions the thoughts or the actions of Mauatua, Jenny or any other of the women.

'We will never leave you, Mr Christian, go where you will.'

Ned Young was the eighth man to join Fletcher. Born on the island of St Kitts in the West Indies, even he agreed he was the worst looking man in the ship, with features that could be frightening when first met. His lilting, vowel-caressing Caribbean accent was made more indistinct by a mouth of teeth that were entirely rotten or well on their way, the result he explained of being brought up on a sugar island. *'You'll be wanting at least one other officer, Mr Christian,'* Ned said.

Ned's mother, or grandmother, was a Negress. He'd never been quite clear about that but he wanted to get away from the Royal Navy and the endless taunts of 'nigger'. He was well educated and seemed well-connected, a more proper companion for Fletcher. *'No. No you can't, Ned,'* Fletcher protested. *'What's the future in that for you?'*

'The future for me? Something better than the prospect of being strung from a yardarm on the other side of the world. I wasn't a mutineer but I did nothing to stop you. Bligh is bound to have it in for me.' Then Ned something likely to encourage Fletcher more than anything else he had heard.

'We will never leave you, Mr Christian, go where you will.'

The details were quickly settled. Fletcher Christian was to have the ship in seagoing condition and the 16 who elected for Tahiti were to be given supplies including arms and a share of everything on the ship. Among those who wanted to settle on Tahiti were

Churchill and Thompson. They were probably the most destructive of the Fort George community and, from their later behaviour, Christian was well rid of them.

There was now action and unity. Water was collected, the sails that had been used as shelters were bent, which means fixing to the yard arms and stays. Very soon Fletcher was to find that leaving Tubuai would be bloodier and more difficult than his arrival.

A party including Hitihiti was sent to collect stock and to find the cow but was ambushed, beaten and plundered. Hitihiti's chiefly status was insulted and the islanders sent a message with the bruised men that they would treat Christian the same way. Fletcher was not intimidated. 20 men were armed and with nine Tahitian men and four of the boys, one of whom always carried the Union Jack, they marched off to retrieve of their roaming stock and to chastise the offenders. The insults to Hitihiti were especially to be avenged.

They were confronted by a war party of about 700 who fought with stones, clubs and spears but with more fury than judgment, retiring with great loss. Some stock was collected but with further ambush being expected Fletcher issued each European with 24 rounds of ammunition and Hitihiti, being an excellent shot, was armed with a musket. Once safely back at Fort George, they found Chief Taaroa with his aged father and young brother, Taaroamiva, who had come to explain that *Bounty*'s stock was the bone, and the flesh, of contention. Tinarau was determined that the animals that roamed into his territory were his property and he was arming more men to ensure he kept them.

Fletcher now needed to show leadership to his own party as well as to block Tinarau. If he had upset anyone on the island it was Tamatoa but this chief had kept his distance. He asked his Tubuaian friends to remain in the fortress to avoid accidental injury. Then he drew up his party into ranks with one Tahitian between two of his men, arming the former with clubs. In silence they marched south through the thick woods towards Tinarau's district.

When they reached a path with dense bush on each side, Burkitt, thinking he heard something stir, stepped to look and received a spear wound in his left side. The Tahitian beside Burkitt levelled his attacker and took his spear. In seconds there was bedlam as Tubuaian warriors rushed in with *'great fury and orrid yells'*. Christian's party quickly faced different defensive directions and fired several times but the Tubuaians continued attacking.

Bounty's men retreated safely to rising ground a short distance to their rear, by which time the Tahitians were armed with captured spears. The Tubuaians attacked the hillock with redoubled energy, not the least perturbed by the growing number of their own dead. The retreat back to Fort George had to be continued, for thick bush above *Bounty's* men was excellent cover for the attacking Tubuaians. Several Tahitians were now wounded and Christian had injured his hand on his bayonet.

They fell back through the bush to a taro ground 200 yards/180 metres away, keeping up a constant rear-guard fire. There they protected themselves behind irrigation banks, a sort of trench warfare. The Tubuaians followed hotly and then halted to throw stones from the edge of the bush.

Fletcher saw that although the furious fighting seemed without order, the Tubuaians were actually organised into parties of 20 to 28 men, each of which had a leader whose orders were followed. One of these rallied his men by venturing from the cover of the bush but even at that distance he was shot dead. Seeing that each man who came into the clear was killed or wounded, the Tubuaians retreated.

Burkitt was growing faint so Skinner, whose musket had been disabled, helped him back to *Bounty*, while the rest waited to be attacked again. It was soon apparent that Fletcher's leadership had beaten the local warriors, who outnumbered them by over ten to one and he was given three rousing cheers.

The Tahitian men and boys immediately began to collect spoils of war but did not get all they wanted. One asked permission to cut out the jawbones of the dead so they could be hung around the ship as a deterrent to other troublemakers. It must have been tempting to Fletcher, thinking of the troublemakers among his own party, but the request was denied, so incensing the man that he had to be threatened with his own death, even so still begging to excise just jawbone for himself. The other Tahitians were happy with spears and clubs and the party met no opposition when they finally gathered the animals they wanted, including the nomadic cow.

Burkitt had had the narrowest of escapes. The spear had struck but not broken a rib and he healed in a short time. Morrison says the affair gave them a very mean opinion of their bayonets for although several men had fallen by them, the blades always broke off and remained in the body. The length of the Tubuaians' spears was such that bayonets were generally no use against them. If they had not had firearms, everyone in Fletcher's party would certainly have died.

On the 14th, the day after the bloody battle, the cow was slaughtered. Butchering it ashore was preferred to manhandling it back on board and it proved excellent meat. For safety, everybody now lived on the ship and while they were feasting young Taaroamiva came aboard to tell them 60 Tubuaian men had been killed (50 according to Heywood), including Tinarau's brother, as well as six women who had been supplying them with spears and stones and many more were wounded. Taaroamiva had been so loyal to Christian that he feared for his life and asked if he and two friends could join *Bounty*. Fletcher agreed and they were doubly pleased when told the ship was bound for Tahiti.

Bounty weighed anchor and made her way relatively easily to the opening, being lighter than before. When clear of the reef, they lay by and *`filld Saltwater to keep her on her legs and at noon made sail, leaving Toobuai well Stockd with Hogs, Goats, Fowles, Dogs and Cats, the Former of which were increased to Four times the number we landed . . .'*

Fletcher Christian's settlement had not failed because of lack of leadership, as he is so fatuously accused. Neither was it the continued belligerence of Tinarau or any unhappiness of the majority of men in his party. Morrison, who was there, says the greatest enemy of Fort George was the determination of the priests of Tubuai that Fletcher Christian would never live on their island.

The priests had virtually all the authority on Tubuai. Once they had seen the Europeans were, as Morrison observes, *'Common Men and liable to accident like themselves'* they could not bear to see *'such Superiority as the Europeans in general usurp over those who differ from themselves'*. Because Fletcher and his party ignored and ridiculed their authority, the priests used every means to prevent the three chiefs making friends with them, believing that if the visitors remained their own power would be lessened, which Morrison adds *'in all probability would have been the case'*.

If there had been fewer in the party, they might have been absorbed into the community with few problems, however Fletcher did make mistakes, largely through expecting Tubuai to be like Tahiti. His ignorance of the divisions and political situation on the island are forgivable, because he was doing something no European had done before.

How would Fletcher know that Tubuaians did not fence in their houses and gardens, as happened on Tahiti? 200 rooting pigs and browsing goats damaged property and gardens badly across all three regions. Then, he unintentionally snubbed Tamatoa by sensibly choosing land that included a spring but that was not in his territory. That was just what the priests needed to divide and then conquer, something not exclusive to South Sea religious extremists.

If these defensive and belligerent priests had lived the next 30 years, they may have reconsidered quite who had the victory. *Bounty* left venereal disease and dysentery. Their decimating effects were compounded when the priests of the London Missionary Society chose Tubuai for their first mission, a greater kiss of death because they infected with ignorance and intolerance. By 1823, the population was reduced from 3,000 to 300.

If Fletcher Christian had not led resolutely firmly on Tubuai, the people who came with him aboard *Bounty* would have been slaughtered, a proposition firmly supported by the increase in the fury and numbers of Tubuaian attacks during his last few days on the island. Any who moaned about Christian's restrictions lived only because of them.

Fletcher Christian's plan was precocious. That he succeeded in persuading such a disparate group to labour to build a moat and enormous earthen fortifications in tropical conditions, while pestered by rats, insects and Tubuaians, was a personal triumph. Frustrated but not beaten, he had lost no men and still had the ship. What was missing was his future.

By September 20th 1789 *Bounty* was again in the lee of Mehetia, where the remaining trade goods, arms and ammunition, alcohol, clothes and practical goods were divided. On the 22nd, she drifted into the lagoon of Matavai Bay and those headed for shore were unloaded before nightfall.

Peter Heywood, quoted by his step-daughter Lady Belcher in her book about the mutiny, remembered how difficult this was. There was a high surf on the day and only one of *Bounty*'s boats was considered safe to *'swim'*. Fearful of travelling in more stable outrigger canoes, even though they admitted the Tahitians handled them better in those conditions than they did their own boat, the Europeans waited endlessly as it made trip after trip. Only when everything else was on shore was the ammunition landed.

Did Fletcher Christian go on shore in Matavai Bay that day? He is quoted as saying he could not face the chiefs to whom he had lied on his last visit. Heywood says he came to say a final goodbye to himself and to George Stewart and it's possible Fletcher did this under cover of darkness. Heywood said Christian took him to one side on the black sands of Matavai, absolved him of any complicity in the mutiny and told him to give himself up to the ship that was bound to arrive in search of *Bounty*. Heywood said he was also entrusted with a message to pass on to Christian's family, relating circumstances that might extenuate, though they could not justify, the crime he had committed.

Heywood kept the meeting secret until well after his subsequent trial in 1792. It was only sensible to distance himself from Christian, who had been such a close friend that Bligh could and did regard Heywood as Fletcher's ally. Heywood apparently lied about his age and was no angel. His name is one of those that appear on the venereal list but he was a gentleman of honour and kept his word, given on the Tahitian sands, to his lost friend. Even when Heywood finally revealed he had had that conversation, he kept the details secret until he had spoken to Fletcher's family.

In the 1930s, it was said that Christian's wild-eyed look and mutinous behaviour might have been caused by syphilis. The shame of this was thought to have been too much for Christian to bear, so, rather than facing society and his mother, he mutinied and took the ship to sail into hidden disgrace. This is a 20th century view of syphilis. In the 18th century diseases of all kinds were accepted differently and few families of the gentry or aristocracy that included adventurous young men would have been free of infectious disease, venereal or otherwise. Neither would they have known the differences between sexually transmitted afflictions.

It is unlikely in the extreme that Fletcher Christian was suffering from tertiary or neuro-syphilis, the stage of the disease that can turn the mind and when death is imminent and inevitable. The most licentious of men is unlikely to have been in this state at 24 for, although this final stage can appear within five years of initial infection, it usually takes ten to 40 years and the longer time is the more usual by far.

Anyway, there is no evidence of Fletcher Christian passing any fatal venereal disease on to Mauatua or to any other Tahitian woman. It wasn't subsequently a problem on Pitcairn Island. Neither did he pass it on to or get it from William Bligh.

CHAPTER 29

MINDS OF THEIR OWN

In 1965 Madge Darby was first to express the theory that Heywood's message concerned a homosexual relationship between Bligh and Christian. The friendship between these two men demonstrably exceeded the norms. Bligh had always favoured Fletcher Christian with special treatment and promotion on *Britannia* as well as on *Bounty*, where he had been allowed unlimited access to his liquor chest. It's likely that these preferments were because Fletcher was better at his job than others and because his professed ability to humour Bligh was thought by his captain to make Fletcher a better friend and companion.

Bligh was certainly a man's man and felt happier in men's company, hence his suitability for a naval career. Betsy had given him only daughters. If he were secretly homosexual, a frustrated and tortured feeling for his young wards, part paternal, part sexual, might explain why he was so vitriolic about Christian, Ellison and Heywood on his return.

Fletcher Christian, ten years younger than Bligh, was the ideal protege, well-born and connected, loyal, charming and a quick learner. With his father dead when he was less than four, he grew up with a resident grandfather but without the interest of an older male member of the family, at least not of the father-figure type whom it is permitted to clasp emotionally and physically. This psychological void, combined with sudden financial betrayal and the practical need for a patron, would certainly recommend Bligh's interest as worthy of encouragement.

Bligh fits well into the role of a surrogate father to Fletcher Christian. The reciprocated and loyal interest of a bright young man from a good family, who was also a talented sailor, would serve only to flatter Bligh. As a single child, Bligh needed, or enjoyed, the feeling of having a surrogate brother or son, too. Such psychological dependencies are more fragile and more explosive than the real thing but do not need to be expressed physically.

In the 21st century it's too easy to think that during long days and nights of sea passage under sail, special relationships might flower into a sexuality two men would never consider when ashore, whatever the opportunity. Yet, the theory this happened to Bligh and Christian ignores the absolute impossibility of such a relationship being concealed on such a cramped and crowded ship as *Bounty*. Bligh was noted as usually having his cabin door open, so if ever it were closed because he and Fletcher were sexually active it would have been obvious to all what was going on. Homosexual sex was also forbidden in the Royal Navy on pain of death, something that ever-zealous Bligh would have heeded absolutely. As well, there is incontrovertible contemporary evidence that a homosexual relationship between the two men did not exist.

Every one of those opposed to Bligh or accused by him and later put on trial in England could have used the charge of homosexuality to damn him and help towards saving themselves. The suggestion would have ruined Bligh forever, spiteful revenge guaranteed for those who wanted Bligh to suffer. Even though their lives were at stake, not one of the accused used the counter-charges of homosexuality against Bligh. This alone should have prevented such theories ever being aired. Madge Derby, Richard Hough and others should have researched more, thought deeper and written less.

Yet there's little doubt in my mind that the intensity of their friendship contributed to its breakdown. Young men must one day grow away from parents or parent figures and the wrench is more difficult for the latter, for whom it is a retrograde step to loneliness. The independence that Bligh gave to Christian on Tahiti was simply because Fletcher was second-in-command and had to have this job. The breadfruit camp was *Bounty* on shore. If Bligh and Christian were still arguing about money, then both would have welcomed the break.

After six months ashore, during which he matured in every way, including sexually, Fletcher came back on board thinking his own thoughts and no longer needed a father or any other substitute, whereas unchanged Bligh would still have expected Fletcher to be dependent. The new relationship would have been mutually agonising. Bligh bullied, insulted and nagged painfully at Christian, hoping to demean him back into his dependent role and soon Christian couldn't stand the sight of Bligh. Something had to give. Fletcher Christian had only one way to salvage his sanity. He followed the common pattern of resolving familial tensions, by doing the equivalent of leaving home.

As soon as Heywood was free after his trial, he did what he promised Fletcher and went to Edward Christian, telling him that Bligh's published narrative was far from the full story, thus stimulating Edward's search for the truth. Fletcher Christian had nobly made it quite clear to Heywood that the blame was all his. He knew he should have been strong enough to weather the daily storms that Bligh sent his way but he was provoked beyond his ability to cope and Heywood made certain Edward Christian knew this.

Fletcher trusted his friend Heywood to let his family know the circumstances were anything but normal, even by the brutish standards of daily naval life. By quietly explaining himself only to Heywood, he showed he was not pursuing a personal vendetta, while, in sharp contrast, Bligh needed to defend his actions publicly and vilified everyone.

The 21st century gives new advantages and tools to help understand Fletcher Christian's revolt against William Bligh and to build a better understanding of the complexities of co-dependent male relationships. We know much more about mental health and what provocation can do in circumstances of isolation and anguish.

Is it possible that Fletcher Christian was not in a normal state of mind on April 28 1789?

In the notes William Bligh made in the open boat after the mutiny, he wrote that Christian was *`subject to violent perspiration and particularly to his hands so that he soils any thing he handles'*. In another description written in Timor, Bligh describes *'violent perspirations'*. The most likely explanation is that Fletcher Christian suffered from hyperhidrosis but there is no evidence to tell us how long this condition had existed, for no other person mentioned it. Hyperhidrosis is the medical term for excessive sweating and it's important to understand that the condition exists independent of other complaints; it does not have to be associated with or a symptom of a physical or mental condition.

It's true that hyperhidrosis can be caused by diseased vital organs but Fletcher had no associated symptoms that would indicate this. Excessive sweating was also associated with opium taking. Drug addiction is unlikely in Christian's case without also believing that Bligh condoned and accepted Christian using the opium in *Bounty*'s stores for recreational purposes. Extreme sweating is also associated with diabetes but this would earlier have led to coma and eventual death in those days.

Hyperhidrosis can affect the palms, the armpits, the feet or the face. Some sufferers are struck in just one of these areas, some in all. Essentially there is no escape. It is chronic and goes on all day and all night. The affected areas are colder than other parts which, together with the constant dampness, makes them seem clammy. Like all sweating, hyperhidrosis gets worse in times of emotional pressure or conflict and thus can be a terrible social handicap, commonly leading to withdrawal and the complications this can cause both professionally and domestically. It is common for sufferers to develop a fear even of accidentally touching the one they love most.

If Fletcher's condition was so extreme that Bligh felt obliged to note he dirtied everything he touched, at a time when shipboard hygiene and nicety was minimal, he was either in a bad way or Bligh was being impossibly prissy. We know Bligh used personal debt to wound Christian and that he would add insult to injury wherever he could, by mocking anything he judged different or less in others. If he ridiculed Fletcher's sweating, it was unforgiveable. Who can bear being humiliated for something they can do nothing about? It's telling that Bligh was the only person to mention Fletcher's sweating palms but perhaps because the condition was impossible to conceal, even if only minor, his sweating palms made excellent evidence for use in identifying Christian if ever he were captured.

The pathology of hyperhidrosis is now understood and can often be helped with surgery to sympathetic nerves but the genesis of most hyperhidrosis remains a mystery, except to say it commonly appears in puberty and it may improve or even disappear in the late 20s and 30s.

Hyperhidrosis is an independent condition, not a symptom of a psychological or nervous disorder as was suggested in Montgomerie's *William Bligh of the Bounty in Fact and Fable* (1937). His conclusion represents a classic mistake, of accepting a generality as a relevant explanation of a specific. It was only when I consulted an acknowledged expert in the field, Professor John Ludbrook, Professor Emeritus at Adelaide University and Professorial Fellow at Royal Melbourne Hospital, that the distinction was clarified.

The social consequences of being a sufferer can certainly lead to some degree of psychological or nervous disorder and it is the order of the appearance of the two that can cause confusion. The clear documentation that Fletcher Christian was admired physically and socially on all his other voyages and by all but a few of those aboard *Bounty* suggests he did not always suffer from the condition or that, if he did, it was minor and others were able to ignore it. The conflict which developed after *Bounty* left Tahiti was ideal for making the condition increasingly severe. In times of stress, sufferers of hyperhidrosis of the face can look and feel as though they have been rained on. The acidity of the sweat pouring into their eyes can make these red with irritation. And isn't this how Christian is described as looking on the day of the mutiny? Apart from a sleepless night, his eyes *'aflame with revenge'* might have been emotion but they were as likely to be red because of reaction to violent perspiration. His long hair would have been sodden and his shirt drenched alarmingly from his armpits. We would all sweat more in the situation. Christian poured with it. Even if he were sane, his hyperhidrosis could have made him look mad.

Combined with the dread seriousness of what Christian was doing and what everyone on board had been enduring, does this further explain why so many of *Bounty*'s crew preferred to watch him take the ship from Bligh rather than dare challenge him? Few want to challenge anyone who seems a mad man.

Was Fletcher Christian actually mad, at least for a while? The unpublished autobiography of Charles Christian gives us further and unexpected illumination from Bligh himself. With hindsight and a growing propensity to tell the truth to listeners who were not officials, Bligh eventually gave a different reason for the mutiny from any given in his published works. Charles wrote of a conversation between Captain Bligh and Major Taubman, the man responsible for putting Fletcher Christian aboard *Britannia.* When Taubman asked Bligh what could possibly be the cause of Fletcher's defection, he replied: *'It was Insanity.'*

`He spoke right,' says Charles. *'But who was it that had drove him into that unhappy state?'* That is the most important question of all but was not the question the court martial of the mutineers was interested in asking.

Bligh always knew the answer, for he had been told by Christian as soon as he was arrested. Charles Christian adds that when Fletcher was a boy, he was *'slow to be moved'*. On board *Bounty, 'Jealousy and Tyranny had produced Ill Usage to so great an Excess . . . and Revenge ensued as an Effervescence from the Opposition of good to bad Qualities.'*

'But who was it that had drove him into that unhappy state?'

Dr Sven Wahlroos was the first to explore Bligh's view that Fletcher was insane on the day of the mutiny had to be faced, after the first edition of this book was published. Dr Wahlroos' *Mutiny and Romance in the South Seas: A Companion to the Bounty Adventure (1989),* combines a chronological account of the mutiny events with an exhaustive encyclopaedia of every person and every place associated with it. A practising psychologist, Wahlroos was the first to use these professional skills to analyse the mutiny. He concludes that everything points to Christian suffering Borderline Personality Disorder. The clinical description, as defined by the American Psychiatric Association, is telling:

Interpersonal relationships are often intense and unstable with marked shifts of attitude over time. Frequently there is impulsive and unpredictable behaviour that is potentially physically self-damaging and the borderline person will often go from idealising a person to devaluing him. *

Clearly, the description might as well be of Bligh but we are focusing on Fletcher Christian. Wahlroos's professional opinion is that thus Christian suffered a *brief reactive psychosis.* He quotes the American Psychiatric Association again to support this:

The essential feature is the sudden onset of a psychotic disorder of at least a few hours but no more than two week's duration . . . suicidal or aggressive behaviour may be present . . . Individuals with Borderline Personality Disorders are thought to be particularly vulnerable . . . situations involving major stress predispose to development of this disorder. *

* Reproduced with permission from the Diagnostic and Statistical Manual of Mental Disorders, Third Edition. Revised. Copyright 1987 American Psychiatric Association.

Clinical psychologist Paul J. Rodriguez agrees with Dr Wahlroos' basic diagnosis and adds that many people diagnosed with Borderline Personality Disorder have commonly experienced childhood neglect, abuse and conflict, including the loss of one or both parents when young. Fletcher Christian's father died when he 3 ½ years old and he later lost his family home through a combination of mismanagement by his mother and profligacy by his older brothers, both of which can be construed as abandonment through neglect of his welfare. Thus, explains Rodriguez, if Fletcher felt he was abandoned by Bligh aboard *Bounty* or believed he was being neglected or demeaned, particularly by a man he once idolised, any inherent instability would have predisposed him to behave in extreme ways, especially during times of stress. On *Bounty* in late April 1789, there seemed to be stress everywhere Fletcher Christian turned and it was very clear that he had been abandoned, betrayed even, by Bligh. Rodriguez says: *Fletcher Christian would have responded to events on Bounty with feelings of deep emptiness and may have had considerable difficulty controlling his anger. He would have dramatically altered his attitude towards the person whom he believed to be abandoning him, from intense admiration to intense devaluation.*

When faced with considerable stress the sort of events which almost anyone would find difficult to endure a Borderline person may suddenly lose touch with reality. This is usually short-lived and can involve deep feelings of detachment from events and people, or grossly inappropriate behaviour.

Brief Reactive Psychosis, the diagnosis by Dr Wahlroos of Fletcher Christian's condition during the mutiny, is now known as Brief Psychotic Disorder with Marked Stressors, a mental state that commonly includes paranoid or grandiose delusions and irrational behaviour. The duration of this sudden-onset condition varies but the sufferer always returns to his or her previous pattern of behaviour. This all fits pretty well with what we know of Christian's behaviour but I shudder to think of how he felt when the Disorder did disappear and he realised what he had done. Was it hours, days or weeks after he arrested Bligh on April 28th?

If Fletcher Christian did suffer from this condition, I would expect a long-standing and inflexible pattern of instability in his relationships, emotions, self-image and control of his behaviour. Instead, every contemporary pre-*Bounty* account describes him as someone particularly affable, charming and socially sensitive, who went out of his way to understand others, especially those of lesser position.

The explanation of this dichotomy, says Rodriguez, is fascinating: *'Individuals with Borderline Personality Disorder may appear charismatic, which masks or makes acceptable behaviour which would otherwise have been judged unsuitable'.*

That is the final piece of the Fletcher Christian puzzle. The very characteristics that might today lead to a greater and earlier understanding of Fletcher Christian are precisely those that led writers and historians astray for over two centuries. His appealing personality and charisma were not signs of great confidence but clues to his fragility, very much making Fletcher neither black nor white but a mixture of both.

So, Fletcher Christian probably was insane on the day of the mutiny. His brother Charles question thus becomes more relevant than ever: *'But who was it that has drove him into that unhappy state?'*

The clear answer is that it was Bligh's treatment of Christian. Regardless of what Christian or Bligh said or did or achieved at any other time of their lives, this time it was their relationship that led to the tragedy. Christian said again and again he had been in Hell for weeks and no one did much to resist Christian when he took the ship from Bligh. Everyone knew that Christian had good reason to be in his state, even if they did not agree with it or fully understand it on the day.

The exact details of what went wrong between Captain Bligh and Fletcher Christian aboard *Bounty* cannot be known, except that men who were there thought the animosity between the two men was long-standing. To this we must add that for most of his naval life Bligh had a reputation for bullying and mentally torturing other men, even those he considered friends and that Christian was by every account a special target for such behaviour by a man who had once been seen by him as a hero and protector.

I simply cannot accept Bligh's protestations that because he had no inkling of the mutiny he could not be held to blame. As commander of *Bounty,* Bligh's duty was to have had that inkling and he lost his ship because of an essential personality defect, an inability to listen to others and to understand they might think differently from him. By offering sympathetic treatment to his former disciple when he needed it most, Bligh might have headed Christian off. Bligh was not interested in helping Christian or acknowledging that his deteriorating mental state, obvious to everyone else on board, had anything to do with him. Instead after sailing from Tahiti, Bligh worked Christian even harder and insulted him even more. Bligh's failure to recognise approaching danger from Christian is not the only time he suffered because of his inability to see another man's point of view. He expected others to think the same way he did and could not understand it when they did not.

These days it might be described as being in denial and he applied such selective thinking in his account of the mutiny, *A Voyage to the South Seas*, published in 1792, Bligh wrote that until the day of the mutiny he considered *`the voyage had advanced in a course of uninterrupted prosperity'*, which anyone who was there would say was delusional. His belief in himself as a faultless administrator and captain was misguided and sorely exposed before and after the mutiny. In Tahiti sails were found to have rotted and, although he blamed others, the responsibility as commanding officer was ultimately his. Much worse was the discovery during the mutiny that one of the ship's cutters was not seaworthy.

Here's what psychologist Dr Sven Wahlroos has to say about Bligh in *Mutiny and Romance in the South Seas*: *Bligh probably did not have any clear-cut mental or emotional illness but he did show prominent compulsive, narcissistic, histrionic and somewhat paranoid tendencies. A major part of his interpersonal problems lay in his almost total lack of understanding of the impact he had on others. His focus was always on himself . . . he felt he had nothing to do with the misfortunes which befell him during his life.*

Of all the many supporting sources for this view, including Beaglehole, Wahlroos quotes just one. In his 1976 book *Captain Bligh,* Richard Humble wrote *'a man who is pathologically unable to accept imperfection is a permanent martyr to himself: he has an enormous cross to bear . . . [Bligh] was not only unable to face up to this [making mistakes], he recoiled from the very idea. From this mental block sprang his tendency to arrogance and diversion of the blame on to others; whatever went wrong it could never be his fault. Inconsistent though he was, he was never inconsistent in this.'*

On April 28 1789 Fletcher Christian paid a terrible price for Bligh's self-deceit and inconsistencies. He had come aboard *Bounty* regarding Bligh as mentor and special friend but was driven temporarily insane because he was abandoned and humiliated in front of others.

Whether or not Fletcher realised he was out of his mind on April 28 1789, the consequences were the loss of everything he had gone to sea to recover, his family name and honour, his fortune and his future.

The result was exile, with men he is unlikely to have chosen and in a part of the world largely unknown.

CHAPTER 30

PURSUIT AND CAPTURE

When Fletcher Christian left Tahiti for the last time, Teina's brother Ariipaea quickly realised the value of *Bounty's* abandoned men and tried to entice as many as possible to Pare, the district of his clan and site of the second breadfruit camp. At first, only Muspratt, Hillbrant, Byrne, McIntosh and Norman were tempted by his offers of land. The others stayed closer to Point Venus. Morrison, who somehow became the acknowledged leader, lived with Poeno, chief of the district of Haapape, the other district that included areas of Matavai and its bay. So did Millward, who had a liaison with one of the chief's wives.

George Stewart and Peter Heywood stuck together, largely remaining aloof from much of what was to follow. They stayed close to Matavai with another chief and Stewart was married in Ma'ohi fashion to that chief's daughter. Heywood was conspicuously industrious about establishing himself as a Tahitian squire. He set up his own household at the foot of a small hill, the top of which afforded him a fine and useful lookout post. The garden and avenue to the house were well tended and planted with decorative and useful trees and plants, an amalgam of Tahitian vegetation and the English cottage garden.

Morrison quickly decided a boat should be built in which they could sail home. He was joined by everyone but Heywood and Stewart, who were prepared to wait for the man-of-war they knew would eventually be sent from England. On November 12 1789 the keel was laid for a 30 foot/9 metre vessel, which the men had to convince their Tahitian hosts was purely for pleasure and for sailing around the coast. If the Tahitians suspected the men wished to sail away, they would have done everything to hinder the construction for they wanted their fire-power for battle. For three months the Englishmen laboured incessantly, exciting the astonishment of the local men and women, not only by their craftsmanship but also by their application to toil for so long a period.

Then, after a Tahitian man had ignored a command to leave the boat builders alone, Thompson shot him and the child he was holding. Thompson quickly moved into the interior, accompanied by Churchill, who shortly afterwards became a chief when his blood-brother or taio died. His new-found status was of little advantage. Thompson discovered Churchill had arranged the theft of his personal muskets, so shot the white chief in the back. Although Churchill was unpopular and his theft of the muskets had been exposed by a Tahitian he had ill-treated, his murder had to be avenged by his subjects. Thompson was pinned to the ground with a branch of wood across his neck and his brains beaten out with stones.

With the two violent men gone, the boat builders worked in peace. By April 30, a year after the mutiny, the boat was fully planked and breadfruit gum was laboriously collected for caulking. On August 5 the vessel, complete with bowsprit, masts, booms and a rudder, was blessed by priests who threw the long leaves of plantain over her. It took half an hour of vigorous pushing to move her the three-quarters of a mile/1.2 kms to the sea. Then she was christened with cider they had made and named *Resolution*, a useful reminder to their hosts that they were associated with *Tute*, Captain James Cook.

The local matting didn't work well enough as sails and Morrison abandoned his attempt to sail *Resolution* to Batavia and on to England.

In September, the forces of Teina and Tu, encouraged by the way their European allies' firearms had carried the day in previous skirmishes, now planned more decisive military action. While engaging their enemy on land in the traditional way, they also attacked from the sea with a fleet of 40 canoes. Somehow, they had persuaded *Resolution* and its crew to join them. The Parean forces triumphed and Teina felt he could now move even further towards dominating all the districts of Tahiti.

One of the spoils of Teina's victory was a *maro ura*, a plaited belt of pandanus leaves decorated with red feathers, a symbol of power and position as significant as the English sovereign's sceptre and orb. Change of ownership whether by inheritance or force was accompanied by a ceremony in which lesser chiefs acknowledged their new allegiance. There were a number of such belts on Tahiti and only a few people swore fealty to each holder. The *maro ura* that the Teina clan won had belonged to a highly important family. This, combined with the courage given him by English musketry, convinced Teina to do something never before attempted. He sent his son Tu around the entire island so that every chieftain could pay homage to him, thereby acknowledging him as first paramount chief or king.

Undoubtedly flattered by the political and military importance with which they found themselves endowed, all the Englishmen attended the first of these ceremonies and to the *maro ura* was now added a new regal symbol of a feather-decorated Union Jack. The boy Tu, the belt and his flag progressed around the island with little trouble but less true submission. Teina proclaimed that all the island's chiefs should then come to his marae to acknowledge the supremacy of Prince Tu. It was unlikely to succeed but Teina wanted to identify those chiefs who were not his allies and friends.

On February 13 1791 the ceremony took place on a newly consecrated piece of ground. Tu was formally invested with the *maro ura.* Three human victims were among the offerings from the loyal island of Moorea. Their eyes were removed as the boy sat with his mouth open, a symbol, it is thought, of ancient cannibalistic rites. The rest of the loyal chiefs followed, bringing sacrifices of one or two bodies according to the size of their territories. The total was 30 corpses, some of which had been killed a month before.

The absence of any chief was taken to be mortally insulting to Tu and the *Bounty* men helped with plans to defeat them. Heywood, Stewart, Cole and Skinner refused to have anything to do with these machinations.

The plan was for all Teina's followers to gather for a celebration in the Papara district, close to that of their enemies. The festivities would hide the positioning of a large attacking force backed by *Resolution*. On March 24 a tremendous banquet was finishing when a ship called *Pandora* anchored in Matavai Bay.

The news quickly spread that she was the ship King George had sent to find Fletcher Christian, *Bounty* and her mutineers.

CHAPTER 31

OBLIVION

Those men who left *Bounty* after the Tubuai adventure were not the first white men to be turned off a European ship to live on Tahiti. Somewhere else on the island was a trouble maker called Brown, who had been left behind by *Mercury* a few months earlier. Fletcher Christian did not meet him. If he had, he would have learned that while he was building Fort George he once evaded discovery only through nightfall.

On the afternoon of August 9 1789, *Mercury*, commander John Henry Cox, sighted Tubuai 11 leagues/61 kms distant. By 8pm the ship was close enough to see lights on shore and fired two guns but had no reply. Having sailed too close to the reef, she had to bear off to avoid being wrecked and then continued on to Tahiti. In daylight she might have sailed around the island, seen *Bounty* and stopped to pay a social call. Fortunately, neither happened and at Tahiti the men on *Mercury* were confused by stories of *Bounty* having returned there under command of someone called *ra'atira Titreano*.

The longer Fletcher Christian stayed at Tahiti, the greater the risk of his discovery or of further revolt. The men he was leaving behind were bound to tell the truth this time and Tu or others might feel obliged to revenge their friend Bligh. At very least, Tahitians could overwhelm Christian's few followers to seize the vessel and its treasures and Christian did not know if he could truly rely on the men who had said they would stay with him.

For all his brave speeches about being left alone to run before the wind that Edward Christian says he made, Fletcher now had companions. His loyal fellow mutineers would want to sail quickly, fearing the ship would be taken from them. They would also want women and to get women to sail away on *Bounty* they would have to be kidnapped. Or that is what the men who write history about men have said.

If you know the reality of women's lives on Tahiti, it's astonishing Fletcher Christian wasn't overwhelmed by women who wanted to sail away on *Bounty*.

I reckon it's equally likely such women as Mauatua urged Fletcher to get on with sailing, before they too were forced from the ship or it was taken from him. Eventually he did sail precipitately and Bligh was later told that he did so because Mauatua discovered a Tahitian plot to take the ship and used this to spur Fletcher into action, wanting nothing to get in the way of her escape from Tahiti. It seems very likely.

Let's look again at what life was like for women on Tahiti. Once the facts are known, it's clearly not true that Fletcher Christian and his eight crew mates had to kidnap women to flee with them.

Upper-class Tahitians, *ari'i,* certainly gave women status, great status sometimes. For women of the other two classes, the landowners, *ra'atira,* and the workers, *manahune,* life was harsh. Brought up to have many of the practical fishing and gardening skills of boys and men, women were then banned from most of their pleasures.

Religion was exclusively for men yet dominated almost every aspect of women's abject lives. Women might only watch or listen from the outskirts to what happened on sacred *marae,* even if it was their son who had been clubbed to death by priests as human sacrifice. The priests' sacred rites, elaborate costumes, drumming, chanting, spells, blood-letting and privileges were only for men, just as women could only be onlookers in the equivalent ceremonies of Europe's Protestants and Catholics.

. . . it's astonishing Fletcher Christian wasn't overwhelmed by women who wanted to sail away on Bounty.

Dancing was for the entertainment of men and some included the exposure of women's most private parts. Then they would have to retire while men ate the roasted pork, the shark and other good things forbidden to women. Women were considered so impure that men did most of the open-fire and pit-cookery because food touched by women was *tapu,* unholy. Men would not wear tapa cloth if a woman had walked over it

Even motherhood was repressed by the priests, so it's likely that most children conceived in Tahiti did not live. William Bligh met women who had been pregnant with six and with eight children, none of whom survived. Children with blood that mixed any of the three classes were aborted, to keep social bloodlines pure. There were three methods of abortion but even a full-term baby was not guaranteed life. When a woman gave birth they squatted, usually held from behind by a man, even if noble enough to justify a special birthing hut decked with scented ferns and censed with fragrant smoke. The men were there to decide if the child was pure blooded enough, pale enough, long enough or perfectly formed enough.

Boys had to be long, strong boned and robust, so they were guaranteed to grow into useful non-dependent beings and then reliable warriors. To ensure that, a boy's head was strapped to a board so that the back of his skull was flattened and forced up into a point.

If a new-born was too dark-skinned, had any defect or was undersized, it would be smothered or have its head bashed in before it had taken its first breath and thus before he or she was considered to have 'joined us'.

Sometimes the murder was just because the new-born was a girl. Daughters were a huge burden for women, because they could eat only food gathered and cooked by their mothers until they left home to get married. Often it was mothers who agreed their baby would not live. Girls that lived had their faces deformed by massage when new-borns, so they had flat noses, and their fingertips were constantly rolled so they became extended and pointed.

Women with new babies were considered impure and so might have only to enter a family home through a new door or even live separately. They could also be divorced at whim. Thus, Tahitian women, facially deformed from birth, were prevented by abortion or murder from normal motherhood, were imprisoned by daughters, risked that any warrior son might be sacrificed by priests, were forbidden to eat or even touch most foods, were shared sexually by a husband's family and friends and prevented from taking part in religion although oppressed by it.

So, universally kidnapped? I don't think so. Grateful for the opportunity to be women on their own terms? I do think so. If you were an 18th-century Tahitian woman, might you not see Fletcher Christian and *Bounty* as a way of escaping its vicious, male-dominated culture, especially to sail away with pale-skinned men who were first thought to be Gods and who had metal and writing and who let you eat the same food as them and let you keep your babies?

Without Pitcairn's founding foremothers, *te tupuna vahine*, from Tahiti, Tubuai and Huahine, we'd know little about Fletcher Christian's mutiny. Bligh and *Bounty* would be barely mentioned and Pitcairn Island would not have been possible. Rather than being kidnapped, it seems more likely that Mauatua and Jenny, at least, actively encouraged Fletcher to sail suddenly and at night, before either Europeans or Tahitians tried to take back *Bounty* and frustrated their chance of freedom.

That these women positively wanted to achieve a new life and a new identity becomes clearer and clearer as we follow Fletcher Christian's search for obscurity. There was a big difference between him and the women. He wasn't following the sort of dream of freedom and identity that the women could formulate but was sentenced to make what he could of a future with no shape. I wonder when he and Mauatua first revealed their beliefs to one another.

During the latter part of the 20th century writers suggested that Fletcher and Mauatua were married by now. The first was Robert Nicholson in his 1965 book, *The Pitcairners*, although he offers no evidence to support this assertion and, when I subsequently met him, he told me he had written this because he thought this is what *might* have happened on Tahiti or Tubuai. In my view there was no marriage and if there had been a Ma'ohi ceremony, it is exactly the sort of occasion James Morrison would have recorded but he did not.

Early on the morning of September 23 1789, two days before he would turn 25, Fletcher Christian cut a cable, leaving a second anchor at Tahiti. *Bounty* slipped through the reef, carrying his eight companions, plus six other men, nineteen women and a baby girl, 35 shipmates in all. Many of *Bounty*'s passengers will have been surprised, as they had been invited on board for a farewell celebration and taken to bed believing the ship would be moving down to Pare the next day but, in the morning, they were outside the reef, sailing past the high peaks of Moorea. One woman immediately jumped overboard and swam for the reef. Others would have followed if they had had more courage. Late in the morning *Bounty* made the atoll of Tetiaroa, where some women jumped when they saw canoes coming that would rescue them.

Fletcher Christian possibly did mean to ignore any protest by the women aboard but things changed, perhaps through a combination of the noise made by the *`much afflicted'* women and because each of the white men had found a willing companion.

Fletcher turned *Bounty* to Moorea, where a canoe put out and six *'rather ancient'* women were rescued by canoe. Any other of the women aboard could have left at Moorea, so it's clear none was kidnapped, not one was forced to leave Tahiti and all they knew. Abandoning Tahiti was what these women most wanted. That's another charge disproved. Fletcher Christian was not a kidnapper.

Once Fletcher Christian put *Bounty* about and sailed away from the turreted twins of Tahiti and Moorea, neither he nor the ship would see them again.

The fates of *Bounty* and those aboard were unknown for almost 20 years.

Sunrise sailing from Tahiti, with the peaks of Moorea on the horizon.

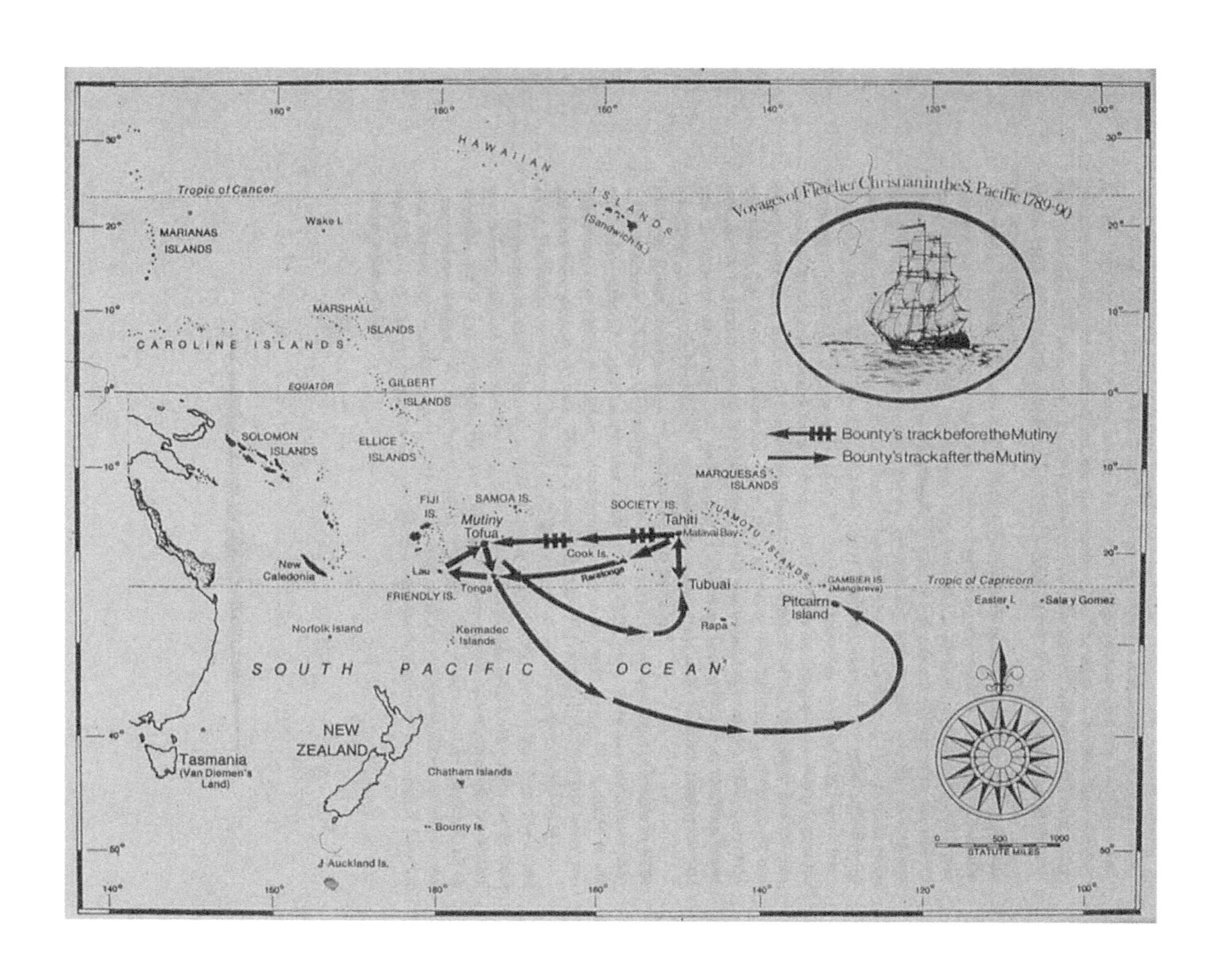

Voyages of Fletcher Christian in the S. Pacific 1789-90
HAWAIIAN ISLANDS
(Sandwich Is.)
Tropic of Cancer
MARIANAS ISLANDS
Wake I.
MARSHALL ISLANDS
CAROLINE ISLANDS
EQUATOR
GILBERT ISLANDS
SOLOMON ISLANDS
ELLICE ISLANDS
Bounty's track before the Mutiny
Bounty's track after the Mutiny
MARQUESAS ISLANDS
FIJI IS.
SAMOA IS.
SOCIETY IS.
Mutiny
Tofua
Tahiti
Matavai Bay
TUAMOTU ISLANDS
Cook Is.
Rarotonga
New Caledonia
Lau
Tonga
FRIENDLY IS.
Tubuai
GAMBIER IS. (Mangareva)
Tropic of Capricorn
Pitcairn Island
Easter I.
Sala y Gomez
Norfolk Island
Kermadec Islands
Rapa
SOUTH PACIFIC OCEAN
NEW ZEALAND
Tasmania
(Van Diemen's Land)
Chatham Islands
Bounty Is.
Auckland Is.
STATUTE MILES

REPERCUSSIONS III

September 23 1789 – January 15 1790
Fletcher's search for a home

January 15 1790 – February 6 1808
Fletcher's fragile Pitcairn paradise

Sunrise over Pitcairn Island, a common first view by voyagers, then and now

CHAPTER 32

PARADISE FOUND

Sailing away on *Bounty* was quite possibly the most thrilling thing imaginable for Mauatua and her companions. Suddenly they were part of a story every bit as fantastic as the legends they had been told around fires about the sailing voyages into unknown waters by their forefathers. Now they were asked to share the seeming paradise in which white men lived like gods on Earth. That there should be human frailties, pain and deceit in the lives of these gods would have been no surprise, not when you knew how Tahitian gods behaved

On that late September day the ship that sailed Fletcher Christian into oblivion would hardly be recognised as one of His Majesty King George's Royal Navy vessels. She bore no resemblance to the smart expedition vessel Bligh had boasted about in his letters to Banks almost two years before. There was only one truly seaworthy small boat, many of the sails had been cut up or given away, booms and spars had been lost at Tubuai. Long months under the tropical sun, thinking the ship would be abandoned, meant her decks needed recaulking. The planks would have shrunk. In storms or a heavy sea, water would drip through the deck, creating mould and stink below, adding to the constant assault of the smell of the animals and of the filthy bilge waters. Ahead was to be no South Pacific cruise of comfort and ease, indeed, considering the voyage *Bounty* made, her crew of mainly inexperienced Ma'ohi men and women had to work day and night under Fletcher's command as they learned to sail *Bounty* and to live at sea. In his search for a home Fletcher Christian guided *Bounty* over nearly 8,000 miles / 12,990 kms of Pacific waters, equivalent to sailing a third of the way around the globe.

Fletcher's *Bounty* crew was Edward Young, John Adams, William McCoy, Matthew Quintal, John Williams, Isaac Martin, John Mills and William Brown. Six Ma'ohi men were clearly on board through choice, too. The noble Tararo was from the sacred island of Raiatea and is said to be a relative of Mauatua. The Tubuaian Titahiti (previously Taaroamiva), younger brother of Chief Taaroa, was on board by choice, as was his friend Oha. Niau was a younger cousin of Tararo but Tahitian, as were Teimua and Manarii.

The route Fletcher sailed *Bounty* was pieced together only in 1958 and even now is so little known that books still show *Bounty* somehow taking four months to sail south-east from Tahiti to Pitcairn, a distance of 1,200 miles / 1930 kms.

Professor H. E. Maude of the Australian National University in

Canberra spent some years on Pitcairn Island during the Second World War and helped to set up its philatelic service. Using his knowledge of Pacific history to locate and interpret new sources of material, much of it records of oral tradition, he finally pieced together the extraordinary journey of *Bounty*, corroborating previously misunderstood or more generally ignored references made to islands and incidents by the Tahitian Jenny, when she told of Fletcher's search for a hiding place.

Remembering the conflicts of Tubuai and now that he had responsibilities for others, including Mauatua, Fletcher Christian had to find an island that was uninhabited, difficult of anchorage and well off the beaten track. The last is the simplest to achieve. The Polynesian triangle shaped by New Zealand, Hawai'i and Easter Island is 10 million square miles/25 million square kms, just slightly less than the continent of Africa and most of it was utterly unknown, even to those who lived there.

Fletcher Christian thought of sailing up to the Marquesas, hoping on the way he might find a suitable island that was close to Tahiti but realised this invited early discovery and instead looked for islands said to have been discovered by Spanish explorers. These were not where they were supposed to be and so he allowed the prevailing westerly winds to sail *Bounty* towards the Tongan archipelago, where he had mutinied five months before.

There are no reports of tension at this stage of the voyage but it must have been bizarre on board as both groups continued to learn each other's customs and language and to cooperate to sail the ship safely. All 12 women and six men from Tahiti, Tubuai, Raiatea and Huahine, some anxious about their bloodline status, had slowly to adopt a new attitude to taboos concerning each other, from accepting the idea of others being over their heads to how they ate. They were used to sitting far enough away from others so as not to hear any chewing sounds and so that they could freely swish their fly whisks. They also dined with small containers of fresh sea water into which they were used to dipping each mouthful. Containers used by men could not be touched by women and anyway men and women should have eaten separately, men first ideally. The confusion caused by relaxation of rules must have been challenging and led to many petty squabbles.

The Ma'ohi men probably had greater problems with the liberation of taboos on what fish and meat women could eat but the difference in personal hygiene must have very difficult for all to bear. Where were the running rivers to bathe in daily and to flush away their waste? Rather than scented breezes they had to endure the stink of *Bounty's* bilge.

I've always thought Christian's trek island to island must have relied equally on traditional navigational skills learned from his

Ma'ohi crew, listening to their centuries' old lore about waves, winds, clouds and stars. If not, he might never have found land at all.

Professor Maude deduced that Bounty passed through the Cook Islands. Several accounts of floating islands with rivers and taro swamps had been collected in 1814 and 1823 and led Maude to the conclusion that Fletcher Christian was the European discoverer of Rarotonga. Bounty introduced the orange to the island and the export of that fruit's juice became a 20th-century mainstay of its economy. There was bartering on Rarotonga and Bounty sailed on to an island Jenny calls Purutea, which seems to be what is now Mangaia. Jenny says a canoe came from the island bringing a pig and some coconuts. One of the canoeists came aboard and was delighted by the pearl-shell buttons on Fletcher Christian's jacket. Christian gave the jacket to the man, who then stood on the gunwale to show it to his friends, when one of the mutineers shot him for no reason and he fell into the sea. With loud lamentations the body was put into the canoe and the islanders who came bearing gifts paddled back to shore with only a dead companion as thanks. Christian was indignant but Jenny said, *'He could do nothing more, having lost all authority, than reprimand the murderer severely'*.

The context in which this statement was made and by whom must be very carefully considered. Jenny gave her interviews around 1818 and whatever her understanding of Europeans was later in her life, it cannot have been very great in 1789. To a Tahitian woman, authority meant autocracy and as Fletcher Christian did not flog the miscreant, or put him in irons the way he might have done even on Tubuai, he would undoubtedly appear to have lost authority in her eyes.

From a European view it might be considered differently. What should Fletcher have done? Killed the murderer? Locked him up? There were not enough capable people on board for him to have done either and I think to have done so would have been dangerous. Considering they were less than half way through their voyage and were anyway quite free of outside regulations, Fletcher had not lost all authority but was running the ship in a sensible pragmatic way, for the greater good.

With superior navigational skills supported by having one of the world's first chronometers on board and with experience as a master's mate that had trained him in best ship-management practice, as well as his professed delight in doing every task on a ship, no-one else was remotely as capable of sailing *Bounty* safely. The other mutineers would know this. Neither Jenny nor anyone else who was on board ever supplied other examples of lost authority. Fletcher was commanding in a way Tahitian Jenny did not understand.

Fletcher sailed *Bounty* on to Tongatabu, the main island of the

Friendly Islands and of today's Kingdom of Tonga. Jenny said they stayed two days, then continued further westwards until they were more westerly than Tofua. He sighted a small low island and proposed to stop there but a brief visit showed it was well populated and thus would not do but they stocked up on eggs, birds and coconuts. This island seems to have been Vatoa or Ono-i-Lau, in the Lau Group in the south of the Fijian islands. By now they had left what we call Polynesia and were venturing into Melanesia, where life was quite different and the inhabitants were almost certainly cannibals.

This marked the end of Fletcher's westward search for a home. It was two months since they had left Tahiti and they had been at sea almost all the time. The ship, although protected from worm by its expensive copper sheathing, would be deteriorating further. The animals and plants on board needed constant water and feed and without the meticulous hygiene and care of Bligh the ship was now more noisome and unhealthier than ever. After so long at sea without the inhibitions of dress required by their ex-captain, the Europeans had probably already adopted the free and comfortable Tahitian *pareu* or loin cloth or like many modern sailors on the Pacific, wore nothing at all.

Fletcher Christian's long searching in the books left in Bligh's cabin at last suggested a solution, somewhere called Pitcairn's Island, 2,000 miles/3200 kms to the east, whence they had come. To get there they would have to loop down towards the icy waters of the vast Southern Ocean before they could find hospitable winds that would guide them to the island.

Choosing Pitcairn and then convincing 27 companions to accompany him there required enormous courage by Fletcher Christian and tremendous faith in Christian by the others, once again dismissing ill-informed claims he was not firmly in command. Pitcairn meant well over a month more at sea and there would be no respite because he didn't expect they would encounter land.

It took even longer than Fletcher hoped. For two exhausting and dispiriting months he tacked *Bounty* into the chill teeth of the south-east trades. It's not surprising some of the dispirited party *'therefore thought of returning to Tahiti'*. Even Fletcher must have been tempted but sailing there would not have been any more pleasant. They were cold and bored and even the wines of Tenerife couldn't temper the fatigue and the monotony. There was no escape from the whine of the wind in the rigging, none from the wretched complaints of the spent timbers on the slow rhythm of the South Pacific. Neither could there be respite from duty, no refreshing uninterrupted sleep. Christian could never relax his vigilance. Sighting the sails of a faster ship, and most were faster, could signal the end of *Bounty*'s voyage.

Bounty headed deep into the cold southern reaches until she was far south-east of Pitcairn and could then turn north-west and up to the warmer waters in which Pitcairn was promised. Days after the dawning of the new year of 1790 there was sudden hope born of unexpected disappointment. When Fletcher Christian sailed *Bounty* to the spot where Pitcairn was promised to be, he found nothing but more water, further horizons of anguish and unkept promises to those on board, yet to Fletcher this was the most encouraging event for months.

The 1767 description he had read of Pitcairn by Carteret, whose 15-year old crew member Robert Pitcairn had been first to spot it, made the island sound perfect for Fletcher's needs. Isolated by unimaginable stretches of water, Pitcairn's smallness makes it even more difficult to find or stumble across than most islands. It has no protective lagoon, so is pounded by violent unhindered surf, which makes landing treacherous. It was said to look fertile, had running water and seemed uninhabited. But where was it?

Fletcher Christian realised at once that, as was common at the time, the island's position had been charted incorrectly. If he could find Pitcairn, he would have sailed *Bounty* and himself right off the face of the known earth and into the oblivion he wished.

He would be invisible.

Knowing it was the island's longitude, its position east or west of the Greenwich meridian, that Carteret had mistaken, Christian zigzagged eastwards along the line of latitude for 210 miles/330 kms. On the evening of January 15 1790, he sighted Pitcairn at last. It was pitifully small to be the object of such excitement, a lonely rock, like the discarded plaything of a forgetful Leviathan.

If there are South Pacific gods, they were determined to extract every measure of patience from Fletcher Christian. After a four-month voyage criss-crossing the South Pacific for almost 8,000 miles/12,875 kms there was to be no simple finale. For 48 hours he and *Bounty* were hurled about in a violent swell that made every thought of landing impossible. Yet, because they saw no signs of life on the thickly-clad peaks that loomed or on the lofty sea cliffs, each hour's delay made the island only more attractive.

A truly safe and calm day for landing at Pitcairn occurs only a few times in the life of most men. At last Fletcher Christian took advice from his Ma'ohi companions that conditions had subsided to the point where an attempt was worthwhile. The wind must have been blowing from the cold south or the east for they rowed through the surf on the west of the island. Here was the only possible alternative landing site to what became known as Bounty Bay on the eastern side and where, incidentally a new quay has now been built, which now makes it easier and more likely for visitors to land.

Christian, Brown, Williams and McCoy, together with three

Ma'ohi men, guided their boat through the sea to land on sharp but slippery boulders, perpetually drenched by surf. They pushed their way through the resistant undergrowth and began climbing the slopes, a glissade of rotting leaf and viscous red mud. For two days they fought their way around the towers and secret passages of this fortress and must have seen most of the island in their uncomfortable, tense exploration. Ancient paths, made by someone or some animal long gone, were overgrown, narrow and easily lost. The mid-summer air would be humid and enervating and there was the constant fear of angry animals or attack by inhabitants.

By the time Fletcher was back onto *Bounty*, he knew Pitcairn was everything that he could wish for. There were paper mulberry trees for making tapa cloth, candle-nut trees for light, pandanus palms for thatching. There was water. Fruit and vegetables grew wild and there were no animals, no mosquitoes. There were coconut trees in abundance and breadfruit, too. With the mangoes and plantains, oranges and sweet potatoes, hogs, goats and chickens aboard *Bounty*, Fletcher and his companions would live like tropical kings. With *Bounty's* cannon mounted atop the cliffs they would also be unassailable.

It was simple to follow the example of the unknown earlier inhabitants and choose to settle on the easier slopes that stretch a short distance along the north-east coast, beginning just above and to the west of the comparatively safer anchorage they would call Bounty Bay. The remains of old gardens had rich red earth, were well drained and had water close. Bounding one side was a high peak, from which a lookout could be kept.

Everything Fletcher Christian had identified as crucial to his survival was here. The island was solitary, almost impregnable and uninhabited. It was also fertile, comfortably more temperate than tropical and its whereabouts was unknown. Pitcairn signalled a new beginning for them all, far beyond practical expectations. It promised the return of all Fletcher cherished, all he had abandoned. Here on Pitcairn with Mauatua as his wife, with land, friends and freedom, he might live with dignity, even as a fugitive from the Royal Navy and Crown.

In January 1790, Pitcairn must truly have felt like Paradise found but it was a fragile Paradise and it would shatter.

In Fletcher and Mauatua's day this was called Pitcairn's Island and after they sighted it on January 15, 1790 it was to be 18 years before another ship found it.

PITCAIRN'S FOREFATHERS – January 1790

Fletcher Christian: Sailed as *Bounty's* Master's Mate, promoted to Lieutenant and Acting Second-in-command; led mutiny, April 28, 1789. Presumably murdered September 23 1793

John Adams: Able-bodied seaman. Died March 5 1829

William Brown: One of *Bounty*'s gardeners, had previously served as a lieutenant. Murdered September 23 1793

Isaac Martin: Able-bodied seaman. Murdered September 23 1793

William McCoy: Able-bodied seaman. Supposedly committed suicide when affected by alcohol distilled on Pitcairn

John Mills: Gunner's mate. Murdered September 23 1793

Matthew Quintal: Able-bodied seaman. Killed by Edward Young and John Adams, probably in late 1799

PITCAIRN'S FOREMOTHERS – *Te Tupuna Vahine* – January 1790

Mauatua/Mrs Christian: (Isabella, Mainmast, Maimiti) Tahitian consort of Fletcher Christian (two sons, one daughter), then of Edward Young (one son, two daughters). She remembered Cook's visits to Tahiti and is said to have left at least one child on Tahiti: see Tamahere in MRS CHRISTIAN – *BOUNTY* MUTINEER (Amazon). Older than Fletcher Christian but by how much is unknown. It's not known when Fletcher began to call her Isabella; Mainmast is a nickname given because of her very upright height and posture. She was never known as Maimiti during her life time as this was adopted by Nordhoff and Hall for their Bounty trilogy. See MRS CHRISTIAN – *BOUNTY* MUTINEER, Author's Notes and was then popularised by the 1936 movie starring Clark Gable and Charles Laughton

Faahotu (Fasto): Tahitian first consort of John Williams. Died childless 1790/91 of a disease of her neck, possibly scrophulous. Williams' demand for another woman started Pitcairn's long descent into strife

Mareva: Shared Tahitian consort of Manarii, Teimua and Niau. Later lived in Adams' household. No children. Died between 1808 and 1814

Puarai (Obuarei): Tahitian, first Pitcairn consort of John Adams. Childless. Fell from a cliff while gathering birds' eggs 1790/91, creating the second shortage of a wife amongst the mutineers

Sully (Sally): Arrived on Pitcairn as a baby, daughter of Teio and an unknown Tahitian father; married Mauatua's second son, Charles, known as Hoppa. Four sons, four daughters. Died March 7, 1826

Teatuahitea (Sarah): Tahitian consort of William Brown. Childless, she died between 1808 and 1814

Teehuteatuaonoa (Jenny): Tahitian consort of John Adams, then of Isaac Martin. No children. Only Ma'ohi woman to leave Pitcairn but not until 1817; death date uncertain

Teio (Mary): Tahitian, first consort of McCoy and later of John Adams, mother of his only son, George. Died March 14, 1829. Arrived on Pitcairn with a baby girl, Sully, who married Mauatua and Fletcher Christian's second son, Charles

John Williams: Able-bodied seaman. Murdered September 23 1793

Edward Young: Acting midshipman. Died of asthma, Christmas Day 1800

Manarii: Tahitian also known as Menalee. Killed by Quintal and McCoy when he hid with them after shooting Teimua

Niau: Tahitian and a younger cousin of Tararo. Shot by Edward Young

Oha: From Tubuai. Shot by Niau in 1790 or 1791

Tararo: Also known as Talaloo, a noble from Raiatea, possibly related to Mauatua. Shot by Niau in 1790 or 1791

Teimua: Tahitian. Shot by Manarii while playing his nose flute for Teraura sometime after Massacre Day

Titahiti: Originally Taaroamiva. From Tubuai, younger brother of Chief Taaroa. Killed by Teraura, then the wife of Edward Young, later wife of Thursday October

---

Teraura (Susannah): Youngest Tahitian woman, first consort of Edward 'Ned' Young, then of Matt Quintal (one son). Married Mauatua's oldest son Thursday October when she was over 30 and he was 16 (three sons, three daughters). Died July 15 1850, so was present in 1838 when Pitcairn passed laws giving women the vote and made education compulsory for girls

Tevarua (Sarah, Big Sully): Tahitian, first consort of Matt Quintal (two sons, two daughters). Died 1799

Tinafanea: From Tubuai, first the shared consort of Titahiti and Oha, then given to Adams. No children. Died between 1808 and 1814

Toofaiti (Nancy, Hutia, Toohaiti): From Huahine. She was first the consort of Tararo but then given to Williams. After Massacre Day she was one of Ned Young's consorts (three sons, one daughter). She died in the Tahiti flu epidemic June 9 1831

Vahineatua (Prudence, Balhadi, Praha Iti): Tahitian, first the consort of John Mills (one daughter, one son) and then of Adams (three daughters). The Pitcairn Island Register says she died on April 29 1831 in Papeete, another epidemic victim: see MRS CHRISTIAN – 'BOUNTY' MUTINEER, page 388

Because both of Mauatua and Fletcher Christian's sons married older Tahitian women, their grandchildren were three-quarters Tahitian, the only such on Pitcairn. When other first-generation Pitcairners (all half Ma'ohi, *half English) married one another, their children were still half-*Ma'ohi *and half-English but if they married a descendant of Fletcher and Mauatua, their children were 6/8ths Tahitian. Thus, Mauatua and Fletcher Christian's descendants were and are still more Tahitian than English if they married Pitcairn or Norfolk Island partners with no outside blood.*

My Christian great-grandfather Godfrey was also more what we now call Polynesian than European. He, my grandfather and my father all married 'out', thus explaining my lack of Tahitian colouring and features – and height!

CHAPTER 33

A PITCAIRN EPIPHANY

The early days of 1790 on Pitcairn, 1200 miles/1900 kms south east of Tahiti, were hard. January is by far the hottest month of the southern summer, and although the temperature on Pitcairn rarely exceeds 90°F/32C in the shade, this is burden enough when you are quickly relieving a vessel of everything movable, including animals, plants, provisions, fittings, tackle and, eventually, even its copper sheathing and timbers. As well, a tense two-way watch was being kept. On land, signs of habitation were apparent everywhere, not recent certainly, but some of the island was so densely forested, so rugged, there was no sure way of knowing who or what might be hiding.

They watched the sea too. While they were unloading the tattered ship, they were at their most vulnerable should they be spotted by another, for the masts and spars of a square-rigger are etched all too obviously against even the roughest lines of nature but to remove them too soon would make escape impossible.

Day after day the 27 men and women were fetching and carrying, filling and emptying the boat that threaded its way over surf and through the treacherous rocks it concealed to the tiny arc of beach. Teio's baby daughter, Sully, one of my great-great-great grandmothers, came ashore in a barrel.

It soon became apparent that a change in weather could prove as disastrous as the crashing sea, at best delaying them for a few days, at worst exposing them to the fury of a cyclone, for this was the season. The longer the ship rode the unredictable South Pacific surf, the greater the risks both of being spied and of losing her, so *Bounty* was run on to the rocks some way to the left of the beach, directly below a 700 foot/215 metre peaked cliff, they named Ship Landing Point. From here a narrow flat shelf gave way to the challenging half-hidden track of red earth that crawled perilously up the cliff-side. Even after this had been cleared of centuries of overgrowth, most men and women needed help to get up or down until new muscles and techniques developed but at the top there were rewards aplenty.

Shelters had quickly been made of palm fronds and *Bounty*'s sails, hidden in the trees well back from the cliff edge and, even with so much to do, it's easy to imagine the moments of leisure Fletcher, Mauatua and their companions snatched amid the anxiety and pressure. After so much deprivation since they left Tahiti on September 23, they could again enjoy true privacy, eat fresh fruits and vegetables, bathe, drink fresh coconut milk or water from the spring that gardener Brown discovered close to the camp. They named it Brown's Water. Pitcairn's seas were full of fish, the rocks encrusted with shellfish and alive with huge saltwater lobsters, the cliffs with sea birds and their eggs. Now that the animals could be fed, fattened and bred properly, they could regularly slaughter a pig or goat for a pit-oven feast. 28 mouths, even if one was an infant, could easily eat a whole animal at one sitting.

How much the traditions of Tahiti, Tubuai, Raiatea or Huahine dominated at this stage is unknown but, judging by Pitcairn in 1980, there was segregation of sexes at eating, even though the taboo that prevented men and women touching the same vessels and implements must have vanished aboard *Bounty*.

The attractions of solid ground and of nature's liberality were not satisfaction enough for everyone. On January 23, less than two weeks after they had arrived, *Bounty* was ablaze, her sun-bleached and warped timbers spurred to greater conflagration by the tar used on her decks and caulking. Was it an accident? This seems unlikely. Whoever fired the ship clearly waited for her to be emptied because no artefacts that would suggest otherwise have ever been recovered from the floor of Bounty Bay. Quintal is discredited for the fire, over-anxious about discovery and retribution, possibly because there was some discussion about preserving the vessel; Fletcher Christian is thought to be entertaining ideas of sailing it away so he could give himself up, something he thought would be honourable.

Fletcher's Brief Psychotic Disorder was well behind him but Pitcairn was the first time since the mutiny that he was free of the heavy burdens of commanding the ship, of responsibility for so many other men and women on Tubuai and at sea. Only now could he consider the consequences of taking *Bounty*, what the repercussions might be and what he should do about them.

Was it really Quintal who set fire to the ship? I no longer accept this as plausible. He always denied it and had just the sort of personality that would make him boast about doing it. If not Quintal, who? If *Bounty's* masts led to early discovery or if Fletcher Christian sailed away in her, inevitably revealing the Pitcairn settlement, *everyone* would be at risk. The eight other mutineers would be hunted down, arrested, tried and executed. The 12 women might be forced to return home, where they too would be punished and then bound again by men's restrictions. Tahiti was now relatively close, so this was well within possibility and there was no guarantee that any who sailed with Fletcher would be safe, at sea or from arrest on land.

The Pitcairners who would be most determined their new way of life was not put at risk were the Ma'ohi women. There are facts that prove this indisputably, facts overlooked for over two centuries of men retailing history as being only what men made it. It's an insight of which I am especially proud, because in an instant it changes everything ever said or thought about the Pitcairn settlement.

It escaped the notice of hundreds of *Bounty* writers and theorists that, after months of searching for a home, not one of the 12 Ma'ohi women was pregnant when they arrived at Pitcairn, something not normally possible. It's certainly not true that South Pacific women ovulate less and that this makes them less likely to conceive. That's merely a way to avoid the facts of abortion and infanticide.

So, no babies were expected when *Bounty* finally discovers Pitcairn. Then, the boss and his wife, Fletcher and Mauatua, produce Pitcairn's first child nine months after arrival, implying an accepted hierarchy, just as there had been on *Bounty*. Who was really in charge on Pitcairn and shaping its future? Quietly and secretly, it was the women.

Acting in concert, the women had taken control of their bodies, aborting any pregnancies until they knew they had a safe, secure and permanent home. That permanence was vital because Ma'ohi believed a person's spirit returned after death to the burial place of its placenta. What would you do with a placenta at sea or anywhere impermanent? It would also have been impractical to be pregnant if the ship were ever endangered, or if stranded on another island, inhabited or not, with a mixed-blood child at the breast. A child's paler skin was not associated with superiority everywhere.

For *Bounty's* women neither permanency nor pregnancy was possible on a ship wandering at sea. These women were determined they would escape the brutality of their previous lives and create the world's first permanent community that mixed their South Pacific blood with that of Europeans. These were prizes they would not give up passively and so I believe it was for their safer, freer future as women and mothers that they burned *Bounty*, just as they had urged Fletcher Christian to sail unannounced from Matavai Bay. No other group on Pitcairn was so committed to making it a safe haven for themselves and their children.

Abetted by constant sea breezes, the fire consumed *Bounty* down to the copper sheathing of her hull and easily burned through the heavy hawsers of rope that had steadied her to trees. She slowly moved off the rocks and then, sentenced by the injuries the rocks had made to her hull, sank remarkably close to land in water less than ten feet/three metres deep. The bones of *Bounty* were not rediscovered until a National Geographical expedition in 1956 and the twisting of the nails and sheets of copper recovered from the sea-bed show the brutality of the flames.

The finality of the fate of those now stranded on Pitcairn was now absolute. They were thousands of miles from anywhere, committed to doing what had only been spoken of, destined only to settle only on Pitcairn whose whereabouts was not known even to maps.,

The continued implementation of Fletcher Christian's ideas of democracy and equality were soon shattered. At an early stage, voting was used by the mutineers to divide most of the island between Fletcher Christian and his eight white companions. Each was allotted both a space to build and separate gardens. The six men from Tahiti, Tubuai, and Raiatea, two of them high-born, were given no land. By this one decision they became servants forever.

It would be unfair to lay the blame for this unfortunate decision only on Fletcher Christian. Now the party was on land his authority was not needed for their safety, so the other mutineers expected a greater say in things and had a majority anyway. An entirely new social order was emerging and dealing with it was difficult for a very young officer and a gentleman, who at sea had always known where he was with these men. As simple seamen they had never had much to lose because they never thought to own much but now saw new horizons. As the landed gentry of a new Pacific kingdom, they had everything to gain by questioning the authority and position of Fletcher Christian and could vote, as he had taught them to do.

The continued implementation of Fletcher Christian's ideas of democracy and equality were soon shattered.

McCoy and Quintal thought of the non-European men as servants from the start and were determined to use them to enhance lives of status and ease. Just as Fletcher was consciously or unconsciously creating for himself the estate denied him at home, his followers were establishing themselves as gentlemen of property, something barely possible and unlikely to have been tolerated in England. The idea of democracy and the equality of men taking root in America and in France had no place here. The taste of equality has always been sweetened by denying it to others and to most mutineers there was no precedent for black men to be equal with white. Certainly, Fletcher Christian would not have seen this in India and the Caribbean.

Mauatua must have become pregnant as soon as they made landfall and she gave birth to the first Pitcairner on a Thursday in October 1790. It was a son and Fletcher named him Thursday October, which might well be something he learned in the Caribbean, where African slaves refused to give their children traditional names until they were freed. It is a clue to the discomfort Fletcher felt.

By this time the settlement had a routine of sorts, with couples taking turns to keep watch from Lookout Point, just to the northwest of the village. Fires were lit only at night and the dogs had been killed, in case the smoke of one, or the noise of the other, advertised domestication where none should be.

Pitcairn was vaguely known to some of the Tahitians as *Hiti au*

Revareva, Land of the Passing Cloud, but was somewhere they thought no-one had visited for centuries. As the settlers explored and marked out gardens and animal pens, they found constant reminders of their predecessors, who it's now known had left some time before 1350. They discovered countless stone chisels and hatchets because the island had every type of material used to make these tools, including shiny black obsidian, which gives the sharpest edge of all. 20th-century researchers found remnants of tool-making workshops and reckoned Ma'ohi sailed huge distances from all over the Pacific to work tools here. A comprehensive and sometimes impressive collection of worked stone artefacts from Pitcairn Island is held but not displayed by *Tāmaki Paenga Hira,* Auckland War Memorial Museum.

The settlers found a marae platform featuring four stone images, each about six feet/1.8 metres high. The idols were crude representations of the human form, as were those also found on a much bigger marae site on the steep western side of the island. Any religious significance was discounted by the new Pitcairners, who used the stones as foundations for buildings, something I saw when I was there in 1980. There were other piles of rocks found inland, whose sea-pounded smoothness indicated they had been dragged up the slopes with great labour. Beneath these cairns there was always a skeleton, indeed skeletons turned up in all manner of places and some of them had a pearl shell under their skulls. There are no pearl oysters around Pitcairn so this probably indicates these men were perhaps from the island of Mangareva.

The three Ma'ohi women who had been allotted to the six men of their own race also wanted only a mixed-race community. There was never a full-blooded Ma'ohi child born on Pitcairn and this can only have been by design of women who wanted nothing of their male-dominated past. Even though every native-born Pitcairner had Ma'ohi blood, an abhorrence of Ma'ohi men became universal. The next generation was brought up to fear them, even though they had never seen what Pitcairners always called black men.

The birth of a son and heir must have reminded Fletcher Christian of the family forbidden him forever, even by correspondence. Was he by now publicly branded a murderer? And if so, of how many men? To his own knowledge, well over 60 Pacific Islanders had been killed as he searched for a home. It might be possible to excuse the battle deaths of these but what about Bligh and the others? Were they alive or dead? Perhaps some of them had survived to tell the world and his family about his mutiny. Such mental agonies, for which there was no resolution, must have been deeply distressing for him and ample reasons for black depression.

There had never been a community like Pitcairn Island before and each day meant either adjusting to the problems or hiding from them, both courses that required effort. The work involved in supporting the community by building, fencing, fishing and farming initially prevented most quarrels on Pitcairn. The six black men worked land for the white men and appeared largely to acquiesce, even though beaten with knotted ropes and threatened with guns. Quintal and McCoy fathered sons, and Mills fathered Pitcairn's first daughter; called Elizabeth she is another of my great-great-great grandmothers.

In 1792, Mauatua bore another son, Charles, named for Fletcher's father and for the brother he met on his last night in Portsmouth. Charles's birth was a Pitcairn Epiphany, an extraordinary event, the resonance of which has been overlooked for centuries, including by me, even though Charles is one of my great-great-great grandfathers.

Charles was born with a club foot, seemingly common in South Pacific islands but Fletcher Christian would never have seen it in Tahiti. Club-footed babies had their skulls crushed before their first breath. When Mauatua decided in a few heartbeats that Charles would live, she traduced and dismissed centuries of priestly domination of women. She rejected all she had been taught was proper for a woman, then to tread a path no Tahitian woman had ever known. Charles being allowed to live with a club foot demonstrated the depth of the revolutionary thinking that Fletcher Christian and *Bounty* had released in Pitcairn's women and gave notice of what every one of its men would have to face.

If he had not known before, Fletcher would now know that the woman he had chosen, or by whom he was chosen, was very special indeed.

Mauatua's decision to let her second son live put every man and woman on notice that Pitcairn was different from anything they knew or expected.

CHAPTER 34

SEPTEMBER 20 1793: MASSACRE DAY

The best accounts of the events on Pitcairn Island between 1790 and 1808, when Fletcher Christian's hideout was again discovered, would be those given by people involved. This effectively rules out all but the adults who arrived on board *Bounty* as most other inhabitants of the island were born after the main turbulence. The oldest children alive on and around Massacre Day were Sully, who was possibly four and Thursday who was about to turn three.

John Adams was the only surviving man in 1808 and said he had been born at Stanford Hill in the Parish of St John, Hackney, of poor but honest parents. His father was drowned in the Thames leaving John and three other children orphans. He was brought up in the poorhouse, where he received rudiments of education and religion and was in his early 40s when the American sealer *Topaz,* Captain Mayhew Folger, found him on Pitcairn. A 20th-century theory said that Adams was Irish and a runaway from trouble in his native village. I have seen evidence that a 'Jack' Adams did go to sea but there is nothing to link him with the Bounty one. An article in the Mariners' Mirror later seemed to show it more likely that the runaway Irishman 'Jack' was the father of the Pitcairner. John Adams had signed on to Bounty as Alexander Smith, and was known to Folger under this name. He may have used an alias because he had deserted another ship but when Staines and Pipon called he used his real name.

Adams was in a real dilemma when Pitcairn was found in 1808. 20 years had passed since he had sailed from England. Now he was a suspected mutineer with new scars, tattoos and wearing beaten bark as clothing, yet men hung on his every word. How much did the visitors know about events aboard *Bounty* and his part in them? What had happened to Bligh and the men in the open boat? Was he party to mass murder as well as mutiny? Was what had happened on Pitcairn likely to be judged under English law? In short, was he liable to arrest and punishment?

He believed that as he was now a practicing Christian and effective pastor of the island, he had atoned for any crime, protesting he was innocent of complicity in the mutiny because he was in his hammock at the time, even though those early visitors probably knew better. Being the only Pitcairner who knew about the outside world, he monopolised and was monopolised by the visitors. That way he knew what was said and to whom and could cunningly give the answers that men expected.

Adams could not be everywhere and some visitors spoke to the women. They managed some general corroboration but the women were forbidden to speak Tahitian and didn't speak English, using instead a confusing mix of 18th-century English and Tahitian. Their children spoke better English but merely repeated they had been told. *Ha'avere* still ruled. This was one of the customs brought to Pitcairn by the Ma'ohi women, a part-serious, part playful habit of lying for the sake of amusement or for deflection of harmful truths. On Pitcairn it was not to enliven boredom as it had been on Tahiti. It was to bury all the women had done to protect their children. There can be no question that Mrs Christian, Mauatua, ensured the game was played for maximum blame on the black men and as little possible mention of Fletcher Christian.

It took Adams and his flock 17 years fully to believe he would not be arrested and taken off the island, something dreaded by the community he led. By the time Captain Beechey arrived aboard HMS *Blossom* in 1825, Adams was certain that any cross-examining about his past was academic rather than incriminatory and began to give fuller and more accurate accounts, while drawing a veil over his own involvement in the horrors that emerged.

The other eye-witness accounts are by Jenny, who had first been with Adams during the Tubuaian experiment and then gone with *Bounty* to Pitcairn as partner of Isaac Martin. She was always fiercely independent and a leader of the women but she had never been a mother and left Pitcairn in 1817 aboard *Sultan*. Subsequently, she talked to a Captain Peter Dillon who translated her account from the jumble of Tahitian and English in which it had been given. The details she remembers are about people rather than dates and places, and thus are all the more believable. They helped me discard much generally accepted history as fable.

Apart from the educated Christian and Young, the rest of the white settlers of Pitcairn Island were from the working classes. Their island refuge provided expectations far beyond those to which they had been born. The pursuit of a self-indulgent way of life while snubbing justice is why they had followed Christian in the first place. The slightest diminution of their new status was likely to spark violent protest.

There is conflicting opinion about the behaviour of Fletcher Christian on Pitcairn. He is either morose and brooding or a happy, active, natural leader. Both pictures were given by Adams. The former was probably what the upright and sermonising 19th-century interviewers wished to hear but Fletcher Christian was a believer in deeds, not words. A busy organiser, digging his garden, building his house, delighting in physical activity of any kind is the more likely picture. Yet Fletcher was human and he had done something quite spectacular. He would have been less than human if he had not sometimes yearned for his home and the freedom he had before he became an outlaw. At such brooding moments, he was said to retire to a cave high in the sheer side of Lookout Point. There he had erected a small watch house and kept a store of provisions. Beechey thought it so difficult to access that a single occupant could hold off a party whatever their size as long as there was a supply of ammunition.

One other source of reliable Pitcairn stories, although second-hand, is John Buffet, who settled on Pitcairn in 1825. He became much respected and trusted and his stories were told when Adams' editing of the truth was no longer possible, so should be given much credence. One of Buffet's most important revelations exposes another possible reason for Adams' later exaggerated loyalty to Fletcher Christian, to whom he always referred as Mr Christian. Buffet was told Adams refused to mend part of his fence through which free-ranging hogs might have damaged Fletcher Christian's garden; their properties abutted. Fletcher warned him that if any pigs did get through the fence, they would be shot. *'Then I will shoot you,'* Adams replied. Immediately the others - not Christian, you will note - bound Adams and sentenced him to be set adrift on the ocean, which would have meant certain death. The story Buffet told said that Fletcher Christian intervened and saved Adams' life.

The real problems of Pitcairn were not petty squabbles about hogs and fences. A struggle for supremacy fermented between the white men and between the white and black men. It was the black women who nurtured Pitcairn Island. The black men were all quickly dead and even if they had passed their knowledge on to those white men who survived the first massacre, the European survivors were commonly drunk or determined to be idle kings of their new castles. It was as well Tahitian girls were brought up the same as boys, taught how to grow and harvest and fish and carve, to swim and to surf on boards, an activity that astonished early visitors. Only Pitcairn's women can have known how best to grow and tend and harvest and fish on a South Pacific island.

When they landed there were 15 men and 12 women; six of the men were black and nine were white. It meant that if each of the Europeans had a wife, the six Ma'ohi men had to share three Ma'ohi women. In fact, five of these men shared two women because Tararo was allotted one of his own, reflecting his high rank. Although sharing women was something Tahitians might ordinarily have done by choice, here it was a further underlining of the lowly status forced on these men. Not owning land was particularly lowering to the two nobles amongst them, Tararo from Raiatea and Titahiti, brother of Chief Taaroa of Tubuai. Slowly their position as nobles, friends and accomplices was eroded. They became slaves. It seems even Pitcairn's women regarded them as second best for not one ever gave birth to a child of their six Ma'ohi partners, a particularly important illustration of how little these women wanted of the men who represented their past lives.

The early days of gardening and building must have provided some sense of coherence, but not easily. Ma'ohi men and women grew up learning about fishing and about gardening for food and it was they who collectively knew where each crop would do best, which banana variety did well up a hill or better on a plain, which fish was safe to eat, which caused hallucinations, even how to make the best fishing lines and hooks. Brown was a professional gardener and Fletcher had been brought up on a farm; Young may or may not have got his hands dirty as a young man in the Caribbean. The rich and fertile soil of Pitcairn would quickly have rewarded them for planting and watering and caring, yet none of the other mutineers was from backgrounds that would include horticulture; anyway, they expected servants and slaves to do everything.

Whatever Pitcairn's foremothers did and learned, the most difficult would be how to handle men who drank alcohol, too generously and too often.

The women had much to learn, too. Whereas Ma'ohi men back home did most of the cooking in pits or by roasting animals and fish in front of open fires, here they had to cook in metal *over* fires while learning to disregard the *tapu* that had restricted them touching men's cooking and eating implements. That would have begun on board *Bounty*, where they would also have been introduced to baking in closed ovens.

Whatever Pitcairn's foremothers did and learned, the most difficult would be how to handle men who drank alcohol, too generously and too often. Unlike the *'ava* they had known on their islands, alcohol changed men's moods, could make them aggressive and brutal. It's not surprising many women still did not want children, yet they continued to assert independence where they could.

These women's determination to make Pitcairn their own is clearly shown by what they abandoned of Tahiti, Huahine and Tubuai. Infanticide was no longer endemic. Baby girls no longer had their face massaged to make flat noses and boys did not have their skulls flattened and pointed. Dancing except for the simplest kinds died out as did all food restrictions. There was no tattooing, mothers did not become slaves to daughters or fear the long hoots of conch shell trumpets that might mean a son would be beaten to death as a sacrifice for priests who forbade women to take part in religion.

What they did with *tapa* cloth is the most telling example of their determination to break with the past. On Pitcairn it was once known as *ahu,* a term seemingly not used elsewhere. Tapa of various colours and of qualities that could range from gossamer to tough floor coverings could be made from the bark of different trees, including breadfruit. Brought up to copy the patterns their mothers and grandmothers and countless other generations had always used, Pitcairn's women universally turned their back on these. Instead, they created patterns of their own, sometimes clearly trying to copy designs of Georgian fabrics presumably carried aboard *Bounty*. This dismissal of powerful feminine traditions of their pasts demonstrates, just as much as allowing a club-footed boy to live, a powerful and united determination to create new identities and futures.

Lengths of Pitcairn's *tapa* were given or sold to visitors with such ease and in such quantity that it's difficult to believe they held sacred or spiritual meaning to Pitcairn's women, dead or alive. The founding foremothers, *te tupuna vahine,* who first made them had fully embraced the Christian religion well before the island was rediscovered and this alone would have negated any otherworldly Ma'ohi connection tapa had to them. The greatest collection is held hidden by the British Museum and there are many who believe their unique designs to be a valuable resource that could be used to raise independent funds for the island. At very least these unrivalled examples of Pitcairn's unique heritage should be displayed, like the stone tools held at *Tāmaki Paenga Hira,* Auckland War Memorial Museum.

There are emotively held views that Pitcairn's early tapa were and are sacred to those settler women and their daughters and that their designs should never be used commercially to the benefit of Pitcairn. So far I have seen no evidence to support this other than opinion, some it from eminent museum curators. I'd welcome being proven wrong because facts are facts and it would be foolish to deny them.

The most poisonous of the several serpents in Pitcairn's apparent Paradise was the determined pursuit of guaranteed sexual intercourse, which soon exacerbated the imbalance between the black and the white men. With a year of arrival, Jack Williams' consort Faahotu died of what seemed to be a scrophulous eruption on her neck. Williams demanded one of the women shared by the Ma'ohi but Fletcher blocked this, suggesting Williams wait for Sully, the infant who had come ashore in a barrel. She was possibly as old as three but Williams was not prepared to wait, even if she was given to him before her puberty.

The white men capitulated and lots were cast for the three 'spare' women shared by the Ma'ohi men. The die chose Toofaiti from Huahine, known as Nancy, who welcomed the outcome. It was the worst possible result because she was the exclusive consort of Tararo. Tararo was deeply wounded that Toofaiti preferred the increased status of having a white husband and weeping with anger and humiliation he took to the hills. Three days later he returned and by force or sweet words took Nancy back and they set up home in a cave below a cliff, still known as Talaloo's Ridge.

Next to lose his wife was John Adams. Puarai was killed when she fell while gathering eggs from a Pitcairn cliff. He was 'given'

Tinafanea, who had been shared between Oha and Titahiti. Three of the six Ma'ohi men had had wives taken from them. The other three, including Hitihiti, now had one between them. Like Fletcher they had been pushed too far and were going to do something about it.

The women knew this but curiously are said to have let their white husbands know only by incorporating clues into the songs they extemporised as they went about their daily tasks: *'Why does black man sharpen axe? To kill white men.'* Fletcher seized a musket and went to find the conspirators, first heading to the large house at The Edge above Bounty Bay, where five of the six Ma'ohi men once lived communally. Oha was the first he met and Fletcher discharged his gun straight at him but it had been loaded only with powder, as a deterrent; Fletcher was not willing to jeopardise his position by killing in anger. Oha misunderstood and mocking Christian's bad aim ran to join Tararo and Nancy in their hideout.

When the four remaining Ma'ohi plotters realised their plan was discovered they quickly extracted a promise of forgiveness from Fletcher and the others in return for an act of treachery. They agreed to kill Tararo and Oha.

First, there was an ineffectual attempt at poisoning. Anxious to break the tension so all could sleep easily, Christian ordered the men to be shot. He chose Niau, the youngest of the men, to do this, saying that if he failed, he would be shot. Once Niau succeeded, Pitcairn was thought safe once more. Nancy returned to live with Williams and a couple of years of relative tranquillity ensued.

In late September 1793 there were five children. The Christians had two boys, McCoy and Quintal each had a son and Mills a daughter. Fletcher's wife Mauatua was due to have another baby shortly and Mills's wife was pregnant again.

The widely accepted narrative blames Quintal and McCoy for pushing the four surviving Ma'ohi to murder because both were cruel and thoughtless towards the men they treated as slaves. If Quintal's man did not prepare his food sufficiently quickly or well, he was severely flogged. When the man was bleeding and whimpering, unable to defend himself, Quintal would rub brine into the lacerations. McCoy was just as free with the clout and lash.

On September 20th 1793 the four black men had had enough and did something about it. That date is still remembered on Pitcairn as Massacre Day.

Posing as a thoughtful servant, Tetahiti borrowed a gun from Martin saying he wanted to shoot a pig for the white men's dinner. Teimua and Niau joined him.

Nine of the ten women left the village to collect eggs in the hills and their European husbands were working in the plantations that were part of their properties. Fletcher worked close to their house at the request of Mauatua, who had not gone with the other women because she expected the birth of their third child.

Williams was the first victim of the black men. When he heard the shot, Martin exclaimed they could all expect a glorious feast that night, believing it was meat and not man that had fallen. The three murderers then asked if Manarii, who was working with Mills in his garden, could help them carry the animal they had supposedly shot. Now a quartet, they moved to the yam plantation where Fletcher Christian was working. As he struggled to remove roots from some newly-cleared ground he was shot from behind and fell down. To finish the job, he was disfigured about the head with an axe, and left for dead. As he was shot or as he lay bleeding in the freshly-turned red earth, Fletcher Christian groaned loudly. McCoy recognised it as the cry of a dying man but Mills contradicted him, saying it was only Mauatua calling her children. Reassured, McCoy went back to work with Mills.

The Ma'ohi were still outnumbered and feared meeting the Europeans if they discovered the truth and banded together against them, so they separated McCoy and Mills. Teimua and Niau hid in McCoy's house and Tetahiti told him that his house was being ransacked. When he burst in, he was shot at twice but neither bullets hit him. He escaped to raise the alarm but he ran into Quintal and the two ran to hide in the bush.

The two would-be assassins of McCoy then shot Mills but as he was only wounded they chopped of his head. Martin was shot next but managed to get to Brown's house. He fell after a second shot and was beaten over the head with a hammer by Manarii until he was still. Teimua had a good relationship with Brown and supposedly shot him with a blank charge, telling him then to pretend to be dead once he fell in his garden. He moved too soon and Manarii beat his brains into Pitcairn's already red earth. Five white men were now dead and the numbers were even. One or two more deaths would make masters of the Ma'ohi.

Adams had been warned of the troubles by Quintal's wife Tevarua and took precautions to secrete himself in the bush with a supply of provisions. He exposed himself precipitately and was shot in the shoulder. As he fell, he was attacked with the butt of the gun and broke two fingers warding off the murderous thrusts. The gun was put to his side but misfired twice. Shocked but stimulated by fear, Adams leapt to his feet and found enough strength to outstrip his would-be assassins.

Finding themselves unlikely to prevail, his pursuers offered Adams protection. Exhausted, Adams accepted and was helped to Mauatua's house, where most women had sheltered after returning to the village. Adams was not the only one the women championed. Young was said always to have been a special favourite of the women and protected from the start. He has equally been suspected of helping plan the massacre, just as he has been blamed for suggesting Fletcher's mutiny, because one of the black men pursuing Adams is supposed to have apologised, saying he had forgotten Young had said Adams was not to be harmed.

When a truce was agreed, there were plenty of women for the remaining eight men. Yet plenitude brings its own complications. The Ma'ohi quarrelled as to who would have which widow. They doubtless had competition from Young and Adams, too. The most troublesome of the victors was the Tahitian Manarii, a man clearly comfortable with murder.

About a week later, he shot Teimua at point-blank range while he played his nose-flute for Young's wife Teraura, Susannah. He had to shoot twice because the first didn't kill Teimua. Susannah had been fond of the murdered musician and as the youngest of the adult women was a keen prize for all the men. Tetahiti made a competitive point of consoling Teraura, so jealous Manarii attacked again. This time the women intervened to protest against his brutality and Tetahiti survived.

Fearful of retribution from black and white alike, Manarii escaped into the mountains to form an unlikely alliance with McCoy and Quintal, who were still too frightened to reappear. There's a thought that both these men should have been the first targets and their survival must have been a blow to whoever planned the killings.

Who did plan the killings? Why did Tevarua tell Adams about it and why was Young's name mentioned by one of the Ma'ohi men? It's hardly likely the black men would have shared their plans yet some of the women clearly knew them.

My view that the women were responsible, wanting to get rid of the drunken white men while plotting together to save Young and Adams, the men who behaved most reasonably and were the least threat to them and to their children. So why was Fletcher Christian murdered? There are no stories of alcohol-fuelled brutality by Fletcher Christian that would justify him being included in the slaughter.

It should be remembered that Mauatua was alive when Pitcairn was discovered in 1808 and would easily have controlled what was said, particularly to protect the ears of the next generation, who had no idea who had spilled the blood that eventually led to their idyllic lives. I do have a developed theory about the reason for Fletcher's inclusion, which is explored in MRS CHRISTIAN *BOUNTY* MUTINEER.

With McCoy and Quintal still alive, no one had won and from now on lives were bargained and spent recklessly. First either Adams or the women got a message to McCoy and Quintal that there would be truce if they disposed of Manarii. They shot him and delivered his severed hands into the village at night to prove it.

Anxious to have the treachery and fear over with, Jenny agreed to take Tetahiti to bed. While they were together, Teraura, still only about 18 years old, attacked him with an axe. According to some stories the slash to his throat was not enough and she had to split his skull with a further blow of the axe. Outside, Young was showing Niau some of the finer points of loading and firing a musket but then shot him.

McCoy and Quintal wanted proof the last of the black men were dead and so body parts of Tetahiti and Niau were sent to them. They agreed to return to the village on October 3rd. Nine men had been slaughtered, five white men and the last four Ma'ohi men. There would never be black children born on Pitcairn.

On the day of the first massacre, a girl had been born to Mauatua and called Mary Ann. Not long after her birth, John Mills II was born. Now there were seven children on Pitcairn and the oldest, Thursday October Christian, turned three a few days after the murders ceased. The situation was now reversed. The Ma'ohi women could choose their men. Adams and Young were each joined by three women, McCoy and Quintal each had two.

In December of the same year, Young began a journal, which only Captain Beechey reported seeing. Most of the details are domestic and banal. Houses were rebuilt for the newly expanded families, new divisions of land were fenced off and gardens were protected from the pigs by digging pits to trap them. It must have been hard work with only four men, none of whom had proven their interest in labour. The women weren't always available because in the next five years 13 children were born. Several of the women were almost constantly pregnant, a definite change from the first years.

If there was any discontent, it appears to have been among the women, some of whom lived promiscuously with the men and changed their abodes regularly. Judging by Young's journal, the changes may have been accounted for by unhappiness and ill-treatment rather than sexual boredom. In fact, Pitcairn Island was now a distinctly unpleasant place to be. The bodies of those who had been murdered remained where they were killed but some had had their heads removed by the women.

On March 12 1794, Young wrote that he had seen Jenny with a skull in her hand, and discovered it was that of Jack Williams. When he insisted it should be buried, the women with Jenny refused, asking him why he wanted such a thing when the other white men did not. Young conferred with the other three and said he thought *that 'if the girls did not agree to give up the heads of the five white men in a peaceable manner, they ought to be taken by force and buried'*. There was no way he could be certain one was not that of a

Ma'ohi. What happened to the five skulls, whoever they had once belonged to, is not known. The incident exasperated some of the women so much it changed all previous resolves. They wanted to leave Pitcairn. On April 14, they were so urgent in their demands that the men began building them a boat. The childless Jenny tore off the planks of her house for the craft and endeavoured to persuade her countrywomen to do the same. It was astonishing that the men should cooperate with such a scheme but in mid-April, the vessel was launched. *'According to expectation'* it upset and although saved from a certain death during their attempt to return to Tahiti, the women became even more despondent and dissatisfied. After this, the defeated escapees were treated badly and frequently beaten by Quintal and McCoy; Quintal bit off most of Tevarua's ear when she returned from fishing with a catch he thought too little.

The day after the abortive escape, the bones of the murdered people were gathered and buried. The impression given in most accounts is that Fletcher Christian, at least, was buried in his own garden, close to where he fell. Although there is a definite Ma'ohi tendency to forget someone once he is dead, it is, nevertheless, an appalling thought that the fathers of the foremothers' young children, lay rotting in the sultry summer of 1793-94.

Quintal was showing signs of mental distress and had seriously proposed that the men should not 'laugh, joke, or give anything to any of the girls', yet on October 3, he gave a party to celebrate the death of the black men a year before.

Only a month later the women were desperate enough to plan to kill the four men as they slept. Once found out, the women were not punished but the men agreed the first woman to misbehave in the future would be instantly put to death. So would each subsequent offender until the *'real intentions of the women'* could be discovered.

Talk meant nothing to the unhappy women, who made a physical attack on the men on November 30. More threats were made but now the women seemed to have the upper hand. Whenever they were displeased they collected their children and some firearms and hid in a remote or fortified part of the island until it pleased them to return. By such mercurial behaviour the women were able to keep their men in a constant state of suspense. More important, they and their children were safe from physical abuse.

Nevertheless, Quintal fathered five children and Young, having had four by Nancy, then had three by Fletcher Christian's widow. McCoy only managed one further child and Adams, after a slow start, fathered four.

By 1796 there was a more sociable atmosphere, with the men entertaining one another in their houses and making life a little more comfortable for their women, possibly because by now there was little or none of *Bounty's* wine and rum remaining. That was not to last. McCoy had once worked in a Glasgow brewery and on April 20 1797 he finally succeeded in distilling raw alcohol from the sweet syrup of the ti-root, *cordyline terminalis*. Drunkenness was once again added to promiscuity.

The raw spirits inflamed the fragile minds of McCoy and Quintal. By the end of the year McCoy lost control totally and threw himself off the cliffs just below Christian's house. By 1799 Quintal was threatening to kill Fletcher Christian's children unless he could have Mauatua as his wife and that same year Tevarua fell or, more probably, jumped from a cliff. The final solution was as bloody as all that had preceded. After Quintal threatened them too, Young and Adams attacked and killed him with axes. Now, there were two white men left.

This is broadly what is believed to be the history of Pitcairn's early days but it's a history collected by men and written about by men, who would have little regard for women, especially men who spent so long at sea that they hardly knew women. In the early 1800s, Queen Victoria's reign was decades away. Jane Austen's books that featured strong-minded young women were published anonymously between October 1811 and December 1815, a world away from that of anyone sailing in the South Pacific. Women simply didn't count in the lives of Pitcairn's early visitors from Europe and America.

It is this misogynistic mind-set against which we must assess early stories of Pitcairn, seeing them as yet more history written by men about men. I believe Pitcairn's women actively encouraged the death of most men, partly for their own relief but mainly for the safety and future happiness of their children. What emotion is fiercer in any community?

Isn't it suspicious that women sang songs rather than telling such as Fletcher or others that they knew the Ma'ohi men were plotting murder? This is classic *'it wasn't me'* strategy, used later somehow to absolve them of any blame. And how about them all being absent from the village on Massacre Day? That made it easier for the black men to kill and for the women to claim they knew nothing about it because they were not there. Once you stop accepting men's versions of history and start asking about the women who were there, Pitcairn's past, including Massacre Day, can be caste entirely differently. The true story of the long-invisible blood of Pitcairn's foremothers is overdue for telling, something I hope I have done in MRS CHRISTIAN – 'BOUNTY' MUTINEER.

These were women who had gone on *Bounty* to be independent women and mothers, who sought a life of fulfilment rather than one truncated by men. Many endured serious physical and mental abuse on Pitcairn, yet nurtured a gentle community of mixed

European/South Seas blood, on an island marooned in scarcely imaginable tracts of ocean. That part wasn't new to them, for they were of the generation that first knew about other lands. Their ancestors had lived isolated for generations.

What was new was the constant terror of brutish drunken men, who would never change their attitudes to women. Why were the women discovered planning to kill the remaining white men? Why did they constantly leave the village with their children to live apart from the men? It would nice for me to know that Fletcher Christian's background made him less likely to be a bully and wife-beater but what if *Bounty's* huge stores of alcohol and another mental breakdown made life dangerous for Mauatua and her children? If nothing else, Fletcher must daily have been profoundly disillusioned that his ideas of a community based on equality had foundered almost as soon as they had arrived on Pitcairn. A loving wife might not have been enough to douse any rage he felt about the failure.

These were women who had gone on Bounty *to be independent women and mothers, who sought a life of fulfilment rather than one truncated by men.*

Imagine your fears as a pioneering Ma'ohi woman when your island was discovered by ships only of men, who had no conception of what life might have been with constantly drunken and demanding mutineers or of what life was really like for women, any women. Imagine the protective responsibility the women felt towards their children, some of whom might remember the last of the murders but couldn't possibly understand why they had happened. Like Adams, not knowing what outsiders might think or do, might not the women have continued with *ha'avere* and lied to protect themselves and their children?

Apart from suspecting that visitors might be judge and jury on them, would Teraura want her younger husband Thursday to know she had murdered Tetahiti with an axe? Would Quintal's children or Edward Young's children benefit by knowing Edward had murdered Quintal because he was drunken brute? The community of those born on Pitcairn looked up to Adams as a faultless leader. Knowing Adams had also wielded an axe to kill Quintal could have removed the security they felt.

Of course those women would hide the truth and intentionally deflect the truth of the past. It's very telling there are no contemporary records of how Fletcher Christian treated Mauatua. Was he too a drunken brute who threatened her and their children? Did his mental state break down again or was he the only one of Pitcairn's settlers who behaved with dignity and fairness before Massacre Day? Why was so little recorded about the man whose mutiny and criss-crossing of the Pacific discovered Pitcairn and who fathered its first son? Perhaps no-one asked the right question but I believe that, if they did, they were certain to get no answer. Why? Because

Mauatua was alive and could control what was said.

Why the truth might matter after so long was much less important than what the truth would do to their children, the oldest of whom was only 18. Some of this new generation might have appreciated knowing men had been killed for their protection but it's not as simple as that. If the women were largely responsible for the deaths of most mutineers, wasn't it better their children were ignorant of this? As I say in MRS CHRISTIAN – *BOUNTY* MUTINEER: 'Could you love a mother who you knew had killed your father?'

The women were safer from suspicion than they might be in the 21st century, when questioning would be more determined. In the early 19th century visitors had a bigger question than anything to do with women, black women at that. What happened to Fletcher Christian?

CHAPTER 35

DRIVEN TO *'THAT UNHAPPY STATE'*

The burning of *Bounty* at Pitcairn Island in January 1790, isolated her complement even more than their dearest or worst hopes. For almost twenty years the community remained unknown as it first writhed bloodily and then succumbed to an all-pervasive form of Christianity that was either inspiring or sickening; it was called both.

Only two ships accidentally found Pitcairn after *Bounty* did. Both thought it uninhabited. One sent men ashore to collect coconuts but did not detect any sign of habitation. Neither did the other ship, which was prevented from landing by violent swells. The next day was apparently the calmest since *Bounty* had arrived but fortunately the intruder had sailed off. The habit of having no fires during the day still kept Pitcairners safe.

On February 6 1808, the Boston sealer *Topaz*, Captain Mayhew Folger, chanced upon the incorrectly charted rock. For reasons never explained but probably connected with the younger inhabitants' mixture of naivety and curiosity, the Pitcairn community made itself known to the ship.

When Thursday October, wearing cock feathers in his hat, led other Pitcairners to board *Topaz* Folger was astonished that he spoke English and claimed to be English, saying all their fathers were English. He didn't know about America and asked if it was in Ireland. He knew Bligh's name, was frightened by a dog and didn't know how hinged doors worked – on Pitcairn doors and windows slid. Folger stayed only for one day during which Adams gave one of his many versions of the island's history and what had happened to Fletcher Christian. Thursday's Tahitian mother Mauatua was still alive, although Thursday said she wasn't. The obfuscations of inaccurate memory and self-protecting *ha'avere* had begun.

That something awesome and Biblical had happened between 1790 and 1808 was certain. Pitcairn was not the expected group of middle-aged men and women and their children. Instead, the island was populated like an embryonic Eden, largely by children and teenagers. Fletcher Christian's older son Thursday October had turned 18 the previous October and was already a father. There was one adult European male, corpulent, pig-tailed, tapa-swathed John Adams. There were nine adult Ma'ohi women, seven from Tahiti and one each from Huahine and Tubuai. The other 14 white and black men who sailed from Tahiti were not to be seen. Everyone told a different story about how and when Fletcher Christian died. This soon made men wonder if he were dead.

The Pitcairners didn't want King George to think them thieves and in exchange for a silk handkerchief Folger was given the Kendall K2 chronometer and azimuth compass, trusting him to return it. After many adventures K2 is now displayed at the Royal Observatory, Greenwich. Mrs Christian gave Folger Fletcher's Chinese porcelain bowl decorated with peonies and pheasants, which went home with him to Nantucket. Now owned by the Nantucket Historical Association, it can be seen in the Nantucket Whaling Museum together with the piece of *Bounty's* copper sheathing I had been given during my 1980 expedition to Pitcairn.

Folger sent news of his discovery back to England via the Consul in Valparaiso but few were interested. France and Bonaparte were causing trouble again. It was not until September 17 1814 that men again sighted the peaks of Pitcairn where none should have been. This was an altogether more serious encounter as these were British naval frigates, *Briton*, Captain Sir Thomas Staines, and *Tagus*, Captain Pipon. The ships had been hunting for an American man-of-war, *Essex*, which had been worrying English whalers.

In 1817 Boston-based *Sultan*, Captain Reynolds, raised Pitcairn and was the last ship that did not previously know of the island. After that the publication of the visit of *Briton* and *Tagus* was causing a sensation and callers became increasingly common. They thirsted for gossip about the mystery of what was vaunted as the world's most perfect community and were also assured of excellent water and good meat, fruit and vegetables, a blessing in remote South Pacific waters. In return, they were expected to behave, to keep their libido, blasphemy and innuendo to themselves and to give such as nails, tools, clothes and religious tracts.

These early visitors besieged the islanders with questions. The fate of Fletcher Christian and *Bounty* had been a favourite topic in the inns of ports and the wardrooms of naval ships for two decades. The enquirers learned much, except for the truth.

The world had been fed with possibilities of Fletcher's escape from self-inflicted exile since as early as 1796. That year the British public was tempted with a slim volume with a long title: *Letters from Mr. Fletcher Christian, containing a narrative of the transactions on board His Majesty's Ship Bounty, Before and After the mutiny, with his subsequent voyages and travels in South America.* On Tuesday September 13 1796, *The True Briton* devoted half a column to a report based on the pamphlet, saying that Fletcher Christian, this *'extraordinary Naval Character'*, had at length transmitted to England an account of himself. It said that after seizing *Bounty* he had visited Juan Fernandez and other islands off the west coast of South America before being shipwrecked while rescuing Don Henriques, Major-General of the Kingdom of Chili [sic]. Suitably grateful, the Spanish Government lucratively employed Fletcher Christian, who was shortly to sail from Cadiz to South America on their behalf.

The True Briton, an evangelical weekly magazine printed in London, says that Fletcher's Letters were published in Cadiz and that, *'we are candidly told by this enterprising mutineer that the revolt . . . was not ascribable to any dislike of their Commander but to the unconquerable passion he [Christian] and the major part of the ship's crew entertained for the enjoyments of Otaheite. . . "It is but justice that I should acquit Captain Bligh in the most unequivocal manner. . ."*

Extracts from the Letters were published everywhere. The gentlemanly defence of Bligh makes one think that perhaps the commander himself, or someone close to him, had encouraged the publication, perhaps as an overdue response to Edward Christian's damaging texts. A copy was sent to Bligh by his publisher, a Mr Nicol. On September 16 Bligh wrote to Sir Joseph Banks: *'Mr Nicol has been so good as to send me down a pamphlet called Christian's Letters - is it possible that wretch can be at Cadiz and that he has had intercourse with his Brother, that sixpenny Professor, who has more Law about him than honour—My Dear Sir, I can only say that I heartily despise the praise of any of the family of Christian, and I hope and trust yet that the Mutineer will meet with his deserts.'*

Clearly the 1794 pamphlets of Edward Christian had wounded Bligh more than he had publicly admitted. Bligh was quite prepared to accept the Letters as authentic and the passion of his correspondence with Banks is thought by Gavin Kennedy to prove the work did not come from Bligh's camp. That is unproven but it does show memories of the mutiny stung and that Bligh's defence was still only to attack others rather than publish facts.

Among the journals that published extracts from the Letters was *The Weekly Entertainer*, seen by William Wordsworth at Racedown, Dorsetshire. The November 1796 issue of this *'agreeable and Instructive repository'* published a letter from him that has excited much conjecture, now if not then: *'Sir, There having appeared in your Entertainer (vide the 255th page of the present volume) an extract from a work purporting to be the production of Fletcher Christian, who headed the mutiny on board the Bounty, I think it proper to inform you, that I have the best authority for saying that this publication is spurious. Your regard for the truth will induce you to apprize your readers of this circumstance. I am Sir, Your Humble servant, William Wordsworth.'*

This was not the only rebuttal of the Letters' authenticity. On September 23, *The True Briton* wrote: *'Letters pretended to have been written by Mr Fletcher Christian having been advertised and an extract from them having been inserted in some of the Public Prints, it is necessary to assure the Public, as we do, from the best authority, that since*

Christian landed at Otaheite in September, 1789, that part of the ship's company who were afterwards brought to England by Captain Edwards, neither he nor the Bounty has ever been heard of. In a matter of so great seriousness, the Public ought not to be trifled with nor imposed upon by idle fabrications and scandalous falsehoods.'

The authority quoted by William Wordsworth was undoubtedly Edward Christian, the family mouthpiece and who had so successfully acted as counsel for William and his sister Dorothy in 1791. The Christian family had not heard from Fletcher Christian and although shamed by his mutiny it was unthinkable he would not write if he were in Spain: his argument was not with them.

On April 15 1815, *The Aberdeen Chronicle*, reporting the visit of *Briton* and *Tagus*, said: *'The mutineers were unheard of for several years. At length, some accounts, which we do not distinctly recollect, represented Christian, the ring-leader of them, to be subsisting by piracy; but this was contradicted, upon the authority of his family, who knew him to be dead at that time, 1804 . . .'*

That was a worrying thing for me to find. How did they know he was dead in 1804? They did not even know where he was, let alone possess details of his mortal state, unless he really had managed to be get back to Europe, returned to his family and died there. Was it true or was the anonymous Aberdonian journalist, lacking access to family information, simply filling an empty column, an unromantic but far more plausible theory?

Not everyone who read or heard about the Letters would know they were thought a forgery and another such narrative, with an even longer title, was successfully published only two years later in 1798. So convincing are the accounts I am sure they led to the belief in 1808 and 1809 that Fletcher Christian was back in Cumberland. There is no contemporary evidence to support this and it wasn't mentioned publicly until 1831, so the rumour was never substantiated, then or subsequently.

The bogus pamphlets were on to a good thing saying Fletcher had escaped and was adventuring in exotic lands. Romantic, dramatic defiance of death by high-profile rebels was what the public wanted. Just as James Dean and Elvis Presley are supposedly alive today, it suited the times of George III and later to think Fletcher Christian might be, too.

In February 1810, well over a decade after the Letters were published, the *Quarterly Review* reported Folger's discovery of Pitcairn Island. Fletcher Christian's children were from then on the object of great interest and represented as *'very handsome, their features strongly partaking of the English: the beauty of one of them, a girl named Mary Ann Christian, for which she is termed "the maid of the South Seas" is said to invite the same admiration which is offered to the most favoured of our own fair country women'*. The news was barely noticed except by Mary Russell Mitford. Helped by Samuel Taylor Coleridge she wrote *'Christina, The Maid of the South Seas: A Poem'*, published in 1811. The first stone of an avalanche of romantic Pitcairn trivia had been dislodged. Late in 1815 a man called Porter published his narrative of a cruise that included the call of *Briton* and *Tagus* at Pitcairn. *The Aberdeen Chronicle* introduced three long extracts saying: *'we are now enabled to complete (the mutineers') history and to describe their present condition'*. The truth was further away than ever.

The history of the Pitcairn community will never be known in full, for from the start there was a deliberate campaign to smudge the past that provided fuel for romantics and purveyors of conspiracy theories. Much of this was done by the surviving women and their art of *ha'avere* as well as by Adams protecting himself against arrest and extradition.

In extreme old age she at last had news of her son Fletcher, whom she could picture only in his early 20s.

The first two visits resulted in different versions of Fletcher Christian's death. Folger stayed at Pitcairn only ten or so hours and we must suspect both the reliability of those who told the tales and those who reported them. There wasn't much opportunity to cross check. Captains Pipon and Staines fared little better.

Two of Fletcher's deaths were murder, one a suicide, one a natural end. Only Adams and Ma'ohi foremothers knew the truth but they were the ones lying. The visitors had been told:

Six years after *Bounty* arrived their servants, the Ma'ohi men, attacked and killed all the English except Adams who was injured

Four years after they arrived, the same revolt of the six Ma'ohi occurred, with only Adams being spared, although injured

Christian became insane shortly after their arrival and threw himself off rocks into the sea

They all lived in tolerable harmony for several years under Christian's government and then he became sick and died a natural death. This was followed by two massacres when the Ma'ohi men killed the Europeans, and were in turn killed by the Ma'ohi women

A further enigma arose about Fletcher Christian's wife. Several stories said Fletcher was shot because he seized the wife of a Ma'ohi after Mauatua died giving birth to their first son. This was given credence by Lieutenant Shillibeer, who interviewed Pitcairners on board *Briton*. Fletcher's son Thursday said he did not know his mother, for she was dead. No Pitcairners contradicted him.

These confused and inconsistent stories about the possibility of Fletcher being murdered would bring little comfort to his brothers Charles and Edward, who were still alive, as was his mother, Ann.

Ann Dixon Christian is perhaps the saddest of all in this complicated story. In extreme old age she at last had news of her son Fletcher, whom she could picture only in his early 20s. Half way around the world he had fathered her only grandchildren, two mixed-blood boys and a girl, each as much Tahitian as Cumberland or Manx. She would never see them or her Tahitian daughter-in-law said to be called Isabella and no one could tell her where Fletcher was, dead or alive. A widow for over 50 years, Ann died aged 90 on March 30 1820 in the house in Fort Street, Douglas, that she shared with her son Charles. Although she had two other famous sons, Edward and Fletcher, the local newspaper noted her only as Charles's mother. He died on November 14 1822 and the same newspaper noted only that he was the brother of Professor Christian, Chief Justice of the Isle of Ely. On December 13 1822 the *Douglas Advertiser* announced a sale of the contents of Charles and Ann's house. It can only be imagined what material from or about Fletcher Christian was lost.

Professor Edward Christian died the following year in 1823, the last of the clan and thus the natural repository of family papers and possessions. I hoped to glean some further clue to Fletcher Christian from his Will and what his family had done about his effects when they resigned themselves to his never returning. Instead of revelation I found something disconcerting that, far from helping pinpoint the fate of Fletcher Christian on Pitcairn, strongly suggested he might, after all, have returned to England.

Like Ann and Charles, Edward had not left a Will that I could trace, a mystery since the Christians from the earliest times were a law-oriented clan. Edward was a Professor of Laws as well as a judge and had a wife and possessions. I had found the Will of his father, Charles. After him neither his wife nor children had made Wills, yet each of them had possessions and close family when they died.

Since 1796, there had been rumours that Fletcher Christian was alive but not in the South Seas. I found myself wondering if he had really managed to get back to England, then living on as secret recipient of his dead family's money and possessions? Peter Heywood is one man who would have agreed wholeheartedly with the possibility but his opinion was not published until 1831 after he and all the Christians were dead.

Sometime in 1808 or 1809, Peter Heywood was walking down Fore Street in Plymouth Dock when his attention was caught by the appearance of a man whose shape so reminded him of Fletcher Christian that he involuntarily quickened his pace. Both men were walking very fast and the rapid steps behind him roused the stranger's attention. He turned, looked at Heywood, and immediately ran off. His face was as much like Christian's as his back and Heywood, exceedingly excited, followed. Both ran as fast as they were able but the stranger had the advantage and disappeared. That Christian should be in England Heywood considered highly improbable, though not out of the realm of possibility for at that time no account had been received of him. The resemblance, the agitation, and the efforts of the stranger to elude him were too strong not to make a deep impression on him. His first thought was to investigate but on reflecting upon the pain and trouble such a discovery could affect Fletcher Christian, he considered it prudent to let the matter drop.

There is nothing sinister about Heywood not following up this intriguing encounter, or in its not being made public for 20 years. The reports from *Topaz* made it clear that the Pitcairners considered Fletcher Christian dead. If he was not, then the *'pain and trouble'* Heywood feared would fall mainly on his old shipmate. Anyway, Fletcher Christian could not have long been in England before finding out that it was Heywood who set into motion the endeavours of Edward Christian to expose Bligh and to publish alternative stories about the mutiny. My opinion is that if the stranger really had been Fletcher Christian, he would have embraced Peter Heywood, rather than fled, and I am sure Heywood thought so, too.

Or had he really seen Fletcher Christian in Devonport?

In the 1820s, some Pitcairn men still wore a clout and the poncho-like tiputa of tapa cloth, decorated with Pitcairn women's unique ochre and yellow designs. Adam's house is on the left and he stands holding children's hands in the centre. The house on the right must be that of Fletcher and Mauatua - are the two young men Thursday and Charles, with their sister Mary looking down on them? Note the gaping dark shape of Christian's Cave on the cliffside of Lookout Point.

CHAPTER 36

THE ENIGMA OF THE ANCIENT MARINER

With real peace possible on Pitcairn as the new century arrived, Young took the opportunity of teaching Adams to read and write better. Fletcher's widow Mauatua had three children by Young who died on Christmas Day 1800, probably of asthma. He was the first Pitcairner to die naturally. Now alone, Adams became an alcoholic, addicted to the spirit distilled from ti-root. During the most dramatic of his alcoholic hallucinations, he saw himself attacked by the Archangel Gabriel and sobered immediately. Once the women had nursed him through the awful exigencies of *delirium tremens* and other effects of withdrawal from alcohol dependency, he turned to the prayer book and Bible among the books rescued from *Bounty*.

With his limited ability to read and only cobwebbed memories of religious instruction in the poorhouse and at sea, Adams cobbled together what he thought a suitably penitential form of Anglicanism and quickly converted the island to a new way of life. The women accepted the new religion quickly because it was the first in which they and their children had been allowed to participate. Adams' rules included two fast days a week, which resulted in frequent fainting among his too malleable flock as they laboured in their fields to grow what they were forbidden to eat.

Life on Pitcairn did indeed become Paradise but even Paradise has loins and by now Thursday October wanted to exercise his. In 1806, when he was about 16, he fathered a child on Teraura, the murderess of Titahiti, who had first been the wife of Young and then of Quintal. The youngest of the women (except for Sully) who had come to Pitcairn, she was over 30 when she married Thursday October to become the second Mrs Christian on Pitcairn and the couple subsequently had six children.

By the time Beechey arrived in 1825, Fletcher Christian's second son Charles had married Sully, the girl who had arrived in a barrel. Fletcher Christian's grandchildren were thus three-quarter Tahitian and Mauatua saw a wedding between her son Edward Young and her grandchild Polly Christian, one of Thursday October's daughters.

In 1825 Adams took the opportunity to ask Beechey to marry him officially to his blind and ailing wife Mary, who had borne him his only son, George. Beechey was the first fully to explore Pitcairn Island's community, past and present, and he described an idyllic existence. The Pitcairners were innocent and beautiful to behold, golden children of Christian love, who did not lie or gossip. If he asked about another islander, they would answer: *'It will do you no good, it will do you no harm'*, and refuse to continue. The men wore a loin cloth and the women a long skirt. Both dropped a tunic called a *tiputa* rather like a poncho over their shoulders. These were wonderfully printed and painted in Pitcairn's unique patterns with red and yellow dyes made from indigenous plants, bark and earth. Otherwise, men and women went naked from the waist up. They wove hats with dexterity and bedecked them with flowers. They anointed themselves with the oils of tropical flowers and entertained with a natural and unaffected Christianity that made visitors ashamed of their Sunday nodding and mumbling.

So breathless and admiring are most, if not all, the visitors' journals that the dearth of hard fact is barely noticed. We know the community lived in bowers in a small group on the cliffs to the west of Bounty Bay. In front of Adams' house was a large lawn, often mistaken for a village square. At the bottom of the clearing was Fletcher Christian's house and beyond that, the house of Thursday October, who shortly changed his name to Friday October, when he learned his father had not compensated for crossing the date line.

Beechey described remains of maraes and the strange carvings in the rocks of the virtually inaccessible cove called Down Rope but does not mention a grave for Fletcher Christian. Mauatua who preferred to be known as Mrs Christian, was not questioned about it. If Adams' flock was really so Christian, should they not have made some attempt to mark the graves of their forefathers? Paganism and idolatry were all behind them, Pitcairn was now a Christian community and Christians honoured the dead. This fact, in particular, convinces me that Adams was not just a benevolent old patriarch but also a hypocritical manipulator of Pitcairn's children. He knowingly allowed Thursday October to say (perhaps even to believe) his mother was dead, yet called Mauatua Mrs Christian. The truth about the early days was certainly protected by the women but by then was particularly to save Adams' neck.

Adams died in 1829, without ever being exposed as a pious fraud. It is impossible to feel anything but affection and sympathy for those he left behind, who had no way of knowing whether his religious teachings were right or wrong. At least they knew the rudiments of reading and writing and the language of their white fathers. They were startled to see dogs and cattle and did not know how to open doors. They shamed naval officers by saying grace before and after meals with genuine sincerity. When Thursday October was once served food by a black servant aboard ship, he left his place at the ship's table saying he did not like black men. The Ma'ohi women and Adams put the blame for Pitcairn's problems on the 'black' men rather than on the alcohol and physical abuse by white men or what its women had done.

Once Adams had gone, laid in a marked grave behind his house, different versions of his stories emerged. In 1831 Sir John Barrow published the first really comprehensive book about the mutiny on *Bounty* and Pitcairn Island. In a footnote he first revealed Heywood's reported sighting of Fletcher Christian in 1808 or 1809. Then, in 1834 Dr Bennet who was the surgeon aboard the whaler *Tuscan,* Captain Stavars, was shown a grave beside a pool said to be that of Fletcher Christian, something mentioned in a long *Narrative of a Whaling Voyage,* published in London in 1840. It has been ignored by, or unknown to, most authors or visitors to Pitcairn, possibly because it is hard to find and because it is not consistent with other records.

There was one person who could have set the record right, Mauatua, the widow of Fletcher Christian. She had remained slim and relatively upright, true to her nickname Mainmast. She had a shock of white hair and was renowned for her stories, told only to her family. The awful truth, at least it is awful to a descendant, is that Europeans could probably not bear to speak to her. She seems to have been a perfect fright in her old age.

In March 1830, a young man aboard the vessel *Seringapatam* wrote in his journal: *An old woman has a world of prejudice to surmount before she can become anything but an object of pity, and often of disgust . . . we recall the stories of witchcraft malice and cruelty imputed to the old and infirm of the female sex. She looked so old and corpse-like that I gladly escaped from her awkward expressions of pleasure at the appearance of my clothes, gun, etc.*

Naturally, I always rather hoped I would never find proof that the young man was referring to my great-great-great-great-grandmother. But he was. In 1841 24-gun HMS *Curacao,* called at Pitcairn and stayed a couple of weeks to help with an epidemic of fever and flu. There are many important records of this visit, including the acerbic journal of the ship's doctor, Gunn. He could not help but note the sadness of a Ma'ohi community that insisted on dressing their children when they swam. I deciphered a previously ignored note in his book, which said that the Christian children were thought to be simple in the head, *'perhaps the influence of their Tahitian mother'.*

That must be seen in early 19th-century context, when black skin was universally associated by Europeans with lower intellect. Only when I was left a privately owned and previously unknown journal of another of *Curacao's* complement that another reason for Gunn's remark became clear. The author of a beautifully illustrated record of this visit to Pitcairn was George Gardner, one of her officers. He noted that the females never ate with the males and called it a relic of the barbarianism, a custom of the uncivilised inhabitants of the South Seas handed down from `. . . *Tahitian parents. There are two of the Tahitian women still living. [One is] Isabella Christian . . . the wife of Fletcher Christian who headed the mutiny. Isabella Christian is the most perfect picture of an old hag I ever saw. She is still surprisingly active. Her age is supposed to be between 80 and 90. She remembers Captain Cook at Tahiti and from what she herself says must then have been a mother. In this tho' there is nothing very extraordinary since they marry even nowadays at the age of 13 or 14.'*

So, Mauatua looked like an old hag. No wonder visitors did not talk to her and thought her children might be mentally deficient. It's also proof she was known as Isabella, although she introduced herself to visitors as Mrs Christian.

Mauatua was certainly over 80 in 1841 and this is the first indication that she might have left children behind on Tahiti, something I used to dramatise her story in MRS CHRISTIAN *BOUNTY* MUTINEER, the companion historical-fiction volume that accompanies this one.

What Gardner and other visitors fail to note is Pitcairn's more recent history, especially that its women had the vote. Between 1832 and early 1838, the Pitcairners had suffered appalling interference by an unauthorised mad man called Joshua Hill, who created competing communities on the island and threatened death to a girl who harvested from the wrong sweet-potato patch. He had been thrown out but the island was vulnerable to drunken crews of whalers and sealers who thought the island fair game for lawlessness because it belonged to neither Britain, France or the United States.

On November 29 1838, the two oldest Mrs Christians, Mauatua and Teraura, helped persuade Captain Russell Elliot of HMS *Fly* to accept Pitcairn as a British Crown Colony. He had no such authority but by his official 'approval' of an extraordinary list of laws and punishments he legitimised the island enough for others to believe it. *Ha'avere* still worked.

The unique list Captain Elliott stamped into law included very clear rules about raising hogs, not killing white birds, who could pick fruit from which tree, the forbidding of claims against the past sins of others and made killing a cat one of the worst temporal sins.

Included in the long list was the island-council's remarkable decision to give the vote to all women over 18 and to make education compulsory for girls. Both were world firsts by a very long way. Considering Pitcairn's isolation and the type of visitor that generally called, the inspiration can only have come from within the island. To me that means Mauatua continued the legacy of long-dead Fletcher Christian, the man who had made her new life possible. Pitcairn's was yet another bloody revolution but the eventual result was something quite new. American and French women didn't get the vote after their 18th-century revolutions.

Mauatua must have been over 80 and after surviving all she had she was dead within a month of Gardner's visit, victim of yet another introduced influenza epidemic. On September 19 1841, she was buried on a site now covered by Pitcairn's village hall. Her son Charles 'Hoppa' followed her to his grave early the following year; his wife Sully had died in 1826.

There were more versions of Fletcher Christian's end to come. One of the least known and most fascinating appeared in a book published in 1898 called *The Mutineer: A Romance of Pitcairn Island*. Turgid with high passion and purple patches in the best manner of the Victorian novel, it purported to be the first true story of what had happened, written by two men, Becke and Jeffery. Becke was an extraordinary Australian adventurer about whom it was said he was *'one of the rare men who have led a wild life and have the culture and talent to give some account of it'*. In Sydney's Mitchell Library and in the National Library in Canberra, I found a great deal of correspondence, both with his collaborator, who lived in England, and with various critics and associates including his agent in London.

Included in the long list was the island-council's remarkable decision to give the vote to all women over 18 and to make education compulsory for girls.

The Mutineer said that Christian survived the gunshot wounds and being hacked at with an axe and then recovered in his cave. When he was well enough, he attempted to put to sea in *Bounty*'s boat to join a sailing ship sighted off Pitcairn. Adams tried to prevent him to protect the secrecy of the community and in the struggle, shot and killed Christian. Before he died, Christian asked to be buried in an unmarked grave as he did not wish people to point out the grave of *'the mutineer'*. No wonder Adams lied, it said. He had killed Christian, albeit accidentally. No wonder everyone else had conflicting stories. They were protecting their patriarch and it was Fletcher's own wish that his grave be not known.

This was such a perfect solution to the mystery of Fletcher's death, even if found in a factional book, that I had to dig deeper. How had Becke come up with this version? The answer is astonishing, he said he got it from Pitcairners. Among his correspondence I found a letter to Egan Mew, QC, of 3 Gray's Inn Place, London. He was a writer for *The Critic*, which had just given *The Mutineer* a wishy-washy review. Becke wrote in his defence: '*I know the descendants of the Bounty Mutineers and the native story of Christian and his life better than any man living. This sounds very egotistical of me but it is true.*

And instead of Christian being . . . the 'full-blooded villain' he was the very reverse. I have been told over and over again by old natives that Christian was the very reverse of a sensual man; that he was intimate with only one Tahitian woman whom he afterward took away with him to Pitcairn; that this woman was seduced by Young; that the other Tahitian men and women would have killed Young, but that Christian, horror-stricken at the bloodshed that had already taken place, carefully protected the man

Pitcairn Island, August 1841, by George Gardner of HMS Curacao, who described Fletcher's widow as 'the most perfect picture of an old hag I ever saw' but did not mention that she and all other Pitcairn women had the vote.

Thursday October, the older son of Fletcher Christian and Mauatua, pictured soon after Pitcairn was rediscovered.

who seduced his wife; and the story of his life in the cave as narrated by Jeffery and myself is true not embroidered fiction.

Furthermore the story of Christian's death by gunshot accidentally received from John Adams/Alexander Smith whilst endeavouring to prevent Christian from putting to sea in the Bounty's boat, is I believe, strictly true.

Anyway, I prefer to believe the native account of the Bounty story to the vague surmises of the many authors who have written on the subject but who only obtained their data from the Court Martial of some of the mutineers or from John Adams' carefully considered statements to Naval Officers. Perhaps you can make an interesting par[agraph?] of this?'

Becke's assertions that he heard these stories first hand journals I found that show he did stay on the island but the exact dates are not clear. Letters that passed between him and his collaborator show that neither was above twisting the story for the sake of drama but Becke passionately believed he had solved the question of Christian's fate.

Is it true, though? Somehow it is too true, and, of course, the *'old natives'* were anything but witnesses to the events, for Becke's stay was probably in the 1850s or 1860s. My research into his background and his book showed me that I could not lightly ignore this version. Neither could I ignore stories that believed Fletcher Christian had escaped and was the inspiration for Samuel Taylor Coleridge's, *'The Rime of the Ancient Mariner'.*

This idea was first suggested by a Mr Porter, who claimed Coleridge's so-called fantasy poetry was based on events and places and people he knew of or had read about. The sufferings of the Mariner can certainly be related to those of Fletcher Christian and it is possible to make a case for the poem being about Fletcher's secret journey back to England. As well as helping Ms Mitford with her poem about Fletcher Christian's daughter, Coleridge once penned a note to himself about Christian's adventures, making the possibility of a link even stronger. Add to this the friendship of William Wordsworth for both the Christian family and Coleridge, and there is enough substance in the enigma for a book, as C. S. Wilkinson proved with his persuasive *The Wake of the Bounty* published in 1953.

Once I had read that, my confusion was complete. I had to do something to sort this out but had no idea what.

Taiyo in full sail in the South Pacific. A mixture of old and new and without a working radio, my expedition's sailing ship made it easier for everyone to identify with Bounty and those who sailed in her.

REPERCUSSIONS IV

1978 - 2019
I follow Fletcher Christian's wake

CHAPTER 37

IGNOMINIOUS BEGINNINGS

Once I realised that in spite of the movies and the many books there was no biography of Fletcher Christian I began to explore the possibilities of writing that book. I called on the only two real explorers I knew. Robin Hanbury Tenison OBE is one of the most important and glamorous of modern explorers and was then leading the Royal Geographical Society's biggest ever expedition to Mulu in Sarawak, for which he later was awarded the RGS Gold Medal. As luck would have it, he was back in London for a short time and I went to see him at the Society's house on Kensington Gore.

'What do you think about a descendant of Fletcher Christian . . .?'

'Following his ancestor's footsteps? Marvellous! Do it! How about becoming a Fellow of the RGS, too?'

Next I went to see Tim Severin, who had just sailed across the Atlantic in an open leather-boat named *Brendan*, showing that the sixth-century Irish monk St Brendan was probably the first European to discover America. Using suppliers from my shop, I had provisioned *Brendan* on the second leg of its voyage from Iceland to the USA. The traditional grains, smoked meats, sausages and cheeses I had chosen, some specially made, were a greater success than modern foods and a considerable help to both health and morale in the cramped and cold boat.

Tim was as enthusiastic as Robin and after his Atlantic trip was envious that mine would be in warm waters. He offered an introduction to Julian Bach, his agent in New York. Julian met me in London's Connaught Hotel the next week. His interest and excitement spurred me on to write a synopsis of *FRAGILE PARADISE,* the biography of Fletcher Christian I planned. Various New York publishers were approached and it seemed likely that either New York Times Books or the Atlantic Monthly Press would commission me. I hoped this would generate a sale to a UK publisher, too. When I was granted an interview by an older customer of my delicatessen, who was considered a top London literary agent, he dismissed the idea saying: 'Who would be interested in the fate of some 18th-century sailor?'

Not an adventurous or curious bone in his body.

I chose to ignore that man and confident of a quick and adequate US sale, I left for Sydney, Australia, to start detailed research in the

Mitchell Reading Room of the NSW State Library, which has a magnificent but (then) largely unindexed collection of *Bounty* related material, including the Bligh/Banks correspondence.

Negotiations did not move as fast as I had hoped and by March 1979, I still had no contract and no money, yet I had discovered enough new information to convince me I was on the right track; I had already found the Bond papers. Thus, I was floored when I was telephoned by New York Times Books at 3.30 in the morning to tell me that an author who had recently written on the subject advised them not to bother with me. The same author later damned FRAGILE PARADISE, as my book was to become, saying it had nothing new to say. That was my first taste of the unlikely and sometimes dishonest passions raised in middle-aged men, who still today feel wounded by what happened aboard *Bounty* more than two centuries ago. I cabled Julian to cancel further negotiations, borrowed money to return to London and spent my remaining $10 during the weekend. On Monday, a telegram waited when I returned from a final visit to the Mitchell Library. Julian had an exceptionally generous offer for the book from Upton Brady of Atlantic Monthly in Boston. I left for London three days later. I could and would write the missing book about Fletcher Christian.

Within six months I knew more about my universally recognised but unknown ancestor than anyone who had written about him. Even though untrained as a researcher or historian (perhaps because of it) it was very easy, revealing how many so-called definitive works are merely patched together of old material at a static desk and how common it is for established writers to get a new commission based on very little but previous success and a 'new theory'.

A book called *The Yesterdays Behind the Doors* was published by Liverpool University Press in 1956 and contains 1,000 years of Christian family history, yet no sources are given, not even those for the family pictures. Every book is but the tip of an iceberg of labour and paperwork, so someone had spent years putting the background of this book together. Perhaps I could find that research? I did, simply by tracing the owner of the estate of the late author. Susan Hicks Beach who was born Susan Christian.

Susan Christian's family papers stuffed all the space beneath a grand piano in a bewitching red-brick Inigo Jones mansion outside Newbury, England. Included were 500 leather-bound pages of Christian family genealogical fact, dating back to 1380, annotated and verified and brim-full with every record of tradition, hearsay or gossip. In neat handwriting, with no mistakes and three colours of ink, impeccable research recorded 21 generations of Christians. 'Vio' Schlaet Christian, a German who was the last Mrs Christian of the Milntown line, had spent over 30 years collecting the material with her daughter Rita Christian Browne. Wills, letters, diaries, paintings, prints, and photographs were all included, hoping that the record would stimulate a member of the family to buy back the estate that they had been forced to sell. For decades they copied by longhand each document they found, tracing signatures and double-checking references. This included material subsequently destroyed by the 1922 fire that ravaged the Irish Public Records Office.

This remarkable cache was owned by young Ewan Christian, who later donated it to the Manx Library in Douglas, Isle of Man. Rita Browne had willed it to his father Lt-Col John Monsell Christian, who had died in 1975. Ewan is now considered the head of the senior British Christian family. With enormous kindness, he let me take exactly what I wanted and promised that no one else should look at a thing until I had published my book. Rather typically, author Richard Hough had looked at the papers and dismissed them as worthless, including its greatest treasure, an unknown memoir by a family member who had never before been mentioned.

Among the trove of letters and notes was the autobiography of Fletcher's brother Charles and digging between the lines of this led, as you will have read, to the discovery of Charles's involvement in mutiny. The original of this work was once in the Manx Museum in Douglas and has long since disappeared but not before it had been copied by Mrs Christian or her daughter, Rita.

With these volumes I could already tell the world much more about Fletcher Christian. Knowing about his ancient family and privileged social background would help a proper assessment of his relationship with William Bligh. With a distinct feeling of dread, admittedly interspersed with excitement, I realised this knowledge further committed me to go to Pitcairn or the book would not be fully credible.

How would I get there, when there was no regular or reliable shipping to the island and how much would it cost? It could take so much time and money that the deadline for writing the book might be jeopardised. The deeper I became committed, the fewer people there were to help because there was no precedent for what I wanted to do. Nigel Winser and Shane Wesley-Smith of The Expeditions Office of the Royal Geographical Society were superb in their moral support but as I was not climbing a Himalayan peak or exploring a rain-forest they could be of little specific help.

Most of all I was certain I dare not go to Pitcairn Island alone. I was bound to be thought biased by previous writers and so I wanted to take others to temper, oppose or complement my impressions. That meant organising my own transport there and back, which meant many thousands of pounds of finance. By now I had also sold the idea of a Fletcher Christian biography to the publishers Hamish Hamilton in London. Like that of Atlantic, their advance was more generous than expected, so I decided I would risk it.

At that time visitors to Pitcairn generally arrived in small sailing vessels as they explored the South Pacific and rarely stayed more than a few days. Even if you organised passage on a rare supply ship from New Zealand or Panama and permission to land, the combination of which might easily take two years to arrange, you could find yourself sailing past Pitcairn because conditions made landing impossible. Having landed, it was common to be trapped for months until the next vessel arrived and was able to take you off. It was a problem solved only by arriving and departing in my own vessel.

I had recently met Curwen relatives, descendants of Fletcher's first cousin, John Christian Curwen and his wife Isabella. David Curwen told me that he had been overwhelmed by the sight of a sailing ship called *Christian Bach,* which was tied up in Greenwich. Inveigling his way on board, he learned it was soon to sail for Australia and intended to call at Pitcairn. *Christian Bach* was a 120-foot/36.5 metre square-rigged brigantine, beautiful, broad and air-conditioned. Each cabin had piped music and a telephone and refrigerator. It sounded just right for me, who cannot swim and had never sailed on anything less than 15,000 tons/15,250 tonnes.

I raced down to Ramsgate where she had sailed, marvelling at the coincidence of her name being a combination of my surname and my New York agent's. Too good to be true, I thought. Indeed, it was.

During weeks of exciting planning it was agreed we would meet *Christian Bach* in Guayaquil, Ecuador, sail to Pitcairn via the Galapagos and Easter Island, include a stop at Tubuai, and finish the journey at Tahiti. Adventure it would be and I'd also see other remote places; there were few cruise expeditions to the Galapagos in the late 1970s. I made a short film for BBC TV, there were pictures published all round the world and *Christian Bach* sailed back to Greenwich, where a press conference was arranged, something I hoped would help raise funds. Nautical friends wagged their heads, suspicious that the shorthanded captain was planning to sail back down the Thames at sunset but with no knowledge of that day's tides.

Christian Bach was soon sinking in the Channel and had to be rescued by the Royal Navy. Once ignominiously returned to port, a writ for non-payment of debts was nailed to her mast. Fortunately, I had not paid over any money but there my luck ended. From then on, the story of the expedition's planning is one of disillusionment, loneliness, fear, sweat and more tears than I knew I had inside me, yet there was no going back. The expedition was officially approved by the Royal Geographical Society but without a schedule, without a real budget, without even a vessel. I could not possibly lose face with the Society, having named the project in honour of Sir John Barrow, one of their Founders.

What became the Royal Geographical Society was formed in 1830 by men who included Barrow, Second Secretary of the Admiralty, who was at the same time working on his book *The Mutiny of the Bounty*, the first serious work on the subject and still one of the most important. It had seemed to be an excellent idea to make the Fellows and the public a little more aware of this man and his interests in 1980, the 150th anniversary of the founding of the Society. It caused no little stir when the prospectus for the Sir John Barrow Commemorative Expedition arrived at the Society. Founders are treated with scarcely less reverence than the Creator but my argument for the expedition's name was accepted and at last I was waiting outside the Committee Room for my official interview, to establish if they would give their stamp of approval.

No trial of endurance in the field can have been much worse. I was seated at the middle of one side of an immensely long and thin bank of tables. There was a smell of leather and pipes in the air. In the distance on my left was the serious white head of Lord Hunt, leader of the successful 1953 Everest expedition, amazed at my confidence in raising a budget of at least £15,000 / almost £60,000 today. Equally far away, to my right, was Nigel Winser of the Expeditions Office, asking good questions about my commitment to long-term interest in Pitcairn, which I could then answer with passion. In between were men of eminence and exploit, who gently probed and questioned with implacable faces. As I answered each question, I tried to look at each man by turning my head so as to answer them equally. It was not unlike being at Wimbledon, except the ball was in my court.

With Robin Hanbury-Tenison's advice and my own wish to contribute something to Pitcairn's future, as well as digging into its past, I had presented a programme we hoped to perform; a definitive map of Pitcairn was something I had hoped the committee would find irresistible. It would be the only one and was to be presented to their Map Room, one of the most famous in the world. Their deliberations and doubts were never revealed, of course, but they did give me their approval. With that I expected my troubles to be, if not over, considerably eased. My expedition was official, and surely that would make other arrangements easier.

CHAPTER 38

TAHITI REGAINED

To complete the voyage and write the book by the end of 1980, I had to be sailing out of Tahiti early in July of that year. By the time *Christian Bach* had foundered and the RGS had given me their blessing, it was the end of March. I had just over three months to find a vessel and raise the money to charter it. Both are notoriously difficult to do and the trap was tightening. Without a vessel, without money or associated budget, I could not raise funding.

The South Pacific was an unknown world to London charter agencies. I discovered Sydney and Auckland were not organised enough to have big chartering services and finally thought my best bet was to try to find a ship on the West Coast of America. Enquiries to quite disparate people came back with the same name, Mary Crowley of Ocean Voyages in Sausalito, across the Golden Gate Strait from San Francisco. There was no time to worry about further expense. Within a week I was in Sausalito and Mary was calming me. She had sailed on expeditions, led expeditions, done everything I had not. She was very sensitive to the problems of putting a group of people together on a ship that was going to be isolated at sea for weeks and what effect such a group might have on Pitcairners. Soon Mary had chosen eight of her vessels that were suitable and willing to do the trip. Comparing one vessel with another is like comparing a buttercup to a Cadillac. You can't. One was big and strong but the crew was probably not right. Another had a wonderful crew but was ultimately too small or too expensive. To me they were all too expensive and I think Mary knew that. She didn't let on and we blithely talked about basic costs of £50/over £200 today, per person per day for ten weeks and agreed that ten people would be about right for the group. I was committing myself to spend (and raise) £35,000/about £150,000 plus air fares and inevitable extras.

One of the ships keen to go to Pitcairn and benefit from the eventual publicity was *Taiyo*, Japanese for ocean and similar to *taio*, the Tahitian word for special male friendships. *Taiyo* had been built in Mexico only seven years before, to combine the lines of an 18th-century brigantine with the advantages of 20th-century technology and comfort. Her hull was steel and the sails on her square-rigged main mast could be managed from the deck rather than requiring crew to scuttle up and down ratlines. She had never been fitted out internally, for the shipbuilders in Mexico had gone bankrupt as she was nearing completion and the government impounded her. Her subsequent history was not encouraging. *Taiyo's* last owner was the pilot of the small aircraft that collided with a jet at San Diego late in 1978. Now she was owned by a consortium headed by Mike Dunn, a Pan Am pilot, was being fitted out in mahogany and teak and looking for a job.

Taiyo was anchored at the Sausalito Marina, just a few minutes from Mary's house. It was duty rather than interest that saw me following Mary to the docks to see her. I had mainly registered the unhappy aspects of the story and words like brigantine, square rig and steel had not meant much. I turned a corner to see *Taiyo* dead ahead and was immediately persuaded. I saw the yard-arms with their furled sails at right angles to the main mast and that was it. Was it some of Fletcher's blood that made my heart leap so? Who knows? I only know that nothing had so moved me for years as that sight. If a price could be agreed, I wanted this ship. It was another immediate and emotional decision, yet if the ship's lines did that to me, I hoped they would do it to others, making it easier to build interest and raise money.

Taiyo would have a crew of four, and accommodation for an additional 12. For $US30,000/about $US90,000 today I could have the ship and six others in my party. Mary took it upon herself to find four fare-paying passengers who, because they would have to be subordinate to the aims of the expedition, would pay $2,500/$7500 each.

Once I started fund raising in earnest, the combination of Royal Geographical Society approval and a beautiful ship worked against me. Most people thought that Royal Geographical Society approval meant they were paying and I was asking others merely to top up the barrel. Some were so taken with the romance of the voyage and the ship that they made promises they should not have. Promises of introductions or contributions came to nothing.

An international airline and an oil company were the most interested potential sponsors and both asked me not to approach other companies. As time passed, I emptied my bank account then borrowed another £5,000/over £20,000. Two weeks before we were due in Tahiti, I still had not paid the final amount to Mary. It took repeated telephone calls to establish that neither the oil company nor the airline was going to help. In one case the decision had been made two weeks before. The men concerned had over-committed themselves and had been too embarrassed to tell me.

This last fortnight was as bitter and solitary as it is possible to imagine. I barely slept as I tried to raise money. I was hardly capable of making a logical decision, yet was faced with deciding whether to cancel the trip and lose the money already sent. I had neither time nor energy to reconsider. The path I was on would have to do.

Just one week from the deadline Upton Brady, the commissioning editor of Atlantic – Little, Brown came up with the balance. There were no sponsors; I was paying for the expedition and for some of its members myself and still had no money for film for the Konica cameras I had been lent. With hours to go flatmate Greg James, now Greg Boki, came to the rescue with £650 / about £2750 today.

Exhausted financially and emotionally I packed for this most important trip of my life in less than an hour. There was no time properly to consider the reference books I should take. I hadn't learned to swim or to dive, knew little about sailing and less about most of the people for whom I would be responsible for the next ten weeks. Against all odds I had an expedition of eleven, including a three-member film team from Marin County, California.

I flew out with Michael Brook, who as a loyal friend and ingenious researcher had been an invaluable support during the past year. When we arrived in San Francisco, there was no one to meet us. We found a cheap hotel, then caught a freezing cable car to Fisherman's Wharf to get drunk and fed economically. I didn't dare use my credit card, the hotel had taken almost all the money we had in our pockets and no one would change travellers' cheques.

I saw the yard-arms with their furled sails at right angles to the main mast and that was it.

Next morning Mary found and mollified us in her incomparable way. Hours were spent agreeing to the details of the film contract and by mid-afternoon we had crossed the Golden Gate Bridge twice and were back at the airport meeting the film crew before flying to Los Angeles to meet more expedition members and then flying on to Tahiti.

In the early hours of July 5 1980, we boarded *Taiyo* in Papeete harbour.

I was pleased there was little moonlight. I sat in the warm tropical night and cried silently with fatigue and relief and no one saw. A fine way for an intrepid explorer to behave.

CHAPTER 39

THE SOUL OF POLYNESIA

Our first South Pacific sun came up explosively, lighting in turn the fabled peaks of Moorea across the lagoon, the knife-edged crags behind Papeete, then the masts of *Taiyo* and the faces of the expedition members. We had stayed up the rest of the night, drinking tepid white California wine, talking and wondering at the great geometric stretch of the Southern Cross above us.

From London there were Michael Brook, Vivien Gay and Jasmina Hilton. Vivien was 30 and a fairly high-powered travel executive, who had promised to be very seasick. Jasmina was a friend of Vivien, half-Iranian, 'a few years older' and an actress. The film crew put together by Mary Crowley was headed by Ted Cochran, a Vietnam veteran who had flown helicopters on Apollo mission retrievals and was a full-time enthusiast for the trip, the ship and the sea. As cameraman, Ted had brought Kim Hoeg, a young Norwegian who had lived most of his life in America. The sound recordist was Marilyn Waterman, a native-born Californian and star graduate of the Film Faculty at Stanford. These three were virtually unknown to each other but their professional and Californian jargons gave them an apparent cohesion.

Richard Hudson was a New Zealander with great knowledge of horticultural and agricultural methods that rely on natural techniques. Elliott Smith was another of Mary's finds, a 40-year old professional photographer from San Francisco, who had worked in the Solomon Islands. From Jackson, Mississippi came Mark Balsiger, a 30-year-old free-lance journalist. From Bedford, Massachusetts, came Andrew Brady, a son of my American editor, admitting to 16 but still 15.

Terry Purkiss was *Taiyo's* English captain, the Australian mate was intriguingly called Tony Fletcher, the Virginian engineer was Hop and Bina, short for Sabine, was the German cook. Terry and Bina seemed very much in love. Three of my shipmates admitted past marriages but were now divorced.

The range of age, sophistication, background and culture was the lifeblood of the expedition, the base on which I would, paradoxically, build firm opinions. By drawing on the varied experiences of all my 14 companions, I felt certain to present an objective and balanced a view of Pitcairn and was more likely to uncover secrets from its past.

First, I had to peel away the layers of normal life from myself and the expedition members. I wanted them to discard their European and American ways of life and understand the simplicity of Polynesia, a term first used in 1756 by a Frenchman; it means many islands and first used to include all the Pacific but was not used by the English until much later and only then for the triangle of New Zealand, Easter Island and Hawai'i. Developing a feel for its music and a taste for its luscious food was important but more vital was to understand properly the deprivations and discomforts of life at sea and on a remote island. Only then might we begin to understand more of Fletcher Christian and his companions and presume to comment on his Pitcairn descendants.

It's been unpleasant to find my instincts were right. Since then, fleeting visits by solo authors have been destructive, self-serving and unbalanced yet widely published.

For all their past importance to the *Bounty* story Papeete and Tahiti were hardly the places to begin our re-education. Air-conditioned shopping centres, television, discotheques, transvestite bars, Chinese restaurants, and superlative croissants meant we might have been in any number of provincial French resorts. I wanted to show the expedition some of older ways, ones shared by *Bounty's* men. On Sunday July 5, while waiting for jet-lag and shyness to dissipate and stores to arrive, we had our first serious meeting under *Taiyo's* striped awning. Captain Terry needed to establish our several degrees of seamanship, which turned out to be low, with Ted, Richard and Mark being the only sailors. The latter two had both also been to Pitcairn and supported Terry when he warned us of the dangers of the waters through which we would be sailing. One ship in seven that sailed into the Tuamotu Archipelago was wrecked because of sloppy navigation or inefficient watch keeping.

I had *Taiyo* for ten weeks and Terry and I agreed we should make a short shakedown cruise from Tahiti and back, sailing north-west to Bora Bora and other islands of the Society Group. Wednesday morning was fixed for departure and the intervening days were spent visiting Matavai Bay, the kaleidoscope of Papeete's early morning food market, filming, and struggling with the problems of stowage aboard *Taiyo*. It was a little aimless for, although we clicked as a group, there was still no sense of reality to the adventure ahead.

At precisely 5 am on July 9 *Taiyo* edged out of Papeete's still lagoon past the ugly concrete docks built by the French and that scar the town's view of Moorea. It was a sickly sunrise and the dreaded *mal de mer* took a wide and sudden toll, which just as quickly disappeared. By noon we were skirting Tetiaroa, the coral atoll Marlon Brando was allowed to buy after he had played Fletcher Christian in the MGM film made on Tahiti in the early 1960s.

Here, just 30 miles from Papeete, was the timeless South Pacific of legend. Most of us were seeing for the first time the startling contrast as the rolling sea of intense blue turned to a still, milky turquoise after thundering over an encircling reef. Behind was the sand, here dazzling and white in the high tropical sun, not black like that of volcanic Tahiti. Then, a jumble of greens, as violently stroked and tangled as a Van Gogh, vivid and thick and punctured with the high heads of coconut palms. As we cruised around the reef looking for the sole, narrow entrance we tried to contact the atoll's main island, which then had a small airstrip and a few thatched huts that could be rented. I was nursing the hope that I could meet Marlon Brando and had written to tell him of the expedition. There had been no answer. Even if he was not at home, his wife, Tarita, who played Mauatua my great-great-great-great-grandmother in the film, might be. When we made radio contact, we were rudely ordered to get off the air. That was the last time our radio ever worked, a re-creation of 18th-century marine isolation that all of *Taiyo's* crew could have done without.

Just as Bligh had done aboard *Bounty*, our captain divided us into three watches, so that each group was on duty for four hours and off for eight, in rotation. He had also arranged two 'dog watches' which purposely interfered with the routine by dividing the 4 pm to 8 pm watch into two, so each watch worked a different time on successive days and nights. When Terry had suggested that the cook and I be excused such duties, I accepted without thinking. Later I spent a lot of time reconsidering. Should I not imitate Fletcher Christian and learn to handle a sailing ship, remember the names of lines and sails, learn how to keep safe and awake on the dreaded graveyard watch, from midnight to 4 am? When I saw how the broken sleep patterns affected the others, I was confirmed in my quick decision. I felt it my responsibility to everyone on board to be as physically and mentally strong as possible. For the expedition's sake I also needed to be well briefed. There were months of thinking and reading to catch up, all that should have been done before I left.

Difficult winds, rough waters, seasickness and tiredness persuaded us to make an early landfall at Raiatea. We slipped through its reef late in the afternoon of the 10th and were unable to express our impressions as we sailed through calmer waters past thin black valleys and luxuriantly draped mountains, flossed with cloud towers of pastel colours. As night fell we tied up at the spartan dock of Utaroa and Marilyn stumbled off to fall on her hands and knees and kiss the solid ground, to the natural consternation of some local children. This was French Polynesia's second biggest town, yet boasted only a dozen or so shops, an endearingly irregular market, and a sprinkling of those untidy, secretive, fascinating jumbles called general stores that germinate throughout the Pacific wherever there are isolated communities and Chinese entrepreneurs. One of these cooked excellent steaks from locally bred cattle and served them with wonderful chips, thick, dry and crisp.

Next morning enquiries around the town produced Mr Chang, who owned the most capriciously decorated nightclub and was also the acknowledged expert on the island's most important site, the ruined marae of Taputapuatea. Before Christian missionaries brought newer and more amazing myths, this was the holiest place in Polynesia, a true Mecca for followers of the cult of Oro. For many centuries brave, pious pagans sailed their carved and festooned canoes from as far as Hawaii and New Zealand to the sacred ceremonies here. The Ma'ohi noble Tararo who went to Pitcairn with Fletcher Christian was from Raiatea and was accorded special status because of this. In Chinese-accented French, Mr Chang told us tales of cannibalism that were not true and stories of human sacrifice that were.

We decided to see Taputapuatea marae for ourselves and amid constant peals of that enchanting easy laughter of all Ma'ohi, we squashed onto wooden benches on the back of open-sided trucks that served as buses.

By evening it was apparent that Polynesia had been absorbed.

The marae was hot, stifling both in temperature and atmosphere. It was so alien to us that most gave up trying to identify or understand the remains and wandered off to enjoy instead the lagoon-side setting. Exactly where sacrificial victims had lain with their heads clubbed in, we were shown how to split open young coconuts and enjoy the cool milk. A girl showed us a rare sight, a bloom of *tiare apetahi,* a gardenia with petals on just one side that grows only in secret glades high on the mountain of Temehani. We didn't stay long, having easily bribed a driver to run a special service to lurch us back the 20 twisting kilometres to the boat. As we sped through settlements scattering children, pigs and dogs, I tried to answer questions about the plants and plantations, teaching others to recognise taro and mango, papaya and tobacco, pineapple, banana, plantain, hibiscus and other flowers. In the market of Utaroa, Bina bargained for fresh fish and sweet potato and hands of bananas. We learned the pleasure of being greeted by each person we passed and soon were wishing the Raiateans *bonjour* too.

By evening it was apparent that Polynesia had been absorbed. We shared the exquisite calm there is in tropical places of just sitting and watching the endless ravelling and unravelling of cotton-wool surf on a reef and then the fast, flashing sunset.

Most of us had discarded our wrist-watches and when Ted Cochran and I bought everyone lengths of printed cloth (the *pareu*) inhibitions went the same way and the uniform was quickly complementing newly-brown bodies. Suddenly we were a group, not a bunch of individuals.

If Raiatea cast the first spell, Bora Bora completely enchanted the entire ship. The island was well on the way to the high point of its Bastille Day celebrations and its normally unnoticeable village had been transformed into a fun-fair. At night golden-limbed men and women entertained with the *tamure*, the famed dance that brings Ma'ohi music and bodies to their most fevered climax. Soaking myself in the spectacle I crept through a fence and, pretending rather more professionalism with a camera than is the case, popped flashes as I sat in the dust an arm's length from the incredible vibrations of hips that varied from lithe to gargantuan. Tiny women, some barely teenagers, displayed agility that defied propriety and huge men conjured thoughts of war and blood as they shook their thighs, rolling their eyes. My Tahitian ancestors had danced like this. Fletcher Christian and his *Bounty* companions would also have sat on the ground, lit by great flares and entertained with pit-cooked banquets. I knew what they had made of it all, but what about my companions?

In spite of rain that made the bars ever more tempting, they watched transfixed for hours, as overwhelmed as I was at the thought of the battering ram effect it must have had on the less worldly minds of 18th-century sailors. I could tell they had seen into the soul of Polynesia and Michael Brook was unusually inarticulate, eventually stuttering: 'I'm just a boy from a small village in Devon, you have to remember.'

Taiyo returned to Papeete on July 15 and when Elliot Smith asked me if I would help persuade a young Tahitian girl to pose for his camera, I soon found myself cajoling her on my behalf as much as his, for this was my idea of a modern Mauatua. Teura was 15, tall with long black hair and with a face that would have launched far more than a thousand canoes. She displayed that seductive mixture of child and woman that only the truly innocent can adopt, sitting sure and erect but giggling behind fingers with bitten nails.

Teura was bought a new *pareu*, put a crown of flowers on her waist-length hair and posed in the flower market against the blooms, buds and leaves that had come from Tahiti's forests and gardens that morning. She confounded us with artless, sparkling smiles and Sphinx-dark stares that no European or American girl of the same age could emulate. She may have been younger than most of the women who went to Pitcairn but feeling I had the blood of a race who could produce such regal and magnificent women made me immensely proud.

The expedition was now ready to sail from Papeete to Tubuai, following Fletcher who established there the first British colony of free men in the South Pacific. Michael climbed up to a yard-arm and threw a garland of frangipani into the water, ensuring we would return. We had shared experiences of the astonishments that Fletcher Christian and his shipmates had seen almost two centuries ago. We had had our first experience of ocean sailing and, like Fletcher Christian, we had stocked up with private stores of treats without which we would have found life at sea impossible. On board were chocolates and mint tea, Camembert cheeses, a tape of bagpipe music, chewing gum, arrowroot biscuits, Coca-Cola and piles of trashy novels.

1,200 miles/1950 kms of South Pacific lay between Tahiti and Pitcairn. I suspected the radio was never going to work. It felt dangerous and also meant I could not make broadcasts back to LBC radio in London, cutting off a source of income.

CHAPTER 40

ECHOES ON TUBUAI

In the last minutes before we cast off from Tahiti for Tubuai, Andrew leapt overboard to kiss and give flowers to a young girl who had flirted with him from the dockside for hours. Once he returned, Terry and I called a meeting. From that moment we formally became The Sir John Barrow Commemorative Expedition and the project's aims were now to take precedence over those of any individual, indeed individuality itself was to be forsaken. Terry forcefully reminded us of the danger of the sea, the importance of safety procedures and of fast compliance with orders. Then, believing a detailed list of instructions would be counter-productive, I explained the rule on which the expedition would be based. I must always know where everyone was when ashore, so that valuable time was not wasted in waiting or searching for each other. Independent action was to be eschewed if the expedition was to be a success, and awareness of this was vital for the safety and enjoyment of all.

It was the first time I had been so direct and it made a noticeable impact. In the next few days, even the sick could be seen reading material about Tubuai and Pitcairn. There was tangible excitement about sailing in the wake of Fletcher Christian. There were also definite rumblings about the food on board.

In Papeete, Vivien, Jasmina and Marilyn had walked away from one lunch preferring to spend money ashore. Once finally isolated at sea, it didn't take much to see that I would have to deal with the unpredictable effects of loneliness and stress that *Bounty's* crew would recognise. The two-week span of a normal holiday away from home was over and I expected disorientation caused by continuing flight from familiarity to become threatening. Sickness struck in the form of a feverish flu, the watch system, although the fairest possible, caused tiredness and with heavy seas combined to produce tension in everyone. Yet working as a crew to sail *Taiyo* to Pitcairn, meant there was no time, energy or audience for tantrums and Captain Terry was justly tough on minor complaints.

I suffered neither the plague nor seasickness, so my status as leader of the expedition was unexpectedly enhanced. I decided to capitalise further on the situation by being both strong and silent, knowing it would be counter-productive, mutinous even, to assume any kind of command at sea. I stayed out of sight, reading *Bounty* material, preparing for Pitcairn. Then, realising I had far more time on my hands than expected, I stopped researching and plundered the ship for the cheap novels I had denied myself for years.

It took five days of rough water and starchy, fatty food to reach Tubuai. As it changed from a grey mist into recognisable trees, mountains and plains, I felt the first sense of real achievement. The discussion on deck quickly demonstrated that Fletcher Christian's decision to settle there seemed perfectly reasonable, even sensible to us all. The most startling feature of Tubuai is how un-tropical it looks if the fringe of palm trees is ignored. Almost immediately behind them rise stark hills of mauve and green and brown.

`My God, we've taken a wrong turning . . . it's Scotland,' said Ted Cochran. Indeed, it might have been. Or Cumberland. Or the Isle of Man. Already it was obvious to us how Fletcher Christian might turn his back on England and expect to find happiness in the South Pacific, because a great deal of it looked just like the homes of his childhood. By the time we paid out our anchor in the lagoon, we had identified Bloody Bay and soon agreed, on purely topographical grounds, that we too would have built Fort George where Fletcher had chosen. It was the most attractive part of the island, provided great opportunity for secrecy and was close to the small lagoon islands or keys where animals could be safely kept.

The little settlement of Mataura, off which we anchored in extraordinarily clear water, has a square with a flagpole and was lined with more of the palm-panelled huts we had seen on Bora Bora. Was Bastille Day celebrated somewhat later down here? No, they were continuing their celebrations to ensure a good send-off for the teams of athletes going to Papeete for the Pacific Games.

Ted and I went in search of an American photographer, Don Travers, who was supposed to live here and who could be helpful. Others went on an informal exploration charged with discovering bars, restaurants, banks and the like. What little village there is spreads languidly along the coast and up the one road that heads inland. Hibiscus, bougainvillea and frangipani crammed the gardens and children smiled before they stared. The Post Office was a walk west and the bank was some way east but no one knew when it was open. We didn't find Don and it was established there were no restaurants other than those temporarily operating on the square. We found the one bar almost by accident; you had to walk through the counter of one of the two Chinese general stores to a shed of corrugated iron and wood, some distance from the village centre. Those less devoted to the refreshment offered by Hinano beer would not have found the establishment. Here, convivially crammed onto wooden benches, we learned there was to be more than one discotheque each night and a Miss Tubuai competition. Once more we were amid Ma'ohi party time and could expect to be royally entertained. We were plied with beer by fishermen, who undoubtedly could less afford it than us, but who gave with such pleasure it was impossible to refuse while able to speak.

That same day I had to cope with tears from one expedition member and then Marilyn's threats to fly home because she didn't think she could cope with more sea sickness. Making the most of the alcoholic cloud under which we all returned to the ship, I continued my diplomatic ploys and arranged to cook for the first time. We sat by candlelight behind the reef of Tubuai and ate sweet-and-sour pork and Terry agreed that I should cook every second day, ostensibly to give the cook a break. This time a Christian had averted a mutiny.

To be fair, cooking for 15 people three times a day was the hardest job on board, exhausting when done seven days a week. Of all Bligh's acute observations on the basic requirements of a viable ship, none is more important than nourishing food and in 1980 it needed to be attractive and varied too. Tomorrow was to be our first working day, so over the conciliatory dinner I repeated my simple request of responsibility to one another.

By 9.30 next morning I was enraged and the day's plans were in pieces. One member of a small advance party had left after breakfast to follow up leads of people with specialised local knowledge and on not finding that contact, had picked up a German itinerant and sped off on the back of his motorcycle. Where or for how long, nobody knew. As I had warned, half the rest were angry, the others followed the flagrant example of irresponsibility and wandered off to follow their own interests. Matters were only solved by the final appearance of Don Travers, the photographer. He was a burly, blond ex-sailor who had stopped at Tubuai on his way to New Zealand, where he had planned to go to university using his GI grant. Instead, he fell in love with Jeanette, one of the most attractive Tubuaian women, married her and settled. He used his knowledge of the people and a mixture of English, French and Tubuaian to help collect the group, to arrange transport for our mountain of film and surveying equipment and to deliver us to the site of Fort George, two miles/just over 3 kms from the town square.

There was little left to see of that once bold place. Without knowing about the wooden plaque chained to a tree by the National Geographic Society, it is easy to drive past the hedge that hides it. Houses are built close to each side and in recent years the fort's brave earth-packed walls had been bulldozed and most of the moat filled. It was said that too many people were drowned there in the rainy seasons. Britain's first free settlement in the

South Seas and that had been home, refuge and battleground of Fletcher Christian's mutineers and their hostages was reduced to a sweet-potato patch.

The first thing I saw was a well in the middle of the garden. It took some time for its significance to emerge. We universally agreed that this spring meant we would have settled exactly where Fletcher had, who must have thought all his dreams had come true. Fresh water is a far greater guarantee of prosperity than simply being friends with the right chief, even if to ignore that was risky. The site also had safe anchorage, handy islets within the reef for the ship's stock, a clear view to the reef entrance and a water source he could control and protect. Already I could see that visiting the places Fletcher had been would let me tell a more accurate version of his life.

The first thing I saw was a well in the middle of the garden.

In 1980 there was enough detail of the moat left for us to make measurements and a map. Louis, a smiling neighbour, gave us permission to survey and photograph as long as we did not damage the young plants. I was amused to recognise two breadfruit trees growing among the sweet potatoes and wondered at their ancestry. What happened next was not amusing but illustrated the ability history has to repeat itself.

Leaving Michael, Mark and Richard to map the fort's site, I pushed through the fringe of toa wood and coconut palm to the narrow, gold-white beach. Not far to my right were the small keys that had once been the pens of *Bounty*'s animals. The beach to my left was being marked into a grid prior to being searched with metal detectors. It had seemed safer to put the errant, unrepentant wanderer to work here, away from the rest of the party, to avoid further interruption of their work. Mark suddenly appeared and told us that Michael had been arrested and driven off *'by some dude who needed a shave but said he was a gendarme'*.

It was difficult to remain cool during the 30 minutes that slowly ticked by before a Land Rover returned carrying Michael, a Frenchman who identified himself as one of the island's two gendarmes, and a surly Tubuaian. I was informed we were trespassing, for Louis, who had given us permission, although of the right family, was from a branch out of favour with that which more properly owned the Fort George site. The large sulky young man was from the aggrieved branch and kept stealing distasteful glances at Michael's bare feet. At some time during the voyage he had allowed Jasmina to paint some of his toenails iridescent pink.

I was expected to see the mayor. First the gendarme had to go home, shave, put on a uniform and apply several handfuls of after-shave lotion in a painful attempt to staunch the flow of blood from his throat, wounded by haste. Then we had to wake the mayor from his post-prandial nap by standing in his lush garden and shuffling.

In separate vehicles we next rattled back along the pot-holed road to see an old man who sat on his haunches and spat and looked at Michael's toes and refused to give us permission to return to the site. He should have been consulted, he was not consulted, and that was that. Further, he was only caretaker of the property on behalf of some 'old ones' who lived in the hills. Perhaps if the mayor sent them a telegram. . .

Our gendarme was as amused as we became, explaining he spent most of his time refereeing such family feuds and the continuing flavour of farce helped me forget the time that was being wasted. He led us on a detour to show us an extraordinary, overgrown marae on which the stone paving, the erect back rests and the petroglyphs were still in place. It was a chilling experience and I was very grateful for his thought to show us. Tubuai was lucky to have such a sensitive and interested man.

Don and Jeanette had joined the others for lunch in a coconut grove close to the disputed garden. As Michael and I tried to make a meal of the few remaining bananas and sun-melted salami, Jeanette told us of her great interest in Tubuai and its history and how she and Don collected stories. Eagerly I explained how much I wanted to identify the site of the battle fought here by Fletcher Christian in which so many islanders had been killed. Surely such an important place would be vividly recalled in oral tradition?

Through the sticky heat of the afternoon, Jeanette drove me to see the island's sages and storytellers. None knew where the battle had been, some didn't even know about it. In fast fading light, we rattled up an overhung track to a collection of huts in a muddy clearing protected from most light and all weather by a high canopy of breadfruit and mango trees. As if on cue, descendants of the biting insects that tormented Bounty's men emerged. Even the smoke of the open cookhouse was no protection. Stamping, scratching and slapping, sidling from the attention of mange-disfigured dogs, we spoke to the oldest man on the island, while his wife wove a basket.

'If it was in the time of my father or his father I would know about it. Before that there was a great sickness and I know nothing'.

So much for continuous oral tradition, I thought, by now more interested in whether the mosquitoes in the South Pacific were malarial. Later, in the circle of light from a paraffin lamp, in the house Don had rebuilt from a shell of coral walls, he, Jeanette and I looked at the books on the island he had collected and learned why the battles with Fletcher Christian and his men had become so unimportant. When *Bounty* arrived, there was an estimated population of 3,000. Less than 50 years later it was only 300. Intestinal sickness and fevers introduced by *Bounty* and then by the missionaries and traders who quickly followed, had massacred the people of Tubuai, their heritage and their memories. The island has still not recovered from the interest of the Europeans. Its population in 1980 was a bare 1,400 and at last increasing. I left them a copy of relevant extracts from *The Journal of James Morrison* and they promised to continue the search for the battle site.

Naturally I am very interested in men who leave a European culture to marry a Ma'ohi woman, without in any way abjuring their backgrounds. Don still loved books and showers; Jeanette nursed her baby while sitting on the floor. They complement one another perfectly and when I asked Jeanette how and why she had learned English, she punched Don in the chest and said: 'Because I wanted him!' Then she laughed that long inimitable fluid laugh of the perpetually happy Ma'ohi woman. Had Mauatua/ Isabella made Fletcher Christian as happy?

Next morning the gendarme told us the mayor had a message saying the `old ones' had also refused permission. Thanks to Don we had found out why. A team of journalists had recently use metal detectors on the site while foolishly circulating stories of a large hidden treasure. Greed flared and came between family factions that had ignored the site for years. Once again, a party led by a Christian was the victim on Tubuai of an internecine war.

I decided to let it rest rather than declare war in the manner of Fletcher. Michael and Mark resolutely walked back to the site with their boards and tape measures and completed the survey in stealth. Later we tracked through the taro swamps, again trying to locate a likely site for the battle and agreed that it could well have been close to the marae our gendarme showed us because the distance from the fort and the terrain fit Morrison's description well.

We spent our remaining short time on Tubuai wishing Fletcher Christian had made things work here. Because of its temperate climate, the vegetation and conditions are less relentlessly tropical than those of Tahiti. There are pastures with indolent cows, as well as pigs wallowing in taro swamps. There are butterflies, lemon trees, roses, flowers, grasses and bushes we recognized. Elliott, Marilyn, Vivien and I walked around to Bloody Bay and identified the marae of Tonohae, imagining the elaborate ceremonies and enormous gift giving that initially welcomed Fletcher. They call the bay Murivai and sitting on its sands we imagined the fright caused by a sailing ship and the horror of the bloody carnage when she hurled death from grapeshot into wooden canoes.

Tubuai's happy blend of familiar and unfamiliar was unexpectedly soothing and the constant smiles and calmness of life here affected us all. We hoped it was a taste of what Pitcairn might have in store for us but before we sailed to discover that Tubuai dramatically demonstrated the brute strength of the sea and the dangers to come.

One of *Taiyo's* two small boats had been damaged as it tried to lace its way through the treacherous coral heads that barred most of the way from our anchorage to the jetty. In case the second boat were also to be damaged, Terry moved *Taiyo* little further east to the almost completed commercial wharf and she was made fast with her stern to the jetty. That night the winds rose to a full tropical storm and with the reef being no protection against the waves, we were more uncomfortable than we had ever been at sea. It was another extraordinary echo, this time of *Bounty*'s experience in Matavai Bay causing her move to Pare. An anchor watch was mounted in the slashing rain.

At 5 am I was wakened by a scream of metal and what seemed to be an explosion. Confused both by sleep and adrenalin, I had to decide whether I was more use on deck or in my cabin; in an emergency at sea a non-swimmer on deck can be a dangerous complication. The tumult was too great to ignore, so I rushed out to find a scurry of half-naked crew and expedition members endeavouring to secure *Taiyo*. An enormous gust had dragged her anchor, so her stern was lifted and then slammed onto the lip of the wharf. With each surge of wind and water, the ship was being further damaged and so was eventually brought parallel to the wharf. By the time the sun came up we were safe, but the horrid squeak of metal against fenders of black rubber was extraordinarily trying, every minute reminding us of the forces we were about to battle. Don and Jeanette drove up early, for they had heard *Taiyo* hammered on to the wharf above the noise of the storm.

Our ship had done more damage than had been inflicted upon her. A little heavy-duty panel beating and a coat of paint would put things right, and we were doubly pleased to be aboard a ship of steel. A wooden or fibreglass vessel would have been wrecked and the expedition ended.

We were a sobered group as vegetables were delivered and stowed and we walked to the store to buy ourselves treats. Arrowroot biscuits, something I had not seen since I left New Zealand in 1965, had become a shipboard rage and could be kept down by the sea sick. We bought cake mixes, more Coca-Cola, shampoo and sun oil to last almost two months. Just after lunch, Tubuai entertained us with a double rainbow but it was an unkind promise of better things that did not come. We sailed into more rain and mountainous seas. The feverish flu had returned, in some cases combined with seasickness. It was a tense ship.

In an attempt to cheer everyone up, I promised to cook lasagne, something I knew I could do extremely well and that was well suited to cold miserable weather. As we passed through the reef and left the lee of Tubuai, I knew I had crucified myself.

Huge pots of boiling pasta, of bechamel sauce and of rich *ragu* plus tumbling canyons of waves are not fit companions for someone cooking on the open sea for the first time. For all that *Taiyo* was beautifully fitted, a basic mistake had been made. Whoever designed the galley forgot that the hob and oven should have been gimballed, freely suspended so that they stayed level whatever was happening to the rest of the ship, just as hammocks do for sleeping men. There was as much lasagne on the walls and floor of the galley as ever got to plates.

I once stamped my foot in utter rage. 'For Christ's sake, stop moving!'

Forbidding even today, Bounty Bay was once even more dangerous with rocky outcrops hindering every approach to land and requiring great skill to navigate.

A 1980 view of Bounty Bay, with the long, steep and sticky red-earth Hill of Difficulty leading up to Adamstown; it is now paved.

CHAPTER 41

A CONTINUOUS PUZZLE

On Saturday 26 July we sailed from Tubuai and our voyage in the wake of Fletcher Christian was fairly much on schedule. On Tuesday we sighted Rapa, an island of startlingly primordial appearance that was to be a respite from the tumbling Pacific swells. Like Pitcairn, it is rarely visited and women far outnumber men, many of whom have gone to Tahiti to find work. For the first time we saw what solitariness could mean to an island community and I was agitated in case I found the Pitcairners as depressed and uncared for as the people of Rapa. There was a sadness on Rapa caused by a rising of expectations that could not be answered by the two or three ships a year that called, just as I believed the case to be with Pitcairn.

We fairly flew out of Rapa with winds gusting up to 30 miles/50 kms, followed by mischievous squalls expelled from the mouth of Rapa's harbour like a series of smoke rings. Lying on my bunk the ship gave the same feeling of power as that of a jet as it hurtles itself into the air. We were rock steady and speeding along and at that rate we expected to be in Pitcairn in five days.

By the middle of Saturday morning, any movement we were making was sideways or up and down. None was forward and for a time we were thrown back towards Rapa. Sunday was no better, so Captain Terry decided we should motor and the inescapable noise and fumes did nothing to improve tempers or health. Each day Pitcairn seemed further away as we successively wallowed and motored, even though perfectly placed to be pushed by the trade winds. Toward the week's end, it seemed we would get to Pitcairn on Saturday, the island's strictly observed Sabbath. I didn't think this wise and told Terry that for all his efforts to get us there quickly, we would have to stay out of sight until Sunday morning. My precautions were unnecessary. On Saturday the wind died altogether and when we were not motoring we drifted backwards. It was galling to be under power in such a magnificent sailing ship but motoring meant more time on Pitcairn and already I feared we might only be able to stay less than two weeks and that certainly would not have been worth the effort. Still with no radio, the Pitcairners had no idea when we would arrive, or if we had changed our mind about coming.

Sunday's sunrise deserved trumpets and Handel. Against a pink and puce sky, bubbling towers of molten clouds poured ever widening rays of light onto a reflective Pacific. Pitcairn was dead ahead, purple, and back-lit. I braced myself against the rail, alone in the bow. My own excitement was made greater by seeing the expressions of joy and achievement on the faces of the expedition members as they came on deck one by one.

As we sailed along Pitcairn's coast, I was amazed at how it towered and wondered why no one had written of the ochre-red earth that showed in great gashes. The cliffs were daunting, aloof, forbidding. Even on this relatively calm morning, I could see enormous spumes of spray as the rollers, unhindered for thousands of miles, crashed into this lonely rock. I stood by *Taiyo's* rail, muttering self-congratulatory expletives, a rather pointless sort of reverie, broken by the start of a partial solar eclipse that ended exactly as the first of two whaleboats full of Pitcairners roared slowly past and told us to continue onto the lee of the island, which that day was opposite Down Rope. Half the passengers leapt from both boats and came on board distributing oranges and bananas. Suddenly it was true. The faces we had seen in National Geographic's *I Found The Bones of The BOUNTY,* the people who shared my ancestors and my family names were on board. I stood back for a moment, considering my new responsibilities. At sea Captain Terry Purkiss was the ultimate authority, now it was me.

When I began to introduce myself, there was some amazement as I was expected to be 'a big one', like the Pitcairners. It made no difference, and soon we were all being plied with questions about the voyage and who we all were. I felt at home. Warren Christian looked like my grandfather William. His brother Ivan was currently the island's Chief Magistrate but they are as unalike as it is possible for brothers to be, one with Ma'ohi features, the other Caucasian. In Tom Christian, the internationally known radio ham, I saw my father's brother Keith. Tom was dark and had black hair, Keith was fair skinned and blonde but they have the unmistakable Christian nose and the same tight curls. Laughing Nola Warren was there, so was her husband Reynold, the Postmaster, and Len Brown, the island's Chief Engineer. When we joined the boats in the lee, most of the rest poured aboard. Glen Clark, Jay Warren, Brian Young, Rex Whiting, the school teacher from New Zealand and his wife Moira, Pastor Stimpson . . .

With a trepidation that was well founded we found ourselves in the famous long boats, sailing past St Paul's Rock into *Bounty* Bay, through the treacherous entrance and then helped onto the jetty. Damp with spray and fuddled with disbelief, we were swept up by this or that family with invitations to stay and within hours I knew the long isolating journey was justified. It had made us ready receptacles for the unique hospitality and way of life of Pitcairn. After the discomforts of five weeks on a ship, bereft of real privacy or fresh produce with no clean clothing or showers other than in cold salt water we were grateful for buckets filled with hot water. Plain honest food at a stationary table with new faces was a banquet.

By next morning everyone had over-eaten enormously. At Tom and Betty Christian's, Andrew Brady and I sat down to a table crammed with two types of steak, fish in coconut milk, chicken, home-made bread, perhaps half a dozen or more vegetable and salad dishes, pilhis (the baked vegetable puddings described by Bligh), and then apple sponge with custard. Others had equally gargantuan stories to tell. All had biting criticism for recent publications about Pitcairn and writers who had described it as unwelcoming and squalid.

It is the publishing of such inaccuracies that has made the Pitcairners wary of strangers and we recognised how their shy self-protection might be misunderstood by some as unwelcoming. One writer who had arrived on the island in just over a week from his home in New York belittled the way they hoarded string, paper, glass and metal. This was typical of how few visitors to Pitcairn grasp the reality of living in such isolation. Keeping everything is simply facing up to the reality of living in isolation and the unreliability of the rest of the world, the pride of the self-sufficient. In a community where there are no servants, the aged proudly look after themselves and even mothers with babies must sew and dig and harvest the year round. There is bound to be some untidiness but there is nothing that approaches the filth created by the pigs in Rapa. There are no pigs on Pitcairn and no sight of the ragged children or disease we saw there, either. Perhaps those untrue things are said of Pitcairn by those who judge solely by contemporary standards of large cities and who have never seen other Pacific islands? Even land-locked communities know they will not survive if they have or grow or keep just enough. Whether it is fruit that rots uneaten, vegetables that moulder undug or great piles of old wood and metal, excesses of these are the only reliable keys to survival, the fact that there is more than needed.

The same misinterpretation of life on an isolated island still happens, when every few years unhappy visitors use Pitcairn to express their personal pain and rage at the world, doing a disservice to journalism, womanhood and to Pitcairners.

In 1980 documents needing official stamps or signatures to help solve an infrastructure or health problem might have taken a year to get to New Zealand and back. It was especially galling to us that, when the rest of the world was playing with the first Walkmen and videotapes, Pitcairn had to use Morse Code as its official communication with the outside world, including that to and from its Governor, the British High Commissioner in New Zealand who never visited.

During our first night on shore, our first night of clean sheets and proper beds, a tropical storm of high winds and heavy rain blew up. In the warmth of a room that did not move, it was a wicked pleasure to feel sympathy but do nothing for *Taiyo's* crew, tossed mercilessly at anchor. By morning all was calm and, apart from our stretched stomachs, we had but one physical complaint, mosquitoes. Almost everyone had suffered dreadfully, and, unfairly, those who had been most seasick were most bitten. The bites were so severe, especially round the ankles, that at least two expedition members were made sick by the discomfort and exasperation. It was only as I read deeper into the books and papers we had brought that I learned Pitcairn had been like Tahiti in *Bounty* days, mosquitoes were unknown, being introduced about a century ago. The Pacific without mosquitoes must indeed have been bliss.

Our first afternoon and morning on Pitcairn were spent wandering, getting land legs and meeting the rest of the islanders. My impressions of these days cling as my most vivid memories of Pitcairn. The storm had churned up the red-clay roads and tracks into a viscous mud. The Pitcairners called it the world's friendliest mud because it sticks so tenaciously. In just a few steps you have thick clumps under each foot. Then the knife Pitcairners carry slices the worst off and you continue. Bare feet were somehow less attractive to this red menace and easier to clean, too, so they were readily adopted. Today the infamous Hill of Difficulty from Bounty Bay up to the village of Adamstown has been concreted, a great advantage.

The single settlement stretches along the cliff that is to the right looking up from Bounty Bay and for only a short distance up the slopes from the main road. In more populous days the single main street was Coconut Grove until it reached the square and church and was then Pitcairn Avenue. Now it is just 'the road'. Although lined with familiar geraniums and gladioli, it is decidedly tropical too and with a continuous puzzle of gardens, half-seen houses, dappled vales, banks of red earth, and the protests of disturbed cocks and hens.

Beside a stately grove of banyan trees, which over the years march across roads and into gardens with the certain arrogance of all magnificent things, are low plots of spiky pineapples, one of the best-tasting varieties in the world. Hibiscus bushes with flowers of flaming coral tangle with vivid blue convolvulus, that spills further onto those epitomes of the tropics, the banana palm, just filtering enough of the strong sun through feathered leaves to light tracks with an unearthly warmth. Suddenly, golden mandarins and oranges glow through dark leaves Douanier Rousseau-like, papayas cluster atop their branchless supports and unripe mangoes hang from thin threads.

A few monumentally fat dogs attempted to yap at we visitors. Each had a personality or physical quirk and there are also many cats but the burdens of tinned food and cat litter are unknown. As in any English village, most houses are ruled by one or two of these creatures. There are rats, too, the only native quadruped but I don't think any of us ever saw one. We were too busy looking for the huge, rattling coconut crab and the spiders said to be five inches/12 cms long. The first time you saw one of the latter, you would swear it was twice as big, especially if that initial encounter was in an outside lavatory, lit only by a torch, late at night. If the rats were really as big as cats, as we had heard, we just didn't want to know!

Above the coconut palms we saw Pitcairn's lovely flashing white fairy terns, which spend idyllic days in flirtation, tumbling and kissing with contagious high spirits. Like Pitcairn's few other birds, the terns are largely songless. Once only a little away from the cliffs, the muffling effect of the thick foliage absorbs even the incessant surf's thunder, so a pause in Pitcairn walks reveals a silence that had a discomforting primeval quality.

. . . if Fletcher Christian did make the difficult climb to the cave and brood there for days, it would have been a very tough man or woman who could have ignored his domineering presence.

Through the treetops and tracks of the village there is one dominating sight. Lookout Point is unavoidable, sometimes menacing and so much bigger than we expected to see on such a small island. It's easy to see the scoop of rock on its face called Christian's Cave and if Fletcher Christian did make the difficult climb to the cave and brood there for days, it would have been a very tough man or woman who could have ignored his domineering presence. Those would have been oppressive days on the island.

As you walk up out of the village, the hills become steeper and much of the foliage changes to the troublesome rose-apple, a scrubby fast-growing tree from Norfolk Island that has become a pest of the first order. It does provide firewood and prevents erosion but is terminally unsightly. In the 21st century rose-apple and every other exotic is gradually being removed, the better to provide more fragrant nectar for the island's exceptional honey. When we first walked across the island and up to the ridge that ripples the length of Pitcairn, an unexpected feeling of vulnerability developed. Although there are countless hollows, glades and other secretive places, the very erectness and isolation of the island and its eerie silence easily gives the sense of it offering no place to hide. Suddenly, it is not safe, not a refuge.

Our lives revolved around the houses, wonderful, welcoming, jumbly houses that, like Topsy, just growed. Most were of weathered boards that had never seen paint - where would it come from? Some had deep verandahs, some were mixes of old and new. I stayed with Tom and Betty in the oldest inhabited house made of massive timbers, pit sawn on the island over 100 years earlier.

On our first afternoon on Pitcairn, 81-year-old Christy Warren waved us into his house. One room endlessly led to another and he seated us at a big table and told us of the hardships of his life, his tremulous tone wavering between pride in the past and sorrow at his present. Now his second wife was dead, he proudly baked his own bread. He showed how clean everything was as he piled the table with cakes and biscuits and served orange cordial; Pitcairn was still teetotal then. He told of unpopular captains who had been thrown off their ships at Cape Horn, of the sailing ships that had visited and stories that made Christy spoken of as the strongest man on island but our evening meals were at 6 and we had to scatter to be on time. Elliott had found his first camera subject but it took the rest of us several days fully to understand the loneliness of Christy.

When we walked about Adamstown, it was not immediately obvious that uninhabited houses outnumbered the inhabited ones. We walked into some up shattered steps and through broken doors, even so feeling it was wrong to do so. They had been abandoned with clothes half out of drawers, postcards stuck into the mirrors of dressing-tables and bibles beside beds. They had been forsaken when their owners went to visit a sick relative or spent a few years working in New Zealand or Australia. New ideas, or the simple problems of return transport, turned the visits into exiles, not all voluntary, and the houses slowly sank back into the earth, noiseless and unnoticed, for their enemies are creeping plants and the voracious borer and white ants. Christy had a loving family and supportive community around him but in his lifetime he had seen the population drop from over a 100 to the current 60. Rather than enjoying the continuation of life that can be a comfort in old age, he had seen a continuous ebbing.

CHAPTER 42

FOOD HANGOVERS

Our objectives on Pitcairn Island fell into two broad categories, those with roots in the past and those concerned with the present. My personal quest was the grave of Fletcher Christian, if such a thing existed. As the Sir John Barrow Commemorative Expedition, we planned to make a definitive map of the island for presentation to the Map Room of the Royal Geographical Society. Projects and information associated with this map would give us a true picture of life on Pitcairn and establish ways that the advantages of the 20th century might assist the Pitcairners, if that were their wish.

Each morning the expedition met in the small, paved square of Adamstown, edged on three sides by public buildings, the Seventh-Day Adventist Church, the Dispensary, the Library, the Post Office and the Court House, also used for Council meetings and films. Earlier visitors had made maps of Pitcairn and credit must be given to the enormously difficult pioneering work they have done, which made our job simpler. Yet, none of these groups had been as interested as we were in preserving the idiosyncratic place names, or they had misunderstood Pitcairn pronunciation and noted names incorrectly. I knew something of the Pitcairners language and accent from the related Norfolk Island language of my grandfather and his sisters. Relying on this and the acquaintance of Richard Hudson and myself with the vowels of the Maori of New Zealand, we hoped correctly to write down the place names. An example of the complications is easily given. One map indicates a place called Jinser Walley. It is easy to interpret the last word for the Pitcairners persist in both the earlier English and the contemporary Tahitian rendering of the letter 'V' and speak of their `willage'; Vivien Gay became Wiwien, even to us. Jinser is not a name but a Pitcairnese pronunciation of ginger. Thus, the proper way to record this place is not Jinser Walley, but Ginger Valley. Another place was commonly mapped as Garnet's Ridge but this is named for the sea birds that cluster there and so is actually Gannet's Ridge. The definite and indefinite article both become `ah'. I'm going to ride 'ah bike', 'give me ah knife', and so on. Thus, Ah Cut, a great notch out of the main ridge is properly recorded as The Cut. Places are frequently preceded by Up or Down, indicating where they are in relationship to the speaker. So, you might refer to some place as Up Hulianda or Down Tedside without the first word being part of the real name.

By deciphering all the place names we could, a reliable record would be made while people still remembered. Names change when they are unwritten and the discovery of original meanings should tell us more of the past. Pitcairn's place names give fascinating glimpses of daily life over almost two centuries. Some are simply possessive, like Jack's Yam and Big George's Coconut and it is generally not too difficult to establish who the eponymous owner was. Others record accidents or incidents once thought important—Sailors' Hide, John-Catch-a-Cow, Where-Freddie-Fall or Bang-Iron Valley, where *Bounty*'s forge and anvil formerly stood. Those are obvious in derivation. Others are not and the delving we did was repaid handsomely.

The most amusing story we unearthed is the origin of the name of a fishing pool, Oo-aa-oo, which when heard could easily be dismissed as some Ma'ohi word, like the Hawaiian Oahu. Until we met Andrew Young. For most of his 81 years, Andrew made it his business to know all there is to know about his beloved Pitkern Island. He tired us out as he strode around the island telling story after story with a waving of a knobkerrie and a deep, firm voice that belied his age. With appropriate actions he told us about Oo-aa-oo. Last century, when Pitcairn women wore long skirts and petticoats, Agnes Christian and her husband Samuel Warren went spear fishing at night, in a rock pool where fish were often stranded when the tide pulled away. Women here are strong and Agnes waded through the pool with her skirts tucked up and with Sam on her shoulders, lighting their way with a flare of candle-nuts strung together and here called a rummer. In a moment of levity, Samuel reached behind him and touched the warm embers of his dying rummer to her rear. Agnes' surprise and perhaps Sam's as he was tumbled into the pool created the onomatopoeic name. There are dozens of such stories, and by listening to them we were hearing and recording the island's history.

A full survey of the island was a task far beyond our ability and time span. Michael Brook had joined a crash course at the Department of Surveying and Photogrammetry at University College, London, where he learned the simple method of surveying the British had used to map India. This was to be employed to map the roads of Adamstown only, for the island is incorrectly described merely as rugged. Like some crumpled piece of tissue paper, it is a three-dimensional maze of steep twists, turns, passes, valleys, ridges and slopes. Andrew Young guided us up to Ship Landing Point, where high winds and vertigo reduced me to crawling on all fours to peer down at Bounty Bay, then to the dust bowl of Hulianda and the arid eroded slope of Red Dirt, where only pandanus palms grow. The tempestuous pool behind St Paul's Rocks (about 250 yards/225 metres long) is almost completely surrounded by a high, jagged escarpment of rock shaped by surf. The sea slams in at one end of the pool, hurtling over barnacled ridges and through fissures, then drains and dribbles out the other until it hits the ocean again in huge vortices. Several men have drowned here, and we spent a long time marvelling at the majesty and danger of the place. We walked through Aute Valley, once entirely cultivated but now a low wilderness, with here and there a well-kept vegetable plot, pineapple plantation or banana grove. There were no signs of the reservoirs once built and which would be just as useful and labour-saving today; rain water tanks were the main source and in droughts even the oldest had to find water elsewhere and to carry it on their shoulders. We sat on high ridges and cliff edges to look at Down Rope and Tautama, where the earliest inhabitants had found material for their tools of stone. From The Cut and Gannet's Ridge we saw gardens, tracks and hills named after Fletcher Christian's three children, Thursday, Charles and Mary.

While Andrew Young guided us on his muscle-stretching tours he also told us a lot about the island's food during the times it was be self-sufficient. He showed us the dandelion greens and other plants they ate or brewed into teas. He told of hard days when fish bones from dinner were boiled to make gruel for breakfast that was poured over potato or yam. I then realised that the unique food of Pitcairn, like their language, a mixture of 18th-century English and Tahitian, had never been recorded. Jasmina quickly volunteered and spent many hours with most of the women and some of the men collecting recipes that had never been written before.

Being in the food business, I should have liked to have done this myself but I had a mountain of talking to do. Every native Pitcairner was a blood relative. I have descent from three mutineers, Christian, Young and Mills, and so could be woven in and out of their genealogical tables in many ways. The continuing marriages between close relatives require new techniques of recording information. Some Pitcairners then had seven lines back to Fletcher alone and we wanted to know about each other.

There were only five Pitcairn family names when we were there. Christian, Young, Warren, Clark and Brown and only the first two are *Bounty* names. There are no Adams and the last McCoy had left recently. Only first names are used and since there was a Glen (Clark) on the island, the children made an immediate distinction by dubbing me Glynn-Christian, said as one word. Having sorted out the present men and women, I then had to turn to the men and women at the top of our family trees. That meant more serious talking. On Pitcairn, this means serious eating and for that there was plenty of opportunity.

The amount and variety of food offered a visitor to Pitcairn Island was astonishing, due as much to the deep freeze as to local industriousness. As Seventh-Day Adventists they are expected to be vegetarians but this is a recommendation rather than a rule and the Pitcairners have always had a special dispensation to eat their goats, although few do. In 1980 the island was still officially teetotal.

Fish features regularly but shellfish and lobster are fairly rigorously opposed by the church. When Andrew Brady and I went out to dine one night, we were presented with a huge dish of rock lobsters, each more than five pounds/2kgs and with wicked giggles were told: 'we're heathens here!' Together with fresh ears of winter corn and juice made of boiled wild strawberries, we dined like kings. There were also other meats, vegetables with coconut milk, apple pie and custard, and the obligatory choice of pilhis.

Pilhis are thought by some to be the only bad things that came out of the *Bounty* mutiny. They are all made from starchy vegetables or bananas grated into a puree on a unique instrument called a *yollo*, an oblong of volcanic rock into which a coarse, criss-cross pattern has been cut or etched. The purees are mixed with coconut milk made by soaking and squeezing coconut meat, then wrapped in banana leaf and baked. Those made with green bananas are perfectly horrid, even to most Pitcairners but those made with yam or sweet potato have a fascinating honey-like flavour and became an expedition favourite. Pitcairners who stayed on coconut-palm free Norfolk Island after the 1856 migration had to use desiccated coconut and their versions are much more delicious. My family's favourite uses very ripe bananas, dessicated coconut and a little flour and bakes it in a roasting pan, then using squares of it to accompany roasts or as an irresistible sweet snack.

Pitcairn's cooking styles are much more varied than might be expected. The Ma'ohi pit ovens filled with hot stones and covered with earth are no longer used but open fires, known here as bolts, are found in many houses, as are closeable stone ovens. The latter are the oldest types of oven known to western man and those on Pitcairn are re-creations of the bread ovens built into the outside walls of the houses left by the mutineers in England. Bread cannot be baked a better or sweeter way and the Pitcairners' wonderful pies and pastries nearly all come from this source. A few kitchens have solid fuel ranges and more are installing modern electric cookers from New Zealand; several families have old and new style kitchens beside one another. Here you need as much capacity as you can get. With 60 people on the island there was an average of one birthday a week. Almost every one is celebrated with a party and a party means all are invited, everyone bringing contributions.

At home with the family or on any of these binges, the Pitcairners served food Ma'ohi-style, which is also the way it used to be done in 18th-century Georgian Britain. Sweet and savoury dishes are put on the table at the same time, so if you have your eye on the passion-fruit ice cream, some favourite flavour of jelly or the coconut rice, you can tuck into that before you tackle breadfruit chips and chicken curry or you can pile all on your plate at the same time. I never got used to seeing Tom enjoying a bowl of fruit, ice cream and jelly from a spoon in his right hand while biting into a slice of bread and yeast extract held in the other. He reckoned the combination was great.

In another echo of ancient Tahiti, twice I ate at a table with men only, while the women stood behind and watched. Pitcairners of the older generation said the custom was because God created men first and thus they must come first in everything. This is actually Tahitian tradition with a new set of Christian clothes, because there men and women ate separately for many reasons, some based on food taboos, some on social status. Even those who ate together did so at some distance from another, so their fly whisks did not inflict injury.

Pilhis are thought by some to be the only bad things that came out of the Bounty mutiny.

For mutual entertainment, we challenged the Pitcairners to one of their famous cricket matches. Visitors to the English cricket grounds of Lord's or the Oval would not have recognised it as such, and we had trouble too. Teams are as big as can be arranged and the rules are simple: the Pitcairners can cheat, their opponents cannot. Ted put beautiful lettering onto a cracked soup tureen that had been retrieved from a wreck at Henderson Island and the Taiyo Challenge Cup Match began Up Hulianda at 11 am on Wednesday August 20. It ended at 4.30 and we lost by several hundred runs in spite of the valiant efforts of our baffled Americans, who had practised secretly for days.

This was the first time a visiting ship had challenged the island for more than ten years. Following tradition, the match was celebrated with a public dinner in the square. Even though the women had been playing exuberantly, extra effort was put into the variety and quantity of food. After so long, many of us still remember the delicious flavours of a lettuce salad dressed with freshly pressed coconut milk.

Expedition members were convinced they woke with food cravings. Some said they trembled if denied cake and mandarins for more than an hour. Many pairs of trousers went to the bottom of suitcases. To do what I had to do, I had to talk with my mouth full, or I should have learned nothing.

Supposedly Fletcher and Mauatua's house; until the 1808 rediscovery of Pitcairn, all houses were single storey

The last traditional Pitcairn House, where the sliding windows and door could still be seen, traditionally said to have been designed because none of the forefathers could forge hinges: it has been demolished.

CHAPTER 43

GRAVES, POOLS AND CAVES

To disperse the miasma of deceit and subterfuge that infects the story of Pitcairn's first 20 years and to get a clearer understanding of what had happened here, I needed to establish details of the original settlement. In pursuit of Fletcher Christian's grave, I hoped also to find that of his wife Mauatua and of their son Charles and his wife, Sully, my great-great-great-grandparents. I had brought photostats of etchings and water colours discovered in libraries and private collections around the world. These would be used to site long-gone buildings and help make sense of half-remembered stories. We stimulated the community to find old photographs too, and the combination of pictures, memories and interest produced results to confound those who say the Pitcairners are not interested in their past.

In the photographic collection of Len and Thelma Brown I discovered Fletcher and Mauatua's grandson, Thursday October II, known as Duddie. He died aged 91 in 1911, and so had been a man of 20 when his grandmother Mauatua had died. In 1980 there were men and women on Pitcairn who remembered him wheezing with asthma and reminiscing. They heard first-hand but 20th-century memories of my Tahitian great-great-great-great grandmother.

The Browns' house seems to sit on the edge of a common ground of the first settlement, for around it are, or were, the forge, and the building where sugar cane juice was reduced by boiling to what is here called molasses, but which is a mother syrup as none of the sugar has been removed. Close by is the sugar-cane crusher itself, like a gigantic, rimless, spoked wheel, said to have come from the West Indies and to be over 100 years old.

Warren and Maisie Christian let me look through the remaining archives of Roy Clark, who had died earlier in 1980. For 70 years Roy recorded and wrote of Pitcairn with talent and detail. Each of his photographs is captioned and most faces are identified, giving sudden life to the names on the branches of the family tree. I was sometimes uneasy to be sitting on a deep verandah sorting these memories, to be enjoying the sight of hens flying into low trees to gorge on bananas, to be listening to Maisie making her famous guava pastries, or to Warren chopping and whittling at the *Bounty* model he was making for Upton Brady, my American editor. It didn't feel like work. Then I would find the face of a long dead relative. Was I looking at the face of Fletcher Christian in an image of one of his descendants or did those noble faces look like the unknown father or brother of Mauatua? There is a similarity between Christian faces over the generations it seems impossible for there not to be echoes of Fletcher somewhere. With a spread of these photographs in front of me, I knew I was as close to Fletcher Christian and Mauatua as it was possible to be but the gap would never be closed.

During the voyage to Pitcairn, close scrutiny of one of the etchings I had brought gave clues to what seemed to be a graveyard different from that seen today. The main clues in the picture are a thatched building, obviously non-existent now but said to have been a church and schoolhouse. There is also a huge banyan tree and under its branches you can just make out an untidy cluster of gravestones. Although this image had been used on a recent stamp, no Pitcairner had noticed the graves. None of the banyan trees that still existed stood in the same relationship to Christian's Cave and Lookout Point as shown in the picture but that could simply be artist's license. Mark Balsiger confirmed suspicions when he reported several people said Up the School when referring to the village square.

It did not take long to find there had been a school on the square just where the Post Office now stands. And the banyan? Yes, it had stood right on one corner, clearly remembered by those who could not recall the old thatched building except for the name. Over years banyans re-establish themselves, marching in new directions through the long suckers they put down. There was one that could believably have moved from that in the old painting.

Disbelief mounted more quickly than excitement as we placed the old gravestones in the 1980 village. Rightly so. The Courthouse, and the one before that, had been built over the first graveyard. We sent Andrew Brady under the Courthouse and he said there was at least one gravestone to be seen in the rubble. An earlier generation had decided to destroy and build over the other stones, rude though they may have been. I was horrified. Land had never been that scarce and these were the graves of their founders! I tried to find ways of explaining or rationalising what had happened. Possibly a sense of guilt made them ashamed of their rough, unmarried ancestors yet respect for the dead and their tombs was part of 19th-century religious morbidity. Only slowly did I realise I was judging and questioning these people by standards other than Pitcairn's. Ma'ohi grief for the dead is transitory and neither do Seventh-Day Adventists mope about death, teaching that the dead ought to be thought of with happiness, as they enjoy their reward in Heaven. Here, at least, the two cultures blended perfectly and on Pitcairn, the living have ever taken precedence over the dead and so building over early graves was not thought impious. It was important to have been reminded early of the folly of mak-

The early church and schoolhouse, with gravestones clearly shown under the banyan tree.

The Meeting Hall in the Town Square, built over the graves of Mauatua, Mrs Christian, and others, probably including that of her son Charles, my great-great-great grandfather ing judgments about Pitcairn based on the mores of other places and earlier times.

The site was almost certainly where Mauatua, my great-great-great-great grandmother and her son Charles had been buried and I was sad not to have been able to mark it, not even with flowers. That night we joined the Pitcairners atop the graves of Mauatua and Charles, to watch the film An Unmarried Woman. It was extraordinarily unnerving.

The longer I stayed on Pitcairn, the more I knew Mauatua had been the most extraordinary of women. Of the nine that arrived with *Bounty*, she alone is remembered and constantly appears in conversation, often as Isabella. I roundly curse those early 19th-century visitors who described her white hair and story-telling ability but who didn't record her stories, if she had been willing to tell them . . .

My host Tom Christian was probably the most sophisticated man on the island, well-travelled and professionally qualified as a radio operator. He was as interested in his heritage and in Pitcairn as I was and between official radio schedules and busy hours talking to the world from his ham-radio shack, he put me on the back of his Honda and we explored and identified places that seemed linked with the *Bounty* settlers. In a book called *The Pitcairnese Language* there is a list of place names collected by the school children of the 1950s, published under Ross, Moverley, et al. It mentions Maimas pool. Could that possibly mean Mainmast another of Mauatua's names? I was determined to find it, even though no one on the island could place it. The water levels on Pitcairn have altered radically over the last century, usually after massive landslides, so a once well-known pool could now be dried up and forgotten.

One morning I went to talk to the island's oldest couple, 85-year-old John Christian MBE and his wife Bernice, 81. This had not been easy to arrange as they were among the most active on the island, forever trundling their heavy wooden wheelbarrow of unique Pitcairn design up the winding tracks and roads that join scattered gardens. After writing my genealogy for Bernice, I joined John as he carved in the dim light of a small window. Almost blind in one eye and troubled with the other, he constructs and rigs sailing ship models that are sold around the world: 'I'm not going to sit around doing nothing, not if I can earn some money!'

There's one model he wouldn't sell, made entirely from the last scraps of a breadfruit tree said to have been brought to Pitcairn by *Bounty*. He let me hold the stoneware water jar from *Bounty* that had come to them through Bernice's family and then I started asking them about place names. Did they know a pool named after Mainmast? 'Of course,' said Bernice immediately, 'that's Down Maimas, over Tedside'.

The western side of the island, steep and forbidding, had never been settled although there are the remains of an important marae there. Once called The Other Side, this had become corrupted to Tedside. It seemed as though Mainmast's pool and Maimas might be the same place. I had also heard Tom refer to somewhere called Mummas and knew that was over Tedside, too. `Could what you call Mummas actually be Maimas?' I asked.

Tom thought. 'Of course, it never occurred to me. Mummas, Maimas, Mainmast!'

We sped over the ridge and were soon sliding across a greasy, dangerous slope under tall old trees below the road through Tedside. It had been years since Tom had been there but he found it. By the time I arrived, breathless and with bleeding hands after the difficult trek, he was hacking grass and moss away from a pool with his machete. We worked for more than three hours in the sun and in the moments we paused I was again aware of the special stillness of Pitcairn, sometimes calming, sometimes frightening, which makes the slightest rustle of branches perturbing and might be someone or something unknown. In the hidden pocket of tall grasses and banana fronds, I understood Pitcairn's how silence must have contributed to the horror of the days of ambush and bloodshed.

When the pool had been drained of filthy water and layers of grey slime, we stepped back to see that the rock had been shaped by man into a large bathtub that seemed so European it could not have been shaped by the men who came here to make stone tools centuries before. Cut by Fletcher for Mainmast? Or was it indeed older and simply discovered by Mainmast? It was easy to see her bathing on a languid afternoon and a beguiling sight it made.

We left the pool to fill with the clear water that dribbled in after running between the rock face and the luscious growth that tumbled down from the road. Two days later I hired the island's tractor to carry a party back and to photograph Tom's oldest daughter, Jacqui by the pool. She hadn't known of the pool and was as touched as Tom and I had been. This seemed to be as close to Mauatua/ Mainmast as I was to get and the identification of this pool gave the Pitcairners further confidence in the expedition and its serious interest in their island. The spontaneous help we had been shyly offered now became universal and we were shown more artefacts and photographs with a mixture of modesty and pride that gave even greater joy to discovery. Reynold Warren showed us a *tu'i*, a hand-shaped stone pestle used for pounding fish, taro or breadfruit. This was noticeably more elegant and worn than the few others we had seen and our engineer Hop, an expert in such things, felt secure enough to identify it as being Tahitian and old and worn enough to have been brought by *Bounty*.

This news was soon passed around Adamstown and several of

Tom Christian beside the excavated pool, discovered after realising that Mummas Pool, meant Maimas Pool. Maimas meant Mainmast, which was a nickname for Mauatua, given because she was so tall and straight backed. Seeing the pool, clearly European in shape, made every Pitcairner and expedition member feel closer to Mauatua and to Fletcher, who is believed to have shaped it.

the long oval wooden dishes that had been used as a mortar with a *tu'i* were rescued from back cupboards. A new interest in the artefacts and memories of our Tahitian forebears was rekindled and Nola Warren and Tom Christian were especially pleased to have the value of such things pointed out.

Pitcairn's younger men and women on Pitcairn were intrigued by our metal detectors that were the first seen on the island and we first took them up to Christian's Cave.

If Fletcher Christian had brooded there for weeks on end, or if he had nursed himself to health after being shot in the back in the 1793 massacre, there might be some evidence of his ammunition stores or of the lean-to Beechey says he built . . . perhaps he had hidden money from *Bounty* and had died before he could use it to escape? Hours of patient work, based on small grids checked with several detectors with differing abilities, found little but spent shells from forgotten goat hunts. For all this painstaking work we found nothing, neither treasure nor other artefacts of the early days. The payoff came as we struggled back and forth with aching chests and tortured leg muscles. We formed definite opinions about the cave and what Fletcher Christian might have done there. Or not.

Christian's Cave is anything but a cave. There are no narrow passages leading to inner chambers, no bats, no glow worms, waterfalls or cathedrals of lime. It is merely a lofty, triangular chip out of the cliff face, some 800 feet/245 metres above sea level It is one of the first spots on the island to be touched by the rising sun, but by afternoon is cold, windy and miserable, a dangerous shock to sweated bodies at the end of a climb.

Access from the village, which sprawls in its camouflage of greens away towards Bounty Bay, today begins in thick foliage close to the school and continues around Lookout Point on uncertain paths made by man and animal, shaded and sun-lit, until emerging on the glare of sparse, daunting slopes. I paused, breathless and slightly dizzy on my first climb, then zigzagged the next long slope on loose shingle and decomposing rock. It is steep enough to make progress on all fours both safer and simpler. 20 feet/6 metres below and to the right of the cave's final access, I had to clamber on great rocks, negotiate a brief but chilling cliff-face ledge and finally scale the short perpendicular climb to the narrow slope of scrub that separates the cave's ledge from the precipice. Satisfaction and relief poured over me in direct relationship to the sweat on my brow, undiminished by the laughter of Pitcairn children and men who had bounced ahead as if jogging in Hyde Park.

A hunted man, armed and provisioned, could undoubtedly defend this place for as long as supplies held out for no more than one attacker at a time could reach the cave and could be shot at will while doing so. At night, it might be a different story. With no water handy and no way to sneak out day or night to gather food, it would have been desperately uncomfortable and exposed.

The tiny figure is me, in the scoop of rock called Christian's Cave, too shallow and far too difficult of access ever to have been refuge for Fletcher Christian after Massacre Day.

It would also have been necessary to stay in one corner of the cave, for snipers could shoot directly into most of it from rock outcrops a stone's throw away.

I heard stories suggesting the cave mouth had been partially obscured by trees earlier in the 20th century and older water colours seem to indicate this may have been so at other times, too. To hide the cave fully from observers, a screen of tall mature trees and thick undergrowth would be needed to ensure secrecy, for even the casual eye is caught by slight movement on the rock face. The idea of there being enough trees to make this scoop a hiding place goes directly against the argued attraction it had to Fletcher Christian, a place to sit and stare at the sea and his new estate, to watch the sun rise behind Bounty Bay and slowly illuminate red gardens and thatched huts. Stories of trees up here seem to have been amplified to fit an unlikely hypothesis.

This is an important point. If Fletcher Christian miraculously survived being shot and then maimed by axes in the October 1793 massacre, he needed a secure hideout while he regained strength and health and prepared to sail away in secret. Could someone shot in the back, probably with a shattered shoulder and useless arm, plus head wounds from a later attack climb to the cave? At night, with infinite care, prodigious recuperative powers and strength, it is just possible, but subsequent dangers inherent in leaving for food and water, crossing cliff faces when sick and injured, would make it a dangerous and painful choice. If he had emerged during the day, he would have been certain of detection; if at night, uncertain of his safety.

Before I went to Pitcairn, I rather fancied the version of Fletcher's death given by Becke and Jeffery in their stirring but stilted *The Mutineer*. Their tale relies on Christian being succoured by his wife and others as he lay in his cave and so calls for a great number of lightly undertaken sorties to and from the hideout, both by Christian and his confidantes. Secrecy and the physical demands of repeated journeys in the time they allotted, even for healthy men and women, makes nonsense of Becke's vehement defence of his book's authenticity. Rather than a story that might be a slight bending of fact to flatter a narrative, he has gone far enough towards fantasy to ensure his story cannot be accepted. This is not necessarily the fault of Becke, who might have been misled by others, including those on Pitcairn, something not unusual. Any visitor to Christian's Cave can see the faults in his version, so either Becke did not go to Pitcairn Island, or he stayed a very short time, or, like others, he and his collaborator, Jeffrey, recomposed the story to enhance the drama.

There is another complication, a story on Pitcairn of an alternative way to the cave, down from the peak of Lookout Point. This would have been by way of vines and tree roots, for there is neither an indication of natural paths nor signs that adjacent landslides might have affected past tracks. Anyway, such aerial access would have made the cave even less secret and infinitely harder to defend and it is hardly a good idea to have the uninvited drop in from above. A man with a bullet-shattered shoulder is most unlikely to have made a cave with such access his goal. So, as a fortress the cave is a possibility. As the desperate hideout of an injured man, it is improbable. As a place where you can retreat into yourself, it is without peer. Most of the expedition used it thus at some time, wishing themselves in other places or other times and there are only gannets and soaring bo'sun birds to interrupt your thoughts.

CHAPTER 44

A POOL OF INFORMATION

Constant high seas put paid to our plans to survey *Bounty*'s wreck and to use our metal detectors among the boulders of the foreshore. A brief dive on the one possible day confirmed there was nothing left but the random piling of ballast bars. Other relics must lie on the shelf of the bay, more swivel guns, more copper sheathing and nails. Some are in Adamstown and in a sad state of decay, despite repeated requests to England and the USA for material and information about preservation. A cannon lies in the grass outside Len and Thelma Brown's. A swivel gun, pristine years earlier, flaked in a shed at Ivan Christian's. There are slowly decomposing balls from both types of gun, hunks of twisted copper and stores of copper nails in most houses. And there are stories of sewing kits, known in the Navy as a hussif, and other pieces that were not submerged but that became victims of the settling of estates or were lost in the depths of stored possessions and the envies of those dead or lately departed.

It's accepted that much of what was once aboard *Bounty* was given, sold or exchanged with the early whalers and sealers who called at Pitcairn. By far the greatest number of these came from ports in New England, including Nantucket and New Bedford. I like to think that dedicated research into old families there could still turn up great *Bounty* treasures, just as Fletcher Christian's Chinese bowl turned up in Nantucket. Oh for the time and the money.

Day after day Tom and I worked to recreate Pitcairn's first village and to establish the validity of any part of any of the versions of our ancestor's death. There was precious little to go on when we got down to it but we agreed the accounts given by the Tahitian Jenny had the greatest veracity.

The site on which Fletcher Christian was supposedly shot is easily shown to those who enquire and has been agreed upon since Pitcairn's first 19th-century visitors. Tom had been shown it as a boy and Old Duddie said his father had shown him the site but his father, Thursday October, was barely three years old at the time of the massacre and, according to Jenny, was playing at his home, which is on the other side of the island. Why is this supposed site of his death so far from Fletcher's house? Why should the site of only Fletcher Christian's murder be remembered?

Young's lost diary said Fletcher was buried close to where he was shot, yet the supposed site of the shooting has never been pointed out as his grave. The plot continued to be used as a garden and I can't believe his family continued cultivating what was literally the soil of their forefather. It seemed likely that site was a successful red herring of Adams and / or the Ma'ohi women, that, like good stories everywhere, became accepted as fact.

When Tom and I stood where Fletcher Christian was supposedly shot on Massacre Day, we rehearsed the version of Fletcher's death given many years later by Jenny, who was then off the island. The day for the murders had been chosen because the women had gone up into the mountains to collect birds' eggs. Here was the first anomaly. If anything could be called 'up in the mountains', this site could. It is only a few hundred yards below the island's long ridge, at the top of John Mills Valley, suggesting it was Mills's land not Fletcher Christian's. As then, it is protected by thick bush and totally insulated from the village, both out of sight and an arduous hilly trek away. It was not Fletcher Christian but the women supposedly up there.

Fletcher Christian is said to have groaned loudly when he was shot, a sound heard in nearby plots and recognised by some as the sound of a dying man. Others men who heard it thought it was Mauatua calling her two children to eat or dismissed the shot as killing the pig that Tetahiti had promised. This all becomes transparently impossible when you are standing where 'tradition' says Fletcher was shot.

Jenny's story clearly implies that most men were gardening close to the village, while the women were as far away as possible up in the mountain. There was no reason in those days for the men to garden in the hills. There was plenty of ground close to home for so few to cultivate and any gardens they did have in other parts of the island were simply divisions of naturally occurring fruit trees. The supposed site is amid such steep and rugged terrain it is unlikely any of the mutineers walked over to discuss an unexpected sound or even that they could have shouted to one another.

The biggest clue to the site being a fraud is the impossibility of Mauatua's voice to have been heard as far away as this. If Fletcher's groan and the dismissal of by other men did happen, it had to be closer to the settlement. Jenny shows this is so by remembering something both intimate to Fletcher and Mauatua and important to a woman. She specifically remembered Fletcher was working close to home that day because Mauatua had not gone with the egg collectors as she was close to giving birth to their third child. That does make sense. On the flatter, more open land around the village it would have been easy for men to saunter over for a chat or to shout at one another. It would also be possible to mistake a sound from the direction of Fletcher's house and garden as coming from Mauatua calling from her door. There was no doubt in my mind that the site traditionally shown as that where Fletcher was shot was an ancient fiction, *ha'avere* with a very long life. Where was the real site?

Ben Christian, Island Secretary for over 18 years, let me borrow the record of land deeds. There was never primogeniture here. Land is divided among a family's children and then subdivided and so on, thus it was possible broadly to piece together Fletcher Christian's original plot of land in the village. Within this swathe at the north-westerly end of the settlement stood Thursday October II's house, the oldest on the island, uninhabited and still showing the sliding shutters used instead of windows and doors. This extraordinary link with Pitcairn's past has inexplicably been pulled down just when better shipping means more visitors to Pitcairn and nothing would be more thrilling for them to see. This site is directly below Tom Christian's house, which stands on the site called Fletcher's, undoubtedly the vicinity of Fletcher Christian's house even if foundations have not been found. Further afield are gardens still owned by Christians and like some gigantic barrier, Lookout Point bounds the north-west extremity. This provides a simple reason for the naming of the cave. It would be Christian's Cave simply because it was on his land, rather than because he adopted it as a den. It would be unlikely to have been called this if it had been on someone else's property.

Fletcher's original gardens are within shouting distance of his house and Mauatua really could have alerted him by a call at the start of childbirth. If he were shot somewhere in this vicinity, his groan might easily have been mistaken for Mauatua, for everyone else was south or south-east of this site and both sounds would seem to have come from the same direction.

Edward Young's diary said that it was not until the following August that the women buried their dead husbands. It is a sobering thought that rivalry and tension were so high on an island so small that the bodies weathered a full sultry Pitcairn summer yet, having been there and seen how easy ambush and attack is, I knew the seeming uninterest was probably sensible self-preservation. We know the Ma'ohi didn't care too much about burial anyway. The women had managed to collect the heads of their men, which was much more important.

When the men killed in October 1793, now all headless skeletons, were buried, it was in a communal grave. Only Fletcher was supposedly buried separately in his garden and I found a solid clue as to where this grave might be, the casual mention in the journal of Dr Bennet, surgeon on a round-the-world whaling voyage aboard *Tuscan*, Captain Stayers. While writing about the Pitcairn he found in 1834, Bennet tells us: *'Fletcher Christian and John Mills were shot on the same day, by the Tahitians: the grave of the former was pointed out to me: it is situated a short distance up a mountain and in the vicinity of a pond.'*

Once more I was searching for a pond, this time certain to be a dry one, for none with water answered the description. Throughout the village I asked about ponds, springs and wells, secretly hoping the mountain mentioned might be Lookout Point, especially as I now believed this to be part of Fletcher's land. As the time to leave Pitcairn was fast approaching, I was anxious in the extreme and steeling myself for failure.

I let my mind cast far and wide, following the vaguest clues. For two days I considered clearing a flat area called Graveyard, towards the centre of the island. This is within such distance of the traditional site of Fletcher Christian's death that I had disproven, that it was a conceivable place for a remote mass grave. Perhaps I was wrong to dismiss that first site? Deeper research into the place name showed it was very old and no one alive had ever heard an explanation of its meaning. Then I remembered the *Bounty* settlers had discovered several graves of their tool-making predecessors. I crossed my fingers and abandoned that site. In the end I trusted my instincts about Adams and Pitcairn's foremothers. I'm sure they would never have allowed such an obvious and dangerous clue to the truth to exist.

Following tangents simply so I could return to my base with a clear head, I went with Tom to his garden below Christian's Cave. It is an important site for artefacts of the industry of those men whose graves had confused me. Tom was going to point out working sites where stone tools had been shaped, and we hoped to find a partly worked or broken tool for me as a memento. We achieved this and, delighted to have the lower half of a broken stone chisel at least 600-years old and that had been polished

This is the dried up shallow pool that fits exactly with Dr Bennet's little known description of where he was told Fletcher Christian was buried.

with use, I suggested to Tom that we take the path that leads to the Cave because some distance along, you come to a dark overhang of rock in which McCoy is said to have set up his still. I hadn't looked at it properly on my peregrinations to Christian's Cave and thought Tom might add some interesting facts.

A few minutes after we plunged into the bush, Tom stopped literally open-mouthed. It was the expression I had seen when he realised that Mummas meant Mainmast. I followed his gaze and quickly understood what had caught his attention. There was as plain a dried-up pool as one could imagine, once one knew. Close to gardens still being used by Christians, it was right in the middle of Fletcher Christian's original plot, within shouting distance of his house, and it was a short distance up a mountain, Lookout Point. Not only did the pool fit Bennet's description, it neatly tied a knot with the threads of my theory and Jenny's accounts.

On available evidence both the Pitcairners and the expedition agreed this was all a fair conclusion. We checked every map, reread every list of place names, listened to more stories but no other pool seemed to have existed that fitted the description. Two days before departure, we had no time for a systematic search for the grave or graves among the banana palms and coffee bushes scattered on the steep hills around the oval pool's crumbling banks. Identifying the pool did not prove Fletcher Christian was buried nearby but discounting one possible murder site and establishing another was a major step forward in the search for Fletcher Christian. There were more hurdles to clear before I was home on that one, not least of which was the emotional one of leaving Pitcairn.

As early as the second afternoon of our stay, expedition members feared the emotional wrench of saying goodbye to the island. Subsequently, the deepening friendships we made and the insinuating charm of Pitcairn drew more into this band. When we joined the Pitcairners in the Courthouse Hall to hear them record their famous songs of goodbye, ringing with rich, reedy harmonies unique to Polynesia, there was not one who could say they did not dread the day. It was put off until August 28, so that Ted and his film crew could record the brief visit of the *Essi Silje*. In 1980 British ships were officially thought unsuitable to call at the British Colony of Pitcairn on their way south to Australasia, so the island relied on the courtesy and personal endeavours of a few Norwegian sea captains to do favours by delivering supplies from the northern hemisphere, an extraordinary situation in which to find a British Colony. We went out to the ship with the long boats and by the time we were delivered to *Taiyo*, salt-splashed and cold, the sun was setting. Some of the Pitcairn women were seasick and so stayed in the long boats, tied up to our ship's side. The rest, most of the island's population, crammed on to our slippery decks for final embraces and tears.

The moment of goodbye, when just we 15 were left on the deck, was ghastly. We stood mute in harsh floodlight from our mast heads as they sang, 'Goodbye' and 'In the Sweet Bye and Bye', and we were grateful we had some preparation for the raw beauty of unaccompanied singing by almost 50 loved people on the open sea. The long boats' diesel motors roared them out of sight into the darkness, then they turned and swept past for a final wave and last glimpses of favourite faces and faltering smiles. The wrench was more than just that of leaving friends.

CHAPTER 45

A DEGREE OF IMMORTALITY

In the 16 days it took to sail back to Tahiti, there was reverie and discussion on *Taiyo*. Little else broke the routine of sleeping, eating and steering except the magnificent attentions of a single albatross. We enjoyed increasing warmth as we bumped north, and pursued suntans with naked determination. Close quarters and the return to duty and discipline brought to festering point the problems of companionship. Yet, none of the potential for explosion brought aboard by 15 strong-willed individuals erupted to such an extent that the expedition's objectives were seriously threatened. There were few blows other than some irregular and infantile bullying but this was as likely to be verbal as physical. It was a triumph for all on board *Taiyo* that emotions were contained. There were disappointments and shortcomings and ugly moments but most of these had happened on land. The equivalent situation on board *Bounty* as she sailed away from Tahiti was never far from my mind but the expedition told me that 16 days of discomfort was nothing to being on board an unhappy sailing ship that would be at sea for many months ahead.

Tahiti was welcome when we tied up in the first few hours of September 13 and headed for the food stalls that alluringly scented the air along the strip of concrete between the park and harbour at the eastern end of town. We devoured steak sandwiches, bags of chips and Coca-Cola. Within days we had melted away from one another, Michael and I flying to New Zealand, Australia, Norfolk Island and Honolulu, to further our research.

During those final days on Tahiti, I spent a stimulating and civilised afternoon with Bengt Danielsson and his wife, Marie-Therese. Bengt was aboard Kon-Tiki in 1947 and since then has become a world respected authority on the anthropology of the South Pacific and the *Bounty* story, as has his wife. We sat in a huge, cool, thatched library and study, one of the scatter of traditional Tahitian buildings on his marvellous estate right on the lagoon. Across the close-cropped grass was an uninterrupted view of Moorea and as we talked of *Bounty* and times past, an outrigger canoe, paddled by a woman with a flower in her waist-length hair, glided across the proscenium of coconut palms. Such diversions and the *mirabelle eau-de-vie* produced by Marie-Therese notwithstanding, Bengt went quickly to the heart of my questions. With typical Scandinavian economy he convinced me that the simpler an explanation was, the more likely it was to be true. He gave me examples which encouraged me not to fear an unfussy solution to many of the vexed *Bounty* questions. It was deeply reassuring to have the keen interest of Bengt and his wife and next day he drove me to see the sites he had identified as the breadfruit camps at Matavai and Pare.

In the cocoons of Air New Zealand DC-10s and the welcome of Mary Crowley's house in Sausalito, I carefully reviewed what we had found about the death of Fletcher Christian. Until the time I began my research, the circumstantial evidence seemed as strong for a dangerous undercover return to his birthplace as for his demise on Pitcairn. Since then I had made two important discoveries.

The first was the 600-year family history mentioned earlier. At the end of the entry about Fletcher Christian, whom the authoresses dub the undoubted black sheep of the family, they write: *'It is . . . extremely unlikely that the Pitcairners should have been deceived in the matter of their leader's death. Moreover, had Fletcher got back to England the only motive worth the risk would have been to see his family and of such an event no tradition has been preserved.'* I find this completely persuasive, considering its provenance. If there had been the tiniest glimmer of gossip from sources in or close to the family, Vio Christian and her daughter, Rita Christian Browne, would have found it and recorded it.

If Fletcher Christian sailed secretly from Pitcairn, he surely would have taken Bounty's *invaluable Kendall chronometer and the azimuth compass.*

It *is* extremely unlikely that the Pitcairners could have been deceived as to the death of their leader. If he had not died on Pitcairn some mention must surely have slipped out in conversation with visitors. More important I believe their first questions would have been about whether or not he had succeeded, had survived. Mauatua at least could not have resisted asking.

No hint of an escape has ever come from Pitcairn. Adams' dissembling created a mystery that led others to infer such a possibility. There is one overriding proof he did not leave Pitcairn that would have been clear to any British sailor, from admiral to ordinary seaman, if they had thought to ask, or if they had known to ask.

If Fletcher Christian sailed secretly from Pitcairn, he surely would have taken *Bounty's* invaluable Kendall chronometer and the azimuth compass. It wasn't on Pitcairn after Folger, the first visitor was given it, so perhaps this explains why the subject didn't occur. By the early 19th-century, even Royal Navy visitors probably didn't know the K2 chronometer had been on board *Bounty*. If they did, no naval man would retail the idea of Fletcher Christian escaping without it.

Heywood's supposed sighting of Fletcher Christian in Devonport is blown out of all proportion by the very thing that was supposed to do the opposite, that it was kept secret until after Heywood's death. It is curious that Barrow believed in the sighting yet did not publicise this until 1831. Barrow wrote *The Mutiny of the Bounty* anonymously but this is not suspicious. Most people did that in those days, when there were terrible laws of slander and libel that could be interpreted too freely for anyone's good.

It is also extraordinary that Sir John Barrow admitted complicity in the 'cover up' of Heywood's story. As Secretary to the Admiralty, he might be assumed to be as responsible as anyone in pursuing those who had incurred its displeasure. Perhaps it helped Barrow's career to be uncontroversial and it was easy to find other explanations for the flight of the man who had surprised Heywood. In Plymouth in the early 19th century there would have been many a man who would run from a naval officer, or any other man, uniformed or otherwise. Impressment although diminishing was still used to man naval ships. There were previous shipmates and captains to fear, as well as cut-throats and footpads.

Perhaps Heywood saw another member of the Christian family, possibly Charles, for there was said to be a widespread disability of the knees among males of the clan, which contributed to the peculiar gait imputed to Fletcher. Anyway, if Fletcher's head and shoulders had been axed on Massacre Day it's very unlikely that anyone would have recognised him.

As to Wilkinson's belief in Fletcher Christian as the inspiration for *`The Ancient Mariner'*, it is as well to consider that the author disarmingly says he has no new information to offer and that he began his research only because he found the signature of an F. Christian in a scrapbook in a Charing Cross bookshop. He was not able to compare this with the mutineer's autograph but because it was associated in the scrapbook with the Losh family, who had assisted Edward Christian, he thought it might be the long-lost key to proving the return of Fletcher. He also mentions the existence in a notebook of Coleridge in the Public Records Office of the scribbled note, *'The adventures of Fletcher Christian'.*

It is absurd to think the latter is anything other than the most basic aide-memoire, the sort of thing any writer does constantly. As to the signature, there are a number of other F. Christians of whom I know and there would have been others besides. I subsequently found two signatures of Fletcher Christian on *Bounty* documents that have survived, which can be seen in the earliest edition of *FRAGILE PARADISE,* but was unable to arrange to compare them with that held by Wilkinson.

Then there is the paper delivered by William Fletcher MP in 1867: *Fletcher Christian and the Mutineers of the Bounty*. William was a relation of Fletcher's mother and I found a handwritten copy (his I think) of the work in the archives of Tullie House Library in Carlisle, together with some letters to him from Lady Belcher, who had Morrison's journal at the time. At first, she scrawled to say she was not well and could not find the work but later advised she had sent it by rail.

William Fletcher does not note any sources so presumably the originator of rumours of Fletcher Christian's return to the area current in 1808 and 1809 is Sir John Barrow, based on Heywood. I spent many days reading every newspaper I could find that exists from the period and from the area but there is no hint of Fletcher's return and I think this is just the type of story that would have been printed. William Fletcher says Fletcher is thought to have been visiting a favourite aunt but all his aunts were dead. Isabella Christian Curwen was alive but a cousin.

The continuing thrust of William Fletcher's story is that Fletcher Christian avoided discovery by hiding in the thick forest that covered the hills around Lake Windermere and the grounds of Belle Isle. It is a melodramatic theory, in the best tradition of the Victorian novel. The repentant mutineer, bullied into abandoning his coloured consort and bastard children, journeys half way around the world in disguise to live out a life of lonely misery within sight of the real Isabella, rich, beautiful but forever unattainable.

The descendants of John and Isabella no longer live on Belle Isle but Windermere still enthralls tourists with the suggestion that as well as being here as a child and young man, Fletcher Christian might secretly have walked where they do after his mutiny, that he might be buried on that romantic island. This is impossible, proven by a discovery of a print in a Cockermouth shop.

When I was there in late 1979, Noreham House on Main Street, which once belonged to Fletcher Christian's grandparents, John XVI and Bridget, had become an antique and junk shop. In a remote corner of the shop I found an engraving of Belle Isle dated 1796. There are almost no trees on the island and most of those to be seen cluster close to the round house. Vitally, there are few trees on the hills that surround Windermere, nothing that would make secure hides for Fletcher. Later research showed it was John Christian Curwen who planted most of the millions of trees that coated Cumberland by 1864, he who had also made Belle Isle into a forest of imported specimen trees.

John started planting on the island in a small way in 1787 and seems continuously to have improved or changed the style of the gardens. The original formal gardens had been demolished in the early 1780s and a raised gravel walk was constructed right around the island at the water's edge. Dorothy Wordsworth gives in her journal for June 8 1802 the final lie to the possibility of hiding in a dense screen of trees on the island. She writes: *'The shrubs have been cut away in some parts of the island. They have made no natural glades: it is merely a lawn with a few miserable young trees, standing as if they were half starved . . . And that great house! Mercy upon us! if it could be concealed . . . Even the tallest of our old oak trees would not reach to the top of it.'*

This pretty much clinches the impossibility of Belle Isle as a hiding place. If it was not covered with bushes or trees in 1802, it

The 1795 engraving of Belle Isle and Lake Windemere that proves Fletcher Christian could not have hidden in trees on the island or the shores, because there were none.

would not be seven years later when Fletcher was said to have hidden there. Every traveller with the will could visit the island and it was also the main summer house of the social John and Isabella Christian Curwens, who introduced boating regattas and swans to the lake.

A fugitive from society on Belle Isle would have to have been as fey as a sprite to avoid detection by the *ton* of Cumberland or the house's servants. By 1809, Fletcher Christian would have been 45 and must have been crippled by his injuries, which would have made him something less than fleet. He could not have rowed himself to the lake shore, or have swum there without great physical difficulty. As well, the supposed axe wounds about his head and neck would have left hideous scars and there is no chance that such an appearance would have been unnoticed or not widely commented upon.

Supposing Fletcher had the guile and luck to escape detection and the physical ability to get to and from Belle Island, he would have little cover there or elsewhere because the shores of Lake Windemere were *not* thick were trees. William Fletcher mistakenly thought what he saw in 1864 had always been like that.

It cannot be said that the Curwens bought silence from those who might have discovered the presence of Fletcher Christian in Cumberland; today's Cumbria properly refers to the combination of Cumberland, Westmoreland and other regions in 1974. As a rich crusader and a distinguished member of the Old Minority of Whig reformers, John Christian Curwen had many opponents who would have paid handsomely to report the harbouring of his criminal relative Fletcher Christian.

There can have been no hiding in Cumberland for Fletcher Christian and little reason so to do. His mother and brother Charles were on the Isle of Man. Edward was in Cambridge or London. The one person, I think, who might be expected to tell us of Fletcher's return is his brother Charles. The cruel, forced anonymity of a younger brother whom he considered to have been driven to temporary madness by the intemperance of another was too meaty a subject for him to abjure. As a further and proximate example of man's inhumanity to man, his pen would have worried at the idea over many pages. If Fletcher were back, he would have gone to Man to see Charles and his mother. If he did, Charles must have alluded to it but he did not.

None of Fletcher Christian's supposed hiding places is verified by contemporary printed or written evidence and none of the stories that incorporate such suggestions has an identifiable ultimate source. There is no trace of belief or proof within the family, either in England or in the South Pacific, that he returned. Indeed, my great-aunts from Norfolk Island would redden and refuse to countenance even the suggestion, let alone the possibility and this was always so within the family.

The seed of his escape was planted as early as 1796, with the publication of the fictional *Letters from Mr Fletcher Christian*. Whoever wrote it certainly knew a lot about South America, and many readers must have believed it to be true because of that ancient need for romantic rebels not to be punished. For all the conflict about the end of Fletcher Christian, his descendants on Pitcairn and Norfolk Island place it firmly on Pitcairn. My corroboration of Jenny's story of Mauatua's call to Fletcher and the identification of Bennet's pool convince me this is the truth.

The ultimate wonder about Fletcher Christian's supposed escape from Pitcairn is not the contradictions but that there should be such a myth. The answer to that is precedent. It seems mankind has always needed heroes of protest to survive. Fletcher Christian was quickly and firmly put into the temporal pantheon because he struck a seminal chord in universal yearnings for adventure and freedom, goals few achieve in reality.

Legends of troublesome but admired men and women supposedly dying but then miraculously discovered to be alive are thousands of years older than the story of Jesus Christ. The ancient Egyptians believed this of their gods and latterly we have James Dean and Elvis Presley, who are not *really* dead. Almost always the survivor has been a protester, law breaker or iconoclast, whose flouting of the rules gave excitement and hope to those less brave. They in turn battle on with their surrogate hero through all adversity, hoping to defeat the system and even that more implacable enemy, death.

The hot pioneering courage and imagination of Fletcher Christian make him worthy of a far more dramatic end than that of perishing five days before his 29th birthday with his head axed into the red earth of Pitcairn, slowly blackening in his own blood and the scorching sun. In the absence of a tombstone or grave to prove his death, the world's romantics have given Fletcher the more enduring memorial of making him into a legend, with a distinct claim to a degree of immortality, from a literary point of view at least.

The Sir John Barrow Commemorative Expedition saw clearly what Pitcairn needed in 1980 and tried to help. At a time when the rest of the world was first playing with videotapes, Pitcairn had only Morse Code for official communications with the Commissioner's office in Auckland or their Governor, the UK's High Commissioner to New Zealand. They had also been advised that they should replace their wooden boats with metal ones but had no information about the performance of a metal boat or ways to research this. They were expected to do what others had decided, petty officials with no understanding of their daily challenges at sea or on land. Their frustration was awful to see but with only dots and dashes as their contact with officials and no way of getting written information in less than many months, they felt demeaned and devalued.

If metal boats could not withstand Pitcairn's surf and currents, Pitcairners would not be able to get supplies or send goods or themselves elsewhere. Bounty Bay offers no harbour, no safe anchorage, no dock for anything but their whaleboats. Ships that do stop must ride the Pacific swells off the island, exactly where depending on that day's weather and the winds, which come without hindrance in any direction. Ships that did stop could only hope Pitcairners get out and then get back safely. The best we could do was to use ham radio to get information to them, a medium that even though being used illegally for such matters, was often the only reliable lifeline for supplies and information for Pitcairners for decades.

Pitcairners worst deprivation was having no independent government representative or law enforcement on the island, not even a policeman of any kind.

There was no-one safe to whom Pitcairners could address questions or complaints in confidence, a right that is basic to life in the UK and on every other British Overseas Territory. I told this to many, including the Pitcairn Office in Whitehall. The response was that the school teacher represented HM Government on the island. There was blank and total misunderstanding that because he was also responsible for handing out desperately needed jobs, no islander dared upset that arrangement.

Lord Tanlaw and Lord Adair asked questions in the House of Lords on my behalf, suggesting the Pitcairners should at least have direct control over some of their budget; at the time it could take 18 months to apply for permission to buy a new motor mower, get the permission, order it and await delivery, delivery that depended on finding a ship and then that weather conditions would allow cargo to be landed.

Little resulted and neither could anyone explain why of all the Commonwealth and the British Overseas Territories only intensely loyal and royalist Pitcairn Island was not invited to the wedding of HRH the Prince of Wales and Lady Diana Spencer, an exclusion that still seems to prevail for occasions of national significance.

After I set up the Pitcairn Island Fund, for which many organised events or donated generously, I received a letter from the Island Council saying that I did not represent Pitcairn and should not be doing such a thing. It was clear this had been dictated by a higher authority who, for reasons of personal power, did not wish Pitcairners to have access to any source of independence. Not wishing further to muddy dirty waters I closed the fund and sent the money to the island, which I believe was used to renovate the store.

Since then I have often offered to work in the UK for Pitcairn Island for expenses only, so that the island is represented by a directly appointed representative on Government committees in the UK and EU. Confusion caused by remoteness and petty officialdom means this has never been accepted and Pitcairn Island is the only one of the 12 Overseas Territories without a spokesperson or representation independent of HM Government in the UK. The island's council was once convinced I was cooking three times a week on BBC-TV only because of my association with Pitcairn Island. The truth was the reverse, that being on TV meant I could talk about Pitcairn, raise money for the Pitcairn Island Fund and such. More recently it was suggested I wanted to work for Pitcairn only because of opportunities to attend glittering receptions. Her Majesty's British Overseas Territories, all of them, deserve bigger minds than this.

Having no independent Government representative or police equivalent on the island greatly contributed to the disaster of the 2004 trials for sexual assault; who could complain to whom? No official in the Foreign and Commonwealth Office seemed to have direct understanding of life on a remote island. Solutions applicable to other parts of the world were applied recklessly, with Baroness Scotland (I believe) heard to shout loudly that 'these men must be punished'. Suddenly almost £4million was found to fund the Gilbert and Sullivan silliness of a new Court of Law being transported to sub-tropical Pitcairn Island, there to sit in wigs and gowns.

A Truth and Reconciliation Commission would have done more, more quickly and been more in keeping with island life and done more to preserve Pitcairn for the future. Under-age sex is a part of every small or remote community in the world including the UK and anyone who thinks otherwise is naïve. The elements of assault and force had to be stopped and punished but did the subject need to be front-page news around the world, forever poisoning the fragile life of Pitcairn Island and giving comfort only to cheap journalists?

Now there are always officials of some kind on Pitcairn, too many and even now too anxious to be associated with a hot social subject that they hope will look good on their CV. At the same time, Pitcairners today feel increasingly able to tell about the methods used on them to get the verdicts required during the Trials; when someone dares write this book I know it will change many points of view but even this overdue publication cannot heal the unnecessary wounds and scars of irresponsible officialdom that lives too far away to be affected by their action's outfall.

The £4 million cost of those trials might have funded infrastructure, including better communications, water storage, renewable energy and more. Perhaps even a small ship manned by Pitcairners, which would not only create jobs but allow the island to profit directly from increased tourism and to trade fresh produce and fish with other islands. These were the objectives of the Pitcairn Island Fund and more latterly of the Pitcairn Settlers Project, which had solid international support for creating a paradigm of organic and sustainable agriculture that would also be a teaching resource for students as well as a profit centre for the island.

Pitcairn Islanders are not perfect but any judgement should be based on who and where they live, under what circumstances. Women writers who visit briefly and then write excoriating articles clearly go there with a personal wound and an agenda that does not include the discovery of reality for Pitcairners. Many years after the trials, men are still described according their past crimes, unheeding of the British creed that these should be put to the back of minds once time has been served. It is a major point of British law that past convictions are never mentioned during a trial. Yes, there are different rules for sex offenders but perhaps there should be differences for a remote community, too. The careers of biased journalists are not damaged by their destructive 'outings' and neither are those of their sensation-seeking editors; the Pitcairners are abandoned to pick up the pieces of their destruction.

A new landing site on the more sheltered Western side of the island now means passengers on cruise ships can be landed 300 at a time. Better shipping means other good people choose to visit for a few weeks, so a better story is emerging, although like all positive news it is unlikely to generate the same oxygen of publicity inflamed by unnecessary trials and bad journalism.

I will never know what Fletcher Christian really looked like. I am not able to sit by the grave of his wife Mauatua or of his son, Charles. I will never know what he thought as he sat on the edge of his cave. The new world he created far below couldn't compare with Ewanrigg or Milntown, Netherhall or Rose Castle. Pitcairn would never give Fletcher the excitements of the backyard ruins of the *'medieval manor house, half castle and half farmstead'* surrounded by red-brick battlements, never recreate the family comforts of Moorland Close.

Fletcher joined the Royal Navy hoping it would regenerate his expectations of superior social and financial position that were lost as a teenager. Now even his expectations were lost.

It's not hard to imagine quite how isolated and, perhaps, a failure Fletcher Christian must have felt, even with the love of Mauatua and their children. I fear his mental stability would have been severely tried, as well as trying for others.

Sometimes I feel I should raise the funds to return to Pitcairn and finally identify the graves of Fletcher Christian and Mauatua. Surely the founders of Pitcairn Island, whose exploits have fired the imagination of millions of men and women for over two centuries, deserve something better than unmarked pits?

Or, is the great lonely rock of Pitcairn the most extraordinary mausoleum that one revolutionary young man and his visionary wife ever shared?

When I sat alone on the ledge of Christian's Cave, with Adamstown below me and Ship Landing Point seen above Bounty Bay in the distance, I was sad that I would never know what Fletcher thought when he did the same and looked down at the new world he had created as just one repercussion of his mutiny.

Elizabeth Mills (front centre) aged about 72, daughter of John Mills and Vahineatua, was the first girl born on Pitcairn Island and is one of my great-great-great grandmothers. Photographed in 1862 on Norfolk Island.

Sarah Christian Nobbs was a granddaughter of Fletcher and Mauatua. The daughter of their second son Charles and of Sully, the baby who was landed on Pitcairn in a barrel, she was said to have looked very like Mauatua. She was a sister of my great-great grandfather, Isaac.

Sons and grandsons of mutineers on Norfolk Island, 1862. From right, Arthur son of mutineer Matthew Quintal and Tevarua, George son of John Adams and Teio, Arthur's son John Quintal and George's son John Adams.

Fletcher and Mauatua's daughter Mary never married. She died on Norfolk Island and rather than being a 'motherly aunt' is more remembered as a rather crabby spinster, by then ashamed of her mutineer father and Tahitian mother as neither is mentioned.

SOURCES AND BIBLIOGRAPHICAL NOTES

The bibliography of the *Bounty*/Pitcairn story is enormous but most works on the subject are variations of someone else's themes or the expansion of slim theories in the hope they will become fat facts. I used primary material in libraries and private collections, well before digitalisation and the internet. I had to go where they were.

The major source for information about the Christian family and the Isle of Man was the private collection of Ewan Christian, which had been used as reference for The Yesterdays Behind the Door, Susan Hicks Beach (University Press, Liverpool, 1956). These remarkable documents have now been generously donated by Ewan Christian to the Manx Museum in Douglas, Isle of Man. The MSS accession number is 9381 and the microfilm references are MIC69 and MIC70.

North Country Life in the 18th Century, Vol. 2 by Edward Hughes (Oxford University Press, 1965) is packed with important Christian family material and the sources for this are all in the Christian Curwen and the Senhouse MSS in the Cumberland Records Office in Carlisle Castle. The papers of William Fletcher MP are across the road, in Tullie House, Carlisle.

Details of Royal Navy life were mainly taken from Sea Life in Nelson's Time (Methuen & Co., London, 1905) by John Masefield. Several articles published in The Mariner's Mirror over the years give excellent insight into the fitting out of *Bounty*.

The background to classical Tahitian life was drawn from the three volumes of Professor Douglas Oliver's Ancient Tahitian Society (University Press of Hawaii, Honolulu, 1974) and cross-checked with several of his sources, which include the manuscripts and published works of Bligh and James Morrison, *Bounty*'s boatswain's mate.

What Happened on the *Bounty* (George Allen & Unwin, translated, London, 1962) by Bengt Danielsson and Rolf DuRietz, clearly presents the complicated story of what happened on Tahiti after *Bounty* left.

For the day of the mutiny I used Bligh's works, Owen Rutter's edition of The Court-Martial of the *Bounty* mutineers (William Hodge & Co., Edinburgh, 1931), The Voyage of the *Bounty*'s Launch John Fryer's Narrative edited by Stephen Walter (Genesis Publication, Guilford, Surrey, 1979), The Appendix by Edward Christian (London, 1794). The latter must be regarded as primary material, not just for what it reports, but also because of the dramatic revelations and corroboration of Edward's subsequent pamphlet A Short Reply to Captain Bligh's Answers (J. Deighton, London, 1975). This second, exceptionally rare and rarely used pamphlet reveals the methodology of the first, explains the danger and disgrace risked by the eminent men who helped collect evidence if they were to be found liars, and even shows that Bligh's servant went independently to the Christian family to give a version of the mutiny different from that of Bligh. A Short Reply also reveals that McIntosh was threatened for talking to Edward Christian and demolishes the methods with which Bligh appeared to get some men (e.g. Lebogue) to retract earlier statements.

The travels of Fletcher Christian after the mutiny were not fully known until the brilliant detective work of Professor H. E. Maude, which was published as 'In Search of a Home' (Journal of the Polynesian Society, Vol. 67, No. 2, June 1958, Wellington). James Morrison gave excellent anthropological detail about Tubuai. The abstracts by Captain Edward Edwards of the journals of George Stewart and Peter Heywood are in his papers, which were recognised in the Admiralty Library, London, by Bengt Danielsson.

For events on Pitcairn Island, each of those mentioned has something to offer and F. W. Beechey's Narrative of a Voyage to the Pacific and Beering's Strait in His Majesty's Ship Blossom . . . in the Years 1825 . . . , 2 Vols. (Henry Colburn & Richard Bentley, London, 1831) is by far the fullest. I decided to accept as authoritative the versions given by the Tahitian woman Jenny; my expedition to Pitcairn Island proved I was right to do so. See what she said in the United Services Journal, London, November 1829, Part 2, pp. 589-593.

The most masterly and thorough interpretative and critical works published about *Bounty* and her men are by the Swedish bibliographer Rolf DuRietz based in Uppsala. As well as a series of small articles and pamphlets he published Studia *Bounty*ana (Dahlia Books, Uppsala, Sweden, 1979, Vols 1 & 2) and had begun a new series titled Banksia. Banksia I (Dahlia Books, Uppsala, Sweden, 1979) is 'Thoughts on the Present State of Bligh Scholarship', and puts into focus the problems of writing about a figure who is so well known. An important work on Fryer was published in 1981.

William Bligh has had many enthusiastic biographers and apologists. None has been more honest or painstaking than George Mackaness, in The Life of Vice-Admiral William Bligh (Angus & Robertson, Sydney, 1931. New and revised edition 1951). He subsequently found and published the startling Bond material published in *Fragile Paradise* the originals are in the National Maritime Museum, Greenwich. As this material was not known to Gavin Kennedy, he did not discuss it in his rich biography Bligh (Gerald Duckworth & Co., London, 1978), so a reassessment of his subject could not be made. Gavin Kennedy's later book, Captain Bligh: The Man and his Mutinies (Gerald Duckworth & Co. Ltd, London, 1989) incorporates the material and is an important reference book.

Sir John Barrow's the Mutiny of the *Bounty* (John Murray [pub], London, 1831) is one of the best works on the subject overall and covers a wider spectrum than my book.

The Heritage of the *Bounty* by H. L. Shapiro (Simon & Shuster, New York, 1936) gives an excellent perspective on Pitcairn's development and David Silverman's Pitcairn Island (World Publishing Company, Cleveland, 1967) is a worthwhile collection of often overlooked sources. The Pitcairnese Language by A. S. C. Ross and A. W. Moverley (Andre Deutsch, London, 1964) includes some excellent essays by Professor Maude and his son and is worth the trouble to track down. The simple style and genealogical charts of The Pitcairners by Robert Nicholson (Angus & Robertson, Sydney, 1965) make it important, but the marriage date he gives for Fletcher Christian is 'entirely his own work' and unsupported. The famous article by Luis Marden 'I Found the Bones of the *Bounty*' in the National Geographic Magazine in December 1957.

The only other book I know written by a descendant of the mutineers is Mutiny of the *Bounty* and Story of Pitcairn Island by Rosalind Amelia Young (Pacific Press Publishing Assn, Mountain View, California, 1894), who was born and brought up on Pitcairn and who is buried there. It gives details and facts not otherwise collected and remains an excellent and entertaining book. Otherwise I find books about Pitcairn Island fail accurately to represent the island and its people.

Naturally William Bligh has been a major source for this book and most of his important papers are in the Mitchell Library, Sydney. His letters and correspondence with Banks are most fruitful, and much work could still be done. These are collected on three reels of microfilm:

MLMS C218 (Reel FM4/1756): MLMS Safe 1/35 (Reel CY 178) Bligh documents and correspondence: MLMS A78 4 (Reel FM4/1748) the Banks Brabourne papers

There is Bligh material in the Dixson Library, Sydney, and in the National Library, Canberra, you will find the Rex Nan-Kivell Collection, which includes interesting secondary material and outstanding illustrative material.

The Dawson Transcripts of Banks letters, which are held in the Natural History Museum are little referred to in connection with *Bounty*. This is where you will discover the letter in which Banks says he expected to be blamed if anything went wrong with the breadfruit expedition. These transcripts also show that the letter of 7 September 1787 in the Banks Brabourne papers, thought to be from Sir Joseph Banks to Sir George Yonge, is the reverse; it is a copy of Yonge's letter to Banks about the former's visit to *Bounty*. Banks had actually sent the original on to Evan Nepean. The complete correspondence (DTC 5:245-9 and DTC 5:259/60) seems to show Bligh was playing these two men off against one another, pretending, for instance, to Yonge that he didn't know where he was to go or to where he was to return.

Dorothy Wordsworth's quote is from Wordsworth's Hawkshead by T. W. Thompson (edited by Robert Woof, Oxford, 1970) a book that contains much detail of meetings between the Wordsworths and the Christian Curwens.

For illumination on Fletcher Christian's hyperhidrosis, I consulted Professor John Ludbrook, MD ChM DSc MMedSc FRCS FRACS, Professor Emeritus, University of Adelaide and a Professorial Fellow, University of Melbourne Department of Surgery, Royal Melbourne Hospital. Professor Ludbrook is a vascular surgeon and vascular physiologist with professional experience of hyperhidrosis.

Much material on Christian's mental condition came from professional psychologist Dr Sven Wahlroos PhD and his masterly Mutiny and Romance in the South Seas: A Companion to the *Bounty* Adventure (Salem House Publishers, Div. of HarperCollins, Topsfield, Massachusetts, 1989). He, in turn, quotes from the Diagnostic and Statistical Manual of Mental Disorders (third edition, revised, Washington DC, American Psychiatric Association, 1987). There is now a fourth edition. Dr Wahlroos' work is a vital reference book, both the only chronological account and a detailed encyclopaedia, and is well overdue for republication. Further insight came from clinical psychologist Paul Rodriguez, BA (Hons) MPsychol MAPS.

For greater clarity on Bligh's command after *Bounty* sailed from Tahiti, I am grateful to my loyal friend and full-time *Bounty* enthusiast Topher Russo, who gave me access to his unpublished Master's thesis, 'Mr Bligh's bad discipline: laxity and recklessness on the high seas' (University of Hawaii, Honolulu, 1994). Similarly, Tasmanian historian Ian Campbell is responsible for new perspectives on Bligh's health and his behaviour during the second breadfruit voyage on HMS Providence in his paper 'Mr Bligh's bad health', which is based on his BA Honours thesis (University of Tasmania, Hobart, November 1994)

First-hand accounts of Bligh as Governor of New South Wales are from Distracted Settlement, edited and introduced by Dr Ann-Maree Whittaker (The Miegunyah Press, Melbourne, 1998).

Undoubtedly there is still more information to be found, which may change my views. I look forward to seeing that material, which is probably amongst the whaling and sealing archives of the United States as are further artefacts from HMAV *Bounty* taken from Pitcairn Island. Much more, I look forward to the day when there is no longer the urge to cast Bligh or Christian as black or white. They are men who are remembered. Few men who are remembered can have been wholly one or the other.

THE AUTHOR

GLYNN CHRISTIAN is a great-great-great-great grandson of Mauatua and Fletcher Christian. He is also the author of MRS CHRISTIAN *BOUNTY* MUTINEER, which dramatically reveals the usually invisible bloodlines of the Ma'ohi foremothers of Pitcairn Island, *te tupuna vahine,* and what they had to endure to protect the future of their children. He is one of the few writers in history honoured to have been able to write about both partners in the pair of his great-great-great-great grandparents, Fletcher and Mauatua Christian

He was born in New Zealand but has lived mainly in London UK since 1965, where he is very well known as a pioneering BBC-TV chef, journalist and prize-winning food writer.

He is the only food writer to have been honoured with a Lifetime Achievement Award from within the food industry, from the Guild of Fine Food in 2008. Among his many books, REAL FLAVOURS – *The Handbook of Gourmet and Deli Ingredients* was voted World's Best Food Book at the Cordon Bleu World Media Awards.

For more information WWW.GLYNNCHRISTIAN.COM

A poignant moment. Ben Christian, then Island Secretary, shows me the bible believed to have been owned by Fletcher Christian. We sit on a retrieved Bounty anchor outside the Courthouse and Village Hall under which I believe lie the grave of Fletcher's wife Mauatua and of Charles, his second son.

INDEX

Bold type indicates close family members of Fletcher Christian

---

All pictures and images not accredited are owned by or are images once owned by Glynn Christian

Plans of converted HMAV *Bounty* courtesy of National Maritime Museum

Print of Thursday October Christian; photo of sons and grandsons of mutineers; photo of Sarah Christian Nobbs; photo of Elizabeth Mills, by permission of Dixson Library, State Library of NSW, Sydney

Illustrations of Milntown, IOM, courtesy of Charles Guard and The Milntown Estate www.milntown.org

Illustration of Rose Castle courtesy of Rose Castle Foundation https://www.rosecastle.foundation/

Illustration of Edward Christian's coat of arms, Gray's Inn © the Masters of the Bench of the Honourable Society of Gray's Inn

Solus·per·Christum
Christian

Printed in Great Britain
by Amazon

34870713R00215